AF323986

# MODERN ASPECTS OF ELECTROCHEMISTRY, No. 46:

# PROGRESS IN CORROSION SCIENCE AND ENGINEERING  I

For other titles published in this series, go to
www.springer.com/series/6251

# Modern Aspects of Electrochemistry

*Series Editors*:
Ralph E. White
Department of Chemical Engineering
University of South Carolina
Columbia, SC 29208

Constantinos G. Vayenas
Department of Chemical Engineering
University of Patras
Patras 265 00
Greece

*Managing Editor*:
Maria E. Gamboa-Aldeco
1107 Raymer Lane
Superior, CO 80027

# MODERN ASPECTS OF ELECTROCHEMISTRY, No. 46:

# PROGRESS IN CORROSION SCIENCE AND ENGINEERING I

Edited by

## Su-Il Pyun

*Korea Advanced Institute of Science and Technology*

## Jong-Won Lee

*Samsung Advanced Institute of Technology*

 Springer

*Editors*
Su-Il Pyun
Department of Materials Science
and Engineering
Korea Advanced Institute of
Science and Technology
Daejon
Republic of Korea
sipyun@kaist.ac.kr

Jong-Won Lee
Energy Group
Samsung Advanced Institute
of Technology
Yongin
Republic of Korea
jongwon277.lee@samsung.com

*Series Editors*
Ralph E. White
Department of Chemical Engineering
University of South Carolina
Columbia, SC 29208

Constantinos G. Vayenas
Department of Chemical
Engineering
University of Patras
Patras 265 00
Greece

*Managing Editor*
Maria E. Gamboa-Aldeco
1107 Raymer Lane
Superior, CO 80027

ISBN 978-0-387-92262-1          e-ISBN 978-0-387-92263-8
DOI 10.1007/978-0-387-92263-8
Springer Dordrecht Heidelberg London New York

Library of Congress Control Number: 2009934519



Printed on acid-free paper

Springer is part of Springer Science+Business Media (www.springer.com)

# Preface

The present volume of Modern Aspects of Electrochemistry is composed of four chapters covering topics having relevance both in corrosion science and materials engineering. All of the chapters provide comprehensive coverage of recent advances in corrosion science.

The first chapter, by Maurice and Marcus, provides a comprehensive review on the structural aspects and anti-corrosion properties of passive films on metals and alloys. These authors look at recent experimental data collected by *in-situ* microscopic techniques coupled with electrochemical methods. A detailed description is given of the nucleation and growth of 2-dimensional passive films at earlier stages, their effect on the corrosion properties of metal surfaces, and the nanostructures of 3-dimensional passive films. On the basis of the experimental data reviewed, the authors present a model for passivity breakdown and pit initiation, which takes into account the preferential role of grain boundaries.

In Chapter 2, Takahashi and his co-workers give a specialized account on the electrochemical and structural properties of anodic oxide films formed on aluminum. In addition to the electrochemical corrosion-related problems of anodic oxide films, the chapter reviews state-of-the-art research of nano-/micro-fabrications based on anodizing treatments combined with chemical/mechanical processes such as laser irradiation, atomic force micro-probe processing and thin film deposition techniques.

Chapter 3, by Janik-Czachor and Pisarek, deals with electrochemical and structural properties of amorphous and nanocrystalline alloys that have been defined as 'meta-stable' alloys by the authors. The chapter addresses the key question of how the structures could be modified to be either more resistive against environmental corrosion or more reactive towards electrocatalytic reactions by hydrostatic extrusion or cathodic hydrogen charging, respectively. An in-depth discussion is given of the passivity of aluminum-based amorphous alloys and stainless steels and the catalytic activity of copper-based amorphous alloys.

Chapter 4, by Di Quarto and co-workers, offers a review of the electronic and solid-state properties of passive films and corrosion layers determined using differential admittance and photocurrent spectroscopy. The authors start with a theoretical analysis of the admittance behavior of passive film/electrolyte junction based on the theory of amorphous semiconductor Schottky barriers. The fundamentals of photocurrent spectroscopy are then introduced with special emphasis on how this technique could be utilized for determining the solid-state properties of passive films.

S.-I. Pyun
*Korea Advanced Institute of Science and Technology*
*Daejeon, Republic of Korea*

J.-W. Lee
*Samsung Advanced Institute of Technology*
*Yongin, Republic of Korea*

# Contents

*Chapter 2*

## ANODIC OXIDE FILMS ON ALUMINUM: THEIR SIGNIFICANCE FOR CORROSION PROTECTION AND MICRO- AND NANO-TECHNOLOGIES

Hideaki Takahashi, Masatoshi Sakairi, and Tatsuya Kikuchi

*Chapter 3*

## ELECTROCHEMICAL, MICROSCOPIC AND SURFACE ANALYTICAL STUDIES OF AMORPHOUS AND NANOCRYSTALLINE ALLOYS

Maria Janik-Czachor and Marcin Pisarek

*Chapter 4*

# PHYSICOCHEMICAL CHARACTERIZATION OF PASSIVE FILMS AND CORROSION LAYERS BY DIFFERENTIAL ADMITTANCE AND PHOTOCURRENT SPECTROSCOPY

### Francesco Di Quarto, Fabio La Mantia, and Monica Santamaria

# List of Contributors, MAE 46

Francesco Di Quarto
*Dipartimento di Ingegneria Chimica dei Processi e dei Materiali, Universita' degli studi di Palermo, Viale delle Scienze, 90128 Palermo, Italy; Tel.: +39-091-6567228; Fax: +39-091-6567280; E-mail: diquarto@unipa.it*

Maria Janik-Czachor
*Institute of Physical Chemistry PAS, Warsaw, Poland; Tel.: +48-22-343-3325; Fax: +48-22-343-3333; E-mail: maria@ichf.edu.pl*

Tatsuya Kikuchi
*Graduate School of Engineering, Hokkaido University, N13, W8, Kita-Ku, 060-8628, Sapporo, Japan*

Fabio La Mantia
*Dipartimento di Ingegneria Chimica dei Processi e dei Materiali, Universita' degli studi di Palermo, Viale delle Scienze, 90128 Palermo, Italy*

Philippe Marcus
*Laboratoire de Physico-Chimie des Surfaces, ENSCP-CNRS (UMR 7045),Ecole Nationale Supérieure de Chimie de Paris,11 rue Pierre et Marie Curie, 75005 Paris, France;Tel.: +33-1-44-27-67-38; Fax: +33-1-46-34-07-53; E-mail: philippe-marcus@enscp.fr*

Vincent Maurice
*Laboratoire de Physico-Chimie des Surfaces, ENSCP-CNRS (UMR 7045),Ecole Nationale Supérieure de Chimie de Paris,11 rue Pierre et Marie Curie, 75005 Paris, France*

Marcin Pisarek
*Faculty of Materials Science and Engineering, Warsaw University of Technology, Warsaw, Poland*

Masatoshi Sakairi,
*Graduate School of Engineering, Hokkaido University, N13, W8,
Kita-Ku, 060-8628, Sapporo, Japan*

Monica Santamaria
*Dipartimento di Ingegneria Chimica dei Processi e dei Materiali,
Universita' degli studi di Palermo, Viale delle Scienze, 90128
Palermo, Italy*

Hideaki Takahashi
*Asahikawa National College of Technology, Shunkoh-dai 2-2,
Asahikawa, 071-8142, Japan; Tel.: +81-166-55-8100; Fax: +81-
166-55-8082; E-mail: takahasi@asahikawa-nct.ac.jp*

# Modern Aspects of Electrochemistry

*Topics in Number 44 include:*

- Basic mathematical models which arise in electrochemical processes regarding galvanic corrosion phenomena
- Near field transducers and their applications to nano-chemistry and electrochemistry
- Symmetry considerations in modern scientific computation
- Analysis of electrochemical systems, including computer simulations and quantum and statistical mechanics
- Techniques to investigate ionic and solvent transfer and transport in electroactive materials
- Monte Carlo simulations of the underpotential deposition of metal layers on metallic substrates
- Models of localized corrosion processes and a review of recent work on the modeling of pitting corrosion
- Conceptual structure of Density-Functional Theory and its practical applications
- Scanning acoustic microscopy's applications to nano-scaled electrochemically deposited thin film systems
- Processes and equations underlying the modeling of current distribution in electrochemical cells

*Topics in Number 45 include:*

- The cathodic reduction of nitrate and electrochemical membrane technology
- Non-haloaluminate ionic liquids
- The properties of nanowires composed of metals and semiconductors.
- Ammonium electrolysis  as a renewable source of fuel
- The usefulness of synchrotron x-ray scattering to a wide range of electrode phenomena

*1*

# Structure, Passivation and Localized Corrosion of Metal Surfaces

Vincent Maurice and Philippe Marcus

*Laboratoire de Physico-Chimie des Surfaces, ENSCP-CNRS (UMR 7045), Ecole Nationale Supérieure de Chimie de Paris,11 rue Pierre et Marie Curie, 75005 Paris, France*

## I.  INTRODUCTION

Ultra-thin layers of hydroxylated oxide (passive films) grown on metals and alloys in aqueous environments provide the surface property of self-protection against corrosion. Passivation is a key to our metal-based civilization and to its sustainable development. Passive films in many cases do not exceed a few nanometers in thickness and effectively isolate the metal (or alloy) substrate from the corrosive environment. However, they are sensitive to breakdown eventually leading to accelerated dissolution of the substrate at localized sites (i.e., pitting) with major impact in practical applications and on economics.

Despite a considerable amount of research on localized corrosion, the complete understanding of passivity breakdown and pit initiation and in particular the reason for its local occurrence remain a vexing problem in corrosion and electrochemical science that must be addressed at the nanoscopic scale. Moreover, the increasing development of materials with nanoscale architecture and

*Modern Aspects of Electrochemistry*, Number 46, edited by S. –I. Pyun and J. -W. Lee, Springer, New York, 2009.

S.-I. Pyun and J.-W. Lee (eds.), *Modern Aspects of Electrochemistry 46: Progress in Corrosion Science and Engineering I*, DOI 10.1007/978-0-387-92263-8_1, © Springer Science+Business Media, LLC 2009

including metallic components requires a detailed understanding and nanoscale control of corrosion processes to guarantee their durability.

The anti-corrosion properties of passivated metals and alloys surfaces have been recently reviewed,[1,6] and are discussed in other chapters of this volume. In this chapter, we address the nanostructure of passive films and its relation with the local properties of protection against corrosion. The relation between the structure of non-passivated metallic surfaces (i.e., in the active state) and their corrosion properties will not be addressed here. Recent *in situ* studies with scanning probe microscopes implemented with electrochemical control addressing this issue at the sub-nanoscopic scale have been reviewed.[7,8] Investigations have focused on the structural modifications of atomically flat model surfaces resulting from the adsorption of anions in the double layer potential region preceding the onset of dissolution and on the dissolution of the metal atoms at specific sites. The dynamics of these processes have been studied in some cases. Dissolution was shown to occur exclusively at step edges leading to the retraction of the terraces by a step flow mechanism. The anisotropic effect of the surface atomic structure of strongly adsorbed layers of anions on the mechanisms of active dissolution of metals was also evidenced. Similar detailed studies remain to be performed on alloys used in practical applications despite some significant advances on the structural aspects of selective dissolution (i.e., dealloying) mechanisms mostly studied on model CuAu alloys.[9-15]

The structure of passive oxide layers is now well-documented thanks to the use of Scanning Tunneling Microscopy (STM) and Atomic Force Microscopy (AFM). Their crystallinity has been demonstrated for Ni,[16–25] Cr,[26,27] Fe,[28–33] Cu,[34–39] Co,[40,41] and stainless steel[42,46] and nickel-based alloys.[47] *In situ* grazing incidence X-ray diffraction (GIXD) experiments have confirmed and complemented this finding for Ni[22,48] and Fe.[49,50] Prior to these studies, it was generally considered that the ultra-thin passive films were amorphous, and that the absence of crystallinity was favorable to the corrosion resistance. This view has now changed, because the structural studies have revealed that in many cases the passive films consist, at the nanometre scale, of crystalline oxide grains with grain boundaries. Non-ordered areas can form between the crystalline grains, particularly during the growth process, i.e., under non stationary conditions. These inter-granular sites of the passive film play a key role in the breakdown of passivity and the initiation of localized corrosion.

The topics addressed in this chapter include the adsorption of hydroxide ions and the initial 2D growth of passive layers, the modified corrosion properties of metal surfaces at this intermediate stage of formation of the passive films, the nanostructure and atomic structure of 3D passive films, the effect of aggressive ions (i.e., chlorides) on the growth and structure of 3D passive films and the nanostructural aspects of passivity breakdown.

## II.  NUCLEATION AND GROWTH OF PASSIVE FILMS

In this Section, data obtained on copper, silver and nickel in alkaline aqueous solutions are reviewed to illustrate and discuss the structural modifications of the metal surface accompanying the adsorption of hydroxide anions and the nucleation and growth of 2D passive layers. The influence of the structure of these 2D layers on the dissolution properties of the metal surface is then addressed for silver and nickel, showing the major influence of crystallization on local properties at the nanometer scale. Such data are relevant to the fundamental understanding of both the corrosion mechanisms when dissolution and passivation compete and the self-repair of depassivated metal surfaces.

### 1.  Hydroxide Ion Adsorption and Growth of 2D Passive Layers

#### (*i*)  Copper

Copper does not passivate in acidic solutions (pH < 5) because copper oxides are not stable and their dissolution is fast. However, at higher pH, Cu(I) and Cu(II) oxide layers are stable and the composition of the passive films has been studied qualitatively and quantitatively with electrochemical and surface analytical methods such as X-Ray Photoelectron Spectroscopy (XPS),[51–54] photo-acoustic spectroscopy,[55] ion scattering spectroscopy,[51] *in situ* Raman[68–58] and infrared spectroscopy.[59] The growth of 3D crystalline

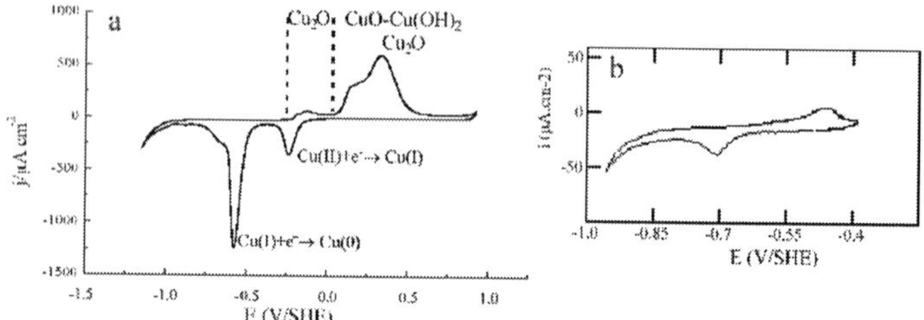

Figure 1. Voltammograms recorded for Cu(111) in 0.1 M $NaOH_{(aq)}$ between the hydrogen and oxygen evolution limits (a) and in the potential range below oxidation (b). The passive film consists of a single layer of Cu(I) anodic oxide and a duplex layer of Cu(I)/Cu(II) anodic oxide depending on the potential regions. (a) Reprinted with permission from *J. Phys. Chem. B* **105** (2001) 4263, Copyright © 2001, The American Chemical Society. (b) Reprinted with permission from *Surf. Sci.* **458** (2000) 185, Copyright © 2000, Elsevier Science.

layers is preceded at lower potential by the adsorption of hydroxide anions as identified by *in situ* Raman.[58,60]

Figure 1 shows typical polarization curves characterizing the 3D growth of the passive film on Cu(111) in 0.1 M $NaOH_{(aq)}$[36] and the $OH^-$ adsorption in the underpotential range of 3D oxidation.[35] It has been shown by ECSTM that $OH^-$ adsorption induces the surface reconstruction of the copper surface and the formation of a 2D structural precursor for the 3D Cu(I) oxides grown at higher potential.[35–38] Figure 2 shows this process for a Cu(111) single-crystal surface in 0.1 M $NaOH_{(aq)}$ on which 3D $Cu_2O(111)$ growth is observed at $E > -0.2\ V_{SHE}$ as described below.[35,37] The sequence of images shown in Fig. 2a was obtained after stepping anodically the potential to $-0.6\ V_{SHE}$, i.e., at the onset of the anodic peak observed at $-0.46\ V_{SHE}$ in Fig. 1b and assigned to the adsorption of $OH^-$ anions according to the following reaction:

$$Me + OH^- \rightarrow Me-OH^{\gamma-}_{ads} + (1-\gamma)\, e^- \tag{1}$$

with Me standing for Cu in the present case.

The initially bare and atomically smooth terraces of the surface (marked m. in Fig. 2a become progressively covered by 2D islands appearing darker on the image (marked ad.) that grow laterally with time and coalesce to cover completely the terraces.

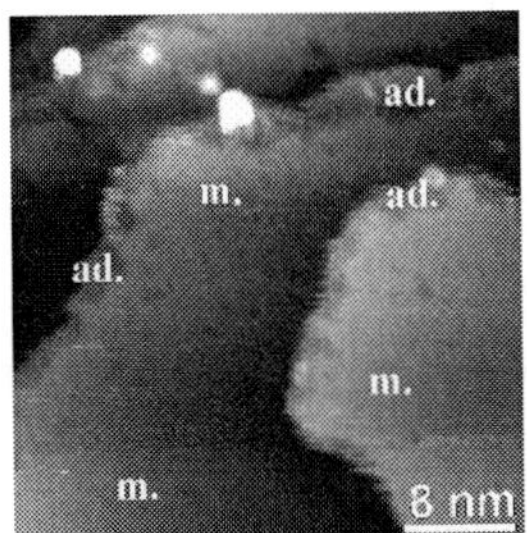


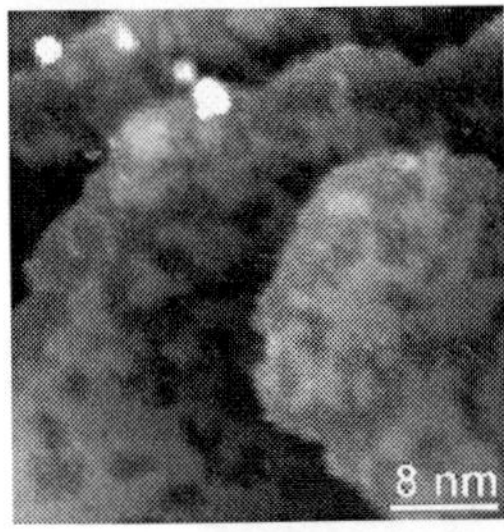


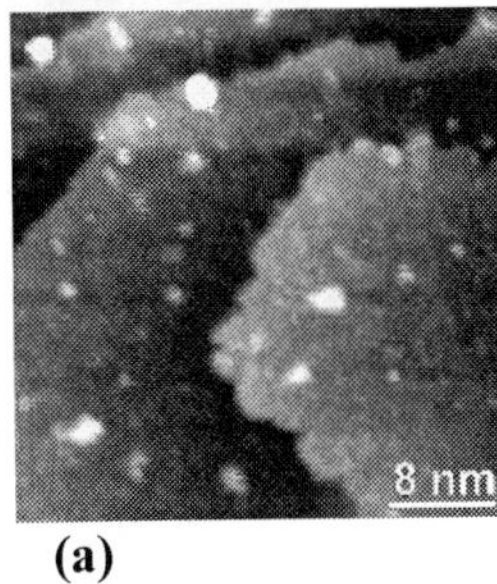


**(a)**

Figure 2. (a) Sequence of ECSTM images showing the growth of the $OH^{\gamma-}_{ads}$ ad-layer on Cu(111) in 0.1 M $NaOH_{(aq)}$ at $-0.6$ $V_{SHE}$. (b) ECSTM image and model of the ordered structure of $OH^{\gamma-}_{ads}$ formed on Cu(111) in 0.1 M $NaOH_{(aq)}$ in the potential region below oxidation. The large and small cells mark the lattice of $OH_{ads}$ species and of reconstructed $Cu_R$ plane, respectively. (a) Reprinted with permission from *Surf. Sci.* **458** (2000) 185, Copyright © 2000, Elsevier Science. (b) Reprinted with permission from *Electrochim. Acta* **48** (2003) 1157, Copyright © 2003, Elsevier Science.

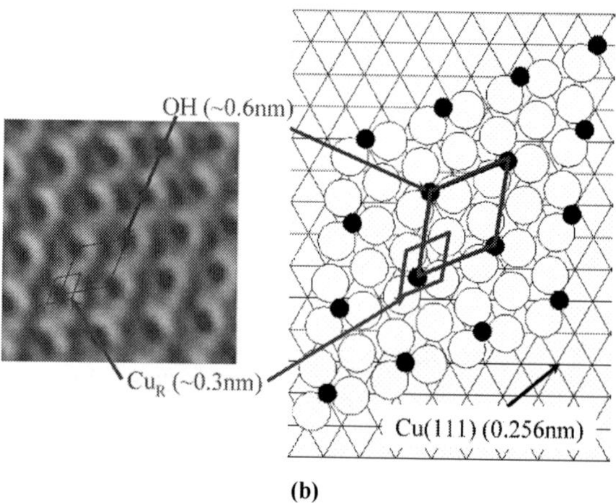


**(b)**

Figure 2. Continuation.

Initial growth of the adsorbed layer occurs preferentially at the step edges, confirming the preferential reactivity of these pre-existing defect sites of the surface, a generic result of Surface Science and Interfacial Electrochemistry studies of solid-gas and solid-liquid interfaces. Figure 2a also shows that the terraces grow laterally due to displacement of the step edges, and monoatomic protruding ad-islands are formed at the end of the growth process. These two features are indicative of the reconstruction of the top-most Cu plane induced by the adsorption of the $OH^-$ anions. The reconstruction causes the ejection of Cu atoms from the topmost plane. The ejected atoms diffuse on the surface and aggregate at step edges. This causes the observed lateral displacement of the step edges. In the final stages of the adsorption process where most of the surface is already covered by the 2D ad-layer, the ejected atoms have a reduced mobility on the OH-covered terraces and aggregate to form the observed monoatomic ad-islands. Some fraction of them may also dissolve in the electrolyte.

Figure 2b shows an atomically-resolved image and the model of the ordered surface according to a $Cu(111)/Cu_R/OH_{ads}$ topmost plane sequence. The reconstruction of the Cu topmost metal plane ($Cu_R$) is confirmed. A hexagonal lattice with a parameter of

0.6±0.02 nm is measured. Each unit cell contains one minimum and four maxima of intensity. The interatomic distance between the maxima (assigned to Cu atoms) is ~ 0.3 nm, which is larger than the interatomic spacing of 0.256 nm in the unreconstructed Cu(111) plane from bulk parameter values, and confirms the reconstruction of the topmost Cu plane into a lower density $Cu_R$ plane. A coverage of ~ 0.2 OH per Cu(111) atom is deduced from the density of the intensity minima, in excellent agreement with the coverage of 0.19 obtained from the electrochemical charge transfer measurements assuming $\gamma = 0$ in Eq. (1). In addition, the position of the $OH_{ads}$ groups corresponds to three-fold hollow sites of the reconstructed $Cu_R$ plane on which they form a (2×2) structure. The structure of the $Cu_R/OH_{ads}$ topmost plane mimics that of the Cu and O sub-lattices in $Cu_2O(111)$, thus forming a structural 2D precursor for the 3D growth of the anodic oxide.

Figure 3 shows the structure of the OH-induced reconstructed ad-layer formed on Cu(001) in the potential range below 3D oxidation.[38] It is assigned to OH-stabilized dimers of Cu atoms ejected from the topmost plane and located above it with the OH presumably adsorbed in bridge sites. The orientation of the dimers along the <100> substrate directions causes the reorientation of the step edges along these directions. By alternating the dimer orientation along the [100] and [010] directions on the terraces, zig-zag arrangements are generated with a main orientation along the <110>directions. Long range ordering of the zig-zag chains is observed in areas limited in width with c(2×6) and c(6×2) superstructure domains. In this case, the local orientation of the dimers along the [100] and [010] directions is thought to initiate a rotation of 45° of the close-packed [1$\overline{1}$0] directions of the 3D $Cu_2O$ (001) lattice with respect to the close-packed directions of the Cu(001) lattice.


## (*ii*) Silver

For silver in alkaline aqueous solutions, the adsorption of $OH^-$ anions at potentials cathodic with respect to the 3D growth of $Ag_2O$ has also been concluded from cyclic voltammetry, *in situ* ellipsometry, *in situ* Raman, *in situ* X-ray Absorption Spectroscopy at grazing incidence (XAS) and *ex situ* XPS data.[61-76] For

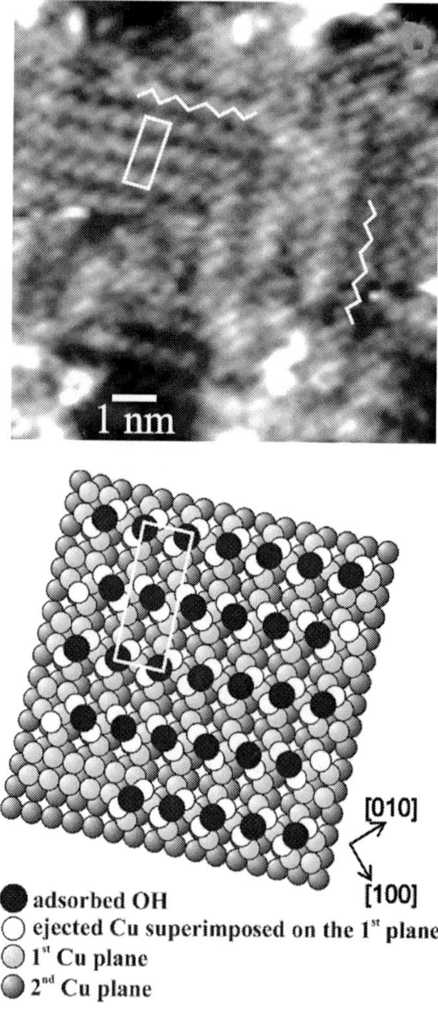



Figure 3. ECSTM image and model of the locally ordered structure of the $OH^{\gamma-}_{ads}$ layer formed on Cu(001) in 0.1 M $NaOH_{(aq)}$ in the potential region below oxidation. Reprinted with permission from *J. Electroanal. Chem.* **554-555** (2003) 113, Copyright © 2003, Elsevier Science.

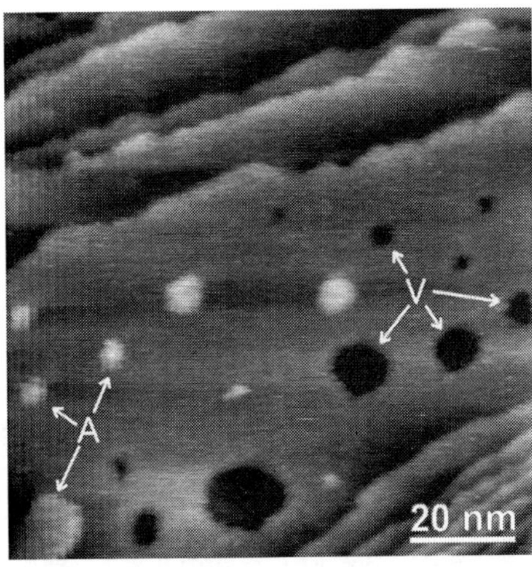


Figure 4. ECSTM image of the Ag(111) surface in
0.1 M NaOH(aq) at –0.3 $V_{SHE}$, showing the absence of
superstructure and high mobility (streaks) on the terraces. Ad-atoms and vacancy islands resulting from the
surface preparation are marked A and V, respectively.
Reprinted with permission from *J. Phys. Chem. C* **111**
(2007) 16351, Copyright © 2007, The American Chemical Society.

Ag(111) in 0.1 M $NaOH_{(aq)}$, OH-induced surface reconstruction
has been evidenced by ECSTM in the potential range preceding
the growth of 3D anodic oxides occurring at $E \geq 0.3$ $V_{SHE}$.[77,78]
However, surface reconstruction requires a potential ($E >$
–0.1 $V_{SHE}$) higher than that (~ –0.45 V) above which anodic peaks
assigned to $OH^-$ adsorption are measured by cyclic voltamme-
try[72,74,76] and the formation of $OH^{\gamma-}_{ads}$ species confirmed by *in situ*
Raman and *ex situ* XPS data.[68,75,76]

The ECSTM images measured in the potential range of
$-0.45 \leq E \leq -0.1$ V reveal atomically smooth terraces and no superstructure (Fig. 4), but the terraces are streaked at high resolution indicating surface mobility. This is assigned to the weak bonding of the $OH^{\gamma-}_{ads}$ species to the Ag(111) surface in this potential
range. This weak bonding and the associated surface mobility

would be consistent with an incomplete charge transfer in the adsorption reaction ($0 < \gamma < 1$).

Figure 5 shows the typical reconstructed terraces observed by ECSTM in the potential range $-0.05 \leq E \leq 0.15$ $V_{SHE}$. Dispersed protrusions (marked D) characterize a 2D reconstructed ad-layer

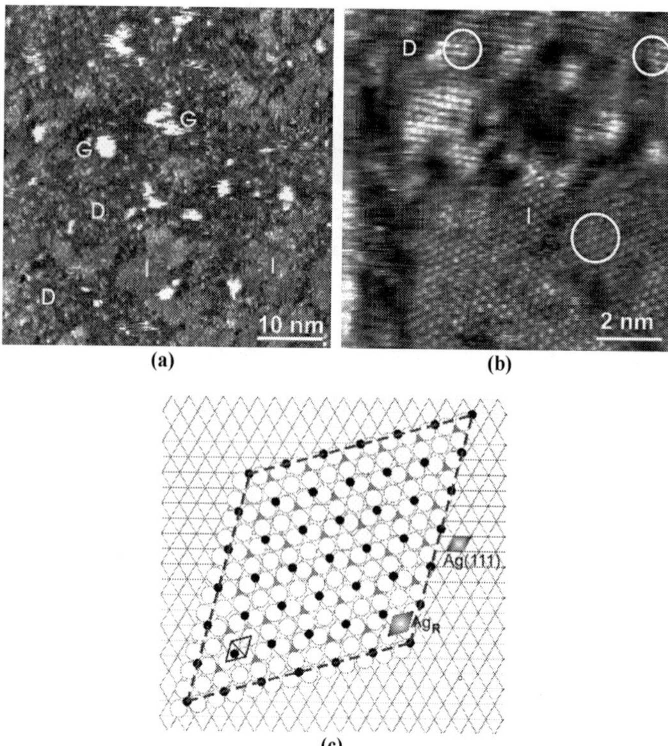

(a)          (b)

(c)

Figure 5. (a, b) ECSTM images of the Ag(111) surface in 0.1 M NaOH(aq) at -0.05 $V_{SHE}$. G, D and I mark grains, the matrix of dispersed protrusions and extended ordered islands, respectively. Circles indicate the atomic lattice. (c) Model of one monolayer of $Ag_2O(111)$ on Ag(111). Unit cells of the reconstructed topmost plane ($Ag_R$) and substrate plane Ag(111) are marked. Black disks and black circles are surface oxygen and silver in $Ag_2O$, respectively. Grey "disks" (partially hidden by the Ag layer) are sub-surface oxygen Reprinted with permission from *J. Phys. Chem. C* **111** (2007) 16351, Copyright © 2007, The American Chemical Society.

only locally ordered in contrast with the well-ordered 2D reconstructed ad-layer formed on Cu(111). These protrusions are assigned to nuclei of a (111)-oriented monolayer similar to silver oxide and consisting of one reconstructed silver layer embedded between two layers of sub-surface species and surface species, corresponding to oxide and/or hydroxide species according to Raman and XPS data.[68,75,76] Indeed, atomic resolution shows that the Ag-Ag distance ($\sim$ 0.32 nm) in the nuclei is enlarged and corresponds to the Ag-Ag interatomic distance in the $Ag_2O(111)$ lattice. This reconstruction of the silver plane in the nuclei of the new surface oxide is indicative of the growth of a 2D structural precursor of the 3D $Ag_2O$ anodic oxide film formed at higher potential ($E \geq 0.3$ $V_{SHE}$).

This model of surface reconstruction of Ag(111) in the underpotential range of oxidation is characterized by a $Ag(111)/O(H)_{sub}/Ag_R/O(H)_{ads}$ topmost plane sequence and differs from that proposed on Cu(111) where no sub-surface layer of adsorbates was proposed, in agreement with the absence of protruding islands on the smooth terraces measured by ECSTM after adsorption. This is attributed to the fact that on Cu(111) the formation of the 2D reconstructed ad-layer coincides with the anodic peak related to the adsorption of hydroxide and does not require a more anodic potential as on Ag(111). On Cu(111), the sub-surface penetration of the adsorbed species also occurs but at a higher potential ($E \sim$ $-0.25$ $V_{SHE}$) corresponding to the onset of the anodic peak marking the 3D growth of $Cu_2O(111)$.

On Ag(111), the $Ag_2O(111)$-like nuclei corresponding to the protruding islands are separated by unreconstructed regions where $OH_{ads}^{\gamma-}$ species are thought to predominate. Long range ordering is promoted by slow potential scans rather than fast potential steps, forming larger periodic islands (marked I in Fig. 5) of the reconstructed ad-layer. Grains ($\sim$ 3 nm wide, marked G in Fig. 5) assigned to 3D oxide nuclei are also observed simultaneously with surface reconstruction. A fraction of the Ag atoms ejected from the topmost plane by surface reconstruction are expected to aggregate in these nuclei due to a limited diffusion on the reconstructed surface. The remaining fraction would dissolve in the absence of scavenging pre-existing step edges, i.e., in the case of extended terraces.

### (*iii*)   Nickel

For Ni in sodium hydroxide aqueous solutions, the adsorption of hydroxide ions takes place at potential below the anodic peak marking the growth of 3D anodic oxides,[79-84] despite the absence of characteristics peaks in voltammograms.[81,84-86] Like on Cu and Ag, ECSTM on Ni(111) revealed OH adsorption-induced surface reconstruction leading to the nucleation and growth of a 2D passive layer,[24,25] but this occurs at a more anodic potential, concomitantly with dissolution as described below.

The ECSTM data obtained at the onset of the active/passive transition in 0.1 M $NaOH_{(aq)}$ evidence the formation of ordered (2×2) islands of limited lateral extension as shown in Fig. 6a.[25] However, this was not observed in 1 mM $NaOH_{(aq)}$. The absence of ordered islands in the more dilute electrolyte is explained by a weak bonding and a high mobility of the adsorbed hydroxide species (presumably $OH_{ads}^{\gamma-}$ with $\gamma \sim 1$). Figure 6c shows a model proposed for the dilute adlayer formed on the Ni(111) surface in this electrolyte. The weakly bonded $OH_{ads}^-$ species are thought to diffuse on the surface and can reside on 3-fold hollow, 2-fold bridge and atop sites as predicted by theoretical calculation of the Ni(111)/vacuum interface.[87] The co-adsorption of water molecules is also considered.

In 0.1 M $NaOH_{(aq)}$, the Ni(111) terraces are also expected to be covered by mobile $H_2O_{ads}$ and $OH_{ads}^-$ species. The higher concentration of $OH_{ads}^-$ species could favor a partial discharge reaction ($\gamma < 1$), strengthening the bonding of $OH_{ads}^{\gamma-}$ with the Ni atoms. Adsorption would be stabilized in the most preferred sites and induce the structural ordering observed by STM. Figure 6b shows the structural model of the (2×2) islands. It is assumed that the threefold hollow adsorption site (corresponding to the nodes of the hexagonal lattice in Fig. 6) is preferred, in agreement with theoretical calculations.[87] If the discharge of the $OH_{ads}^{\gamma-}$ species is incomplete ($0 < \gamma < 1$), the repulsive interactions between $OH_{ads}^{\gamma-}$ must be screened by co-adsorbed species such as $H_2O_{ads}$ in order to grow the (2×2) domains. Given the radii of $H_2O$ (0.15 nm) and $OH^-$ (0.13 nm), up to two water molecules can be co-adsorbed

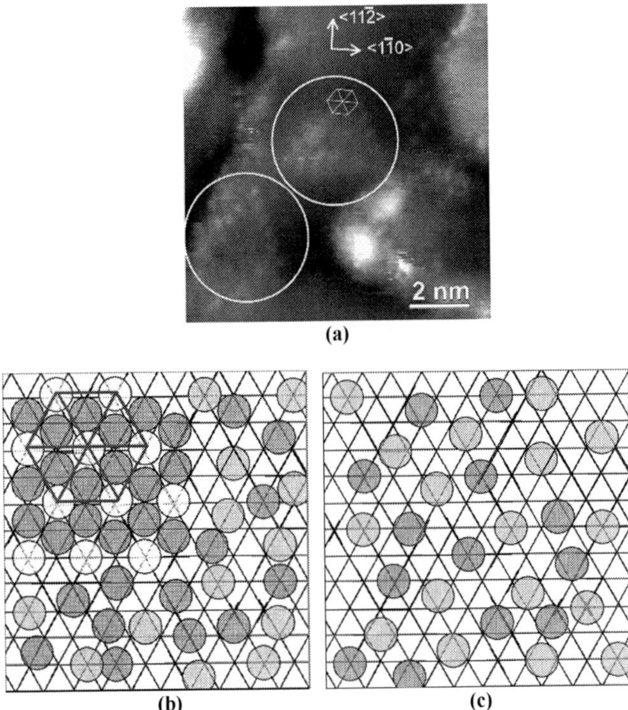


Figure 6. (a) ECSTM image of the Ni(111) terraces in 0.1 M NaOH(aq) at $-0.55$ $V_{SHE}$. Unreconstructed islands (circled) having a (2×2) superstructure (marked by the hexagon) are formed. (b) Model of the adlayer in 0.1 M $NaOH_{(aq)}$. The (2×2) ordered islands (marked by the hexagon) consist of strongly bonded $OH^{\gamma-}_{ads}$ species co-adsorbed with $H_2O_{ads}$. (c) Model of the disordered dilute adlayer of less strongly bonded and mobile $OH^-_{ads}$ co-adsorbed with $H_2O_{ads}$ in 1 mM NaOH(aq). Dark grey, bright grey and white disks mark $H_2O_{ads}$, $OH^-_{ads}$ and $OH^{\gamma-}_{ads}$, respectively. Reprinted with permission from *J. Electrochem. Soc.* **153** (2006) B453, Copyright © 2006, The Electrochemical Society.

with one $OH^{\gamma-}_{ads}$ species per (2×2) unit cell. The water molecules could contribute to stabilize the (2×2) arrangement by forming hydrogen bonds with the co-adsorbed $OH^{\gamma-}_{ads}$ species. The high

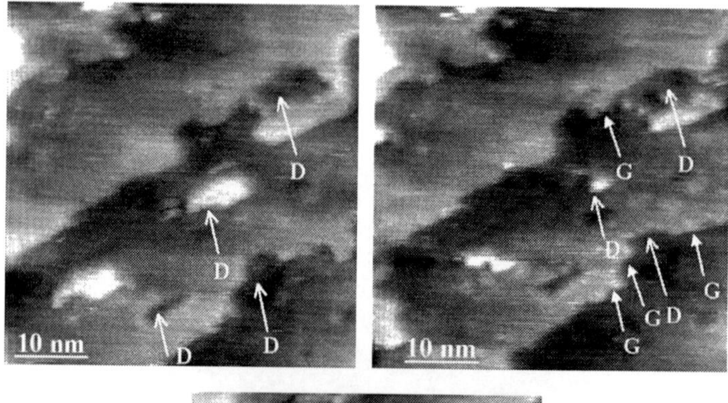

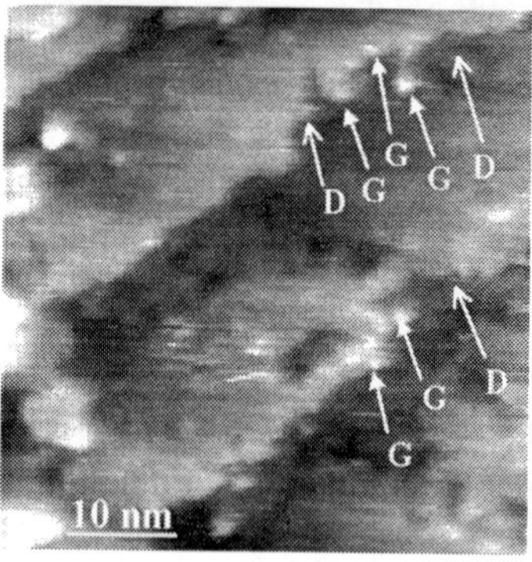

Figure 7. ECSTM images of the Ni(111) surface recorded in 1 mM $NaOH_{(aq)}$ at $E$ = −0.3 $V_{SHE}$, showing the step flow process generated by dissolution (sites marked D) and the nucleation of 2D grains of the passive film (marked G) at the step edges. Reprinted from *J. Solid State Electrochem.* **9** (2005) 337, Copyright © 2005, Springer.

packing density of this adlayer, with a nearest neighbor distance between co-adsorbed species of 0.29 nm, combined with a relatively weak bonding of the $OH_{ads}^{\gamma-}$ species is thought to be responsible for the limited lateral extension of the ordered domains.

Figure 7 illustrates the nucleation of the 2D passive film observed on Ni(111) in 1 mM $NaOH_{(aq)}$ (pH ~ 11) at $E$ = −0.3 $V_{SHE}$, i.e., also at the onset of the passivation peak. A slow dissolution of the nickel surface is observed. The dissolution takes place at the step edges, leading to the consumption of the metal terraces by a step flow mechanism, as observed on numerous substrates including nickel.[7,8,19,88] The nucleation of the passivating oxide occurs preferentially at the step edges at this potential. It is characterized by the formation of 2D grains with a lateral size of ~ 2 nm. The ongoing dissolution at the nearby step edges not yet passivated leads to the formation of isolated 2D islands of the passivating oxide.

At $E = -0.28$ $V_{SHE}$, dissolution of the metal terraces by step flow is still observed but the 2D growth of the passivating oxide is faster. It forms extended islands of 2D nanograins blocking the dissolution at the step edges. These 2D islands extend laterally with time to completely cover the surface. This leads to the formation of a passivating oxide constituted of 2D nanocrystals (~ 2 nm) (see Fig. 8). The nanocrystals have an hexagonal lattice with a parameter of $0.31 \pm 0.01$ nm assigned to 2D nanocrystals of $Ni(OH)_2(0001)$ possibly in strained epitaxy on $Ni(111)$.[24] Similar strained epitaxial 2D films have been observed on $Ni(001)$ in 0.1 M NaOH(aq.).[18] On $Ni(111)$ in sulfuric acid solution (pH ~ 3),[23] crystalline 2D islands with a lattice parameter consistent with $Ni(OH)_2(0001)$ have also been reported in the initial stage of

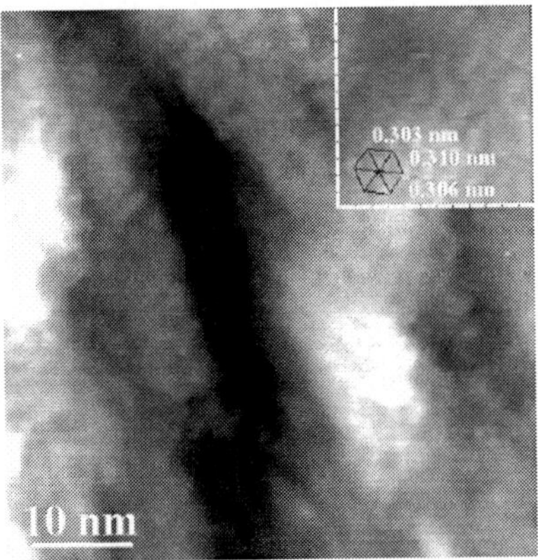


Figure 8. ECSTM images of Ni(111) covered by a 2D $Ni(OH)_2(0001)$ layer in 1 mM $NaOH_{(aq)}$ at $E = -0.28$ $V_{SHE}$. The insert shows the atomic lattice Reprinted from *J. Solid State Electrochem.* **9** (2005) 337, Copyright © 2005, Springer.

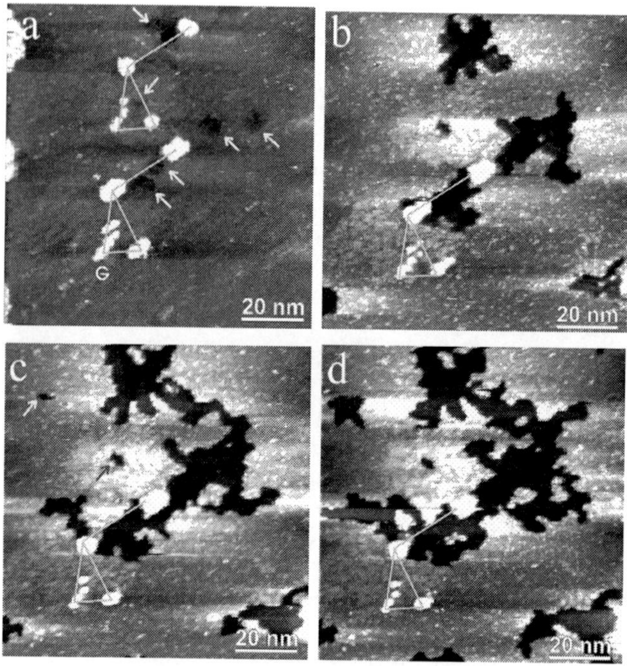

Figure 9. Sequence of ECSTM images of the Ag(111) surface covered by the 2D Ag$_2$O(111)-like ad-layer in 0.1 M NaOH$_{(aq)}$ at +0.15 V$_{SHE}$, showing the nucleation (pointed by arrows) and propagation of 2D pits. 3D oxide grains are marked. Reprinted with permission from *J. Phys. Chem. C* **111** (2007) 16351, Copyright © 2007, The American Chemical Society.

growth prior to complete coverage of the surface, suggesting that in both acid and alkaline solutions, nickel hydroxide predominates in the sub-monolayer regime of growth of the passive film.

## 2.   Influence of 2D Passive Films on Dissolution

### (*i*)   Silver

As described in Section II.1(*ii*), the 2D reconstructed ad-layer formed on Ag(111) has an heterogeneous structure with ordered Ag$_2$O(111)-like islands separated by unreconstructed areas where OH$_{ads}^{\gamma-}$ species likely predominate. Figure 9 shows a dissolution sequence observed when increasing the potential at $E = 0.15$ $V_{SHE}$, still below the potential $E \geq 0.3$ $V_{SHE}$ required for 3D oxide growth. The ECSTM data reveal the effect of the surface heterogeneity on the dissolution process.

Nucleation of 2D etch pits is observed on the terraces (Fig. 9a). It is assigned to the dissolution of the topmost metal atoms at the defective sites of the $Ag_2O(111)$-like ad-layer. The local chemical nature of these sites is not determined but it seems possible that the sub-surface relocation of OH adsorbates required to form the reconstructed ad-layer weakens the bonding of the topmost Ag atoms with their Ag nearest neighbors and leads to the formation of $[Ag(OH)_2]^-$ complexes that have been observed to dissolve by XAS for this system.[71] Once initiated, a pit grows by a 2D mechanism of propagation due to a preferential dissolution at the newly created step edges (Fig. 9b,c,d). Preferential dissolution at pre-existing step edges has been observed on numerous metals including silver,[7,8,89,90] and the 2D pit propagation observed on Ag(111) shows the prevalence of this mechanism also in the presence of the 2D $Ag_2O(111)$-like ad-layer.

Due to the poorly ordered structure of the ad-layer, the dissolving step edges are not aligned along well-defined crystallographic directions, in agreement with the step roughening effect induced by non-ordered adsorbed surface structures.[8] The lateral propagation of the dissolution reaction in the topmost atomic layer of the terraces is irregular since it is governed by the local resistance of the ad-layer. This leads to channels that isolate ordered fragments of the topmost terraces that are observed to resist to dissolution (Fig. 9c,d). The other sites where dissolution is blocked correspond to the 3D grains initially present at the surface and assigned to the Ag(I) oxide nuclei. This illustrates, at the local scale, the blocking property against dissolution provided by the growth of the anodic surface oxide.

Figure 10 shows a continuing dissolution sequence observed at $E = 0.15$ $V_{SHE}$ and illustrating the nucleation of 3D pits. As the dissolution of the topmost terraces progresses laterally, the second plane of the substrate at the bottom of the 2D pit is exposed to the electrolyte and reacts to form an heterogeneous ad-layer similar to that grown on the topmost plane (Fig. 10a). Again, 2D etch pits can nucleate at the defects of this newly formed ad-layer and propagate laterally in the second metal plane depending on the local

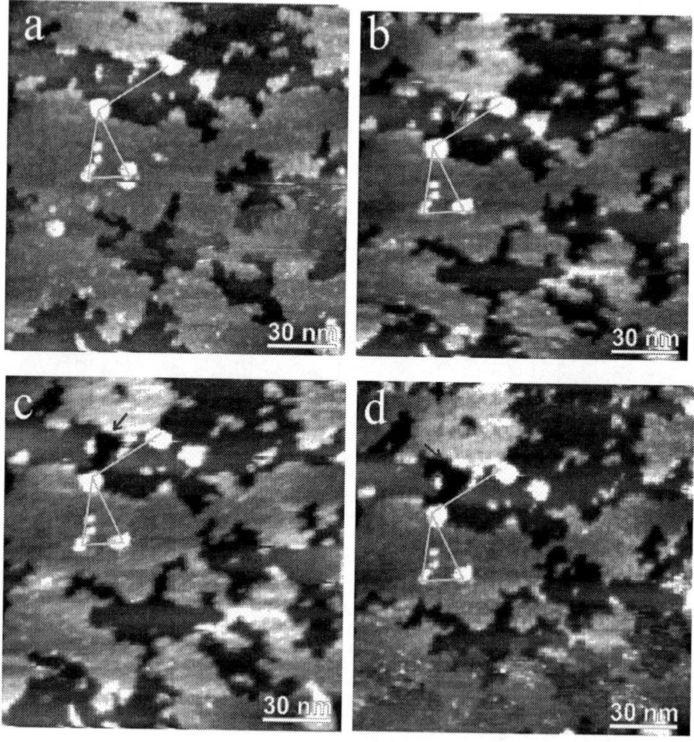

Figure 10. Sequence of ECSTM images of the Ag(111) surface in 0.1 M NaOH$_{(aq)}$ at +0.15 V, showing the initiation (pointed) and propagation of 3D pits. Reprinted with permission from *J. Phys. Chem. C* **111** (2007) 16351, Copyright © 2007, The American Chemical Society.

resistance of its heterogeneous structure (Fig. 10b,c,d). The images in Fig. 10c,d show that, at 0.15 V, the dissolution of the second substrate plane is stopped by the resistive fragments of the topmost plane. This corresponds to the encounter of a dissolving step edge in the second plane with a resistant step edge in the topmost plane and shows that dissolution does not propagate underneath the upper resistant plane. Consistently, dissolution of the second plane is also blocked by 3D oxide nuclei (Fig. 10d). This contributes further to the irregular growth of the pits in these conditions of polarization.

## (*ii*)  Nickel

Figure 11 shows a sequence of ECSTM images for Ni(111) in 0.1 M NaOH$_{(aq)}$ after prolonged polarization at –0.55 V$_{SHE}$ at the onset of the active/passive transition, also showing the effect of the heterogeneity of the 2D passive layer on the resistance to corrosion.[91] Resistant terraces and corroding terraces, marked R and C in Fig. 11a, respectively, characterize the surface in addition to the grains preferentially observed at the step edges and assigned to the 3D passive film nuclei. High resolution images show that the R terraces are covered by the ordered 2D layer of Ni(OH)$_2$(0001)

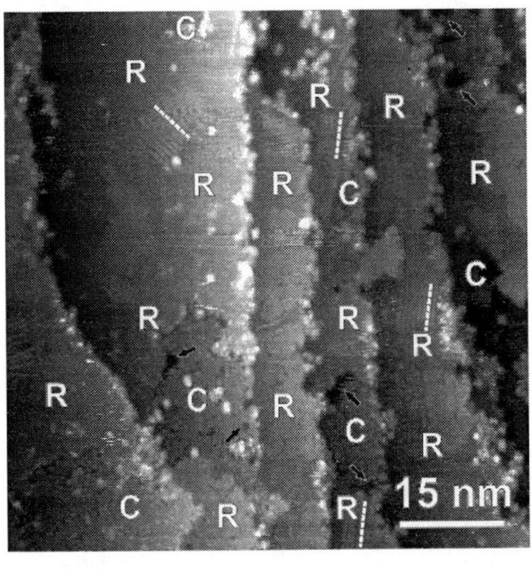


(a)

Figure 11. Sequence of topographic ECSTM images of the Ni(111) surface in 0.1 M NaOH$_{(aq)}$ recorded after prolonged polarization at -0.55 V$_{SHE}$; (a) 2800 s, (b) 2980 s and (c) 3100 s, showing the effect of the nanostructure of the 2D passive layers on the corrosion resistance. Resistant terraces (R) and corroding areas (C) are indicated. Arrows point the sites of initiation of dissolution whereas circles mark their lateral growth. Reprinted with permission from *Electrochim. Acta* (2008) in press, Copyright © 2008, Elsevier Science.

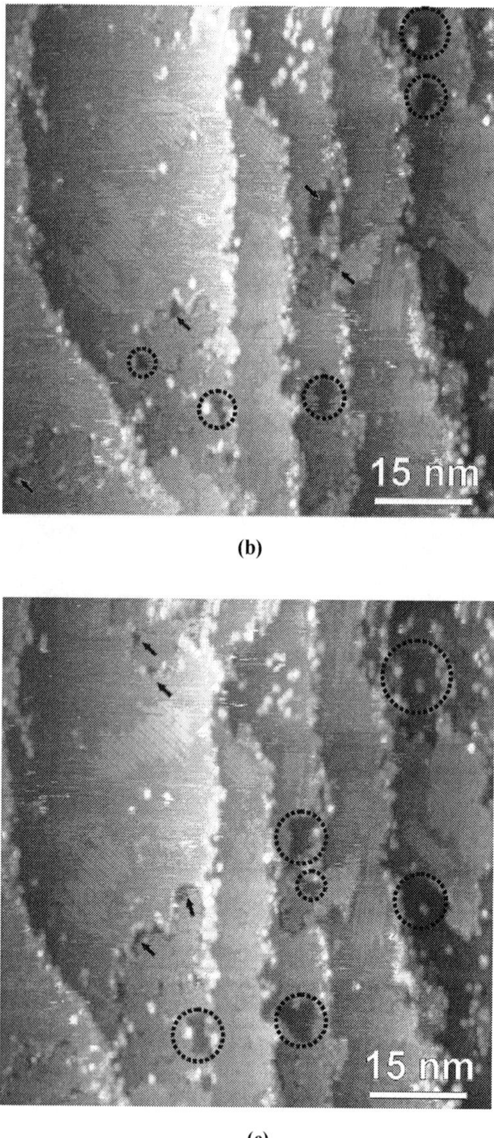

(b)

(c)

Figure 11. Continuation.

formed in this potential range and described above in Section II.1(*iii*). The measured height difference between the resistant (R) and corroding terraces (C) is ~ 0.07 nm (i.e., less than 1 atomic plane), indicating that the corroding areas are not bare but likely also covered by adsorbed particles assigned to $Ni(OH)_x$ species. However, a significant difference is the absence of superstructure indicating the absence of structural order in the corroding areas.

The time sequence in Fig. 11 evidences that the formation of the 2D pits is blocked at the step edges protected by the passive film nuclei. However, such pits measured to be from 0.24 to 0.35 nm deep nucleate on the terraces inside the C areas and at the boundaries between the R and C areas (as pointed by arrows), and grow laterally by dissolution of the terraces in the C areas (as indicated by circles). This is a direct evidence of the less resistive property of the non-ordered monolayer of $Ni(OH)_x$ particles in the corroding areas and at the boundaries with the ordered monolayer. The 2D pits are assigned to the dissolution of the $Ni(OH)_x$ particles, mobile on the metal terraces after detachment from the non-ordered adlayer formed in the corroding areas.

It is also observed that the C areas propagate into the R areas. This propagation suggests that, under the influence of the scanning tip, $Ni(OH)_x$ fragments can also detach from the boundaries of the ordered monolayer to aggregate with the non-ordered monolayer.

These data show that the dissolution sites, commonly observed to be the pre-existing step edges at moderate over-potential in the active region,[7,8] are transferred to the substrate terraces in the presence of passive film nuclei blocking the step edges. Thus, from a step flow process depending on the local structure of the pre-existing step edges, the dissolution mechanism becomes dependent of the local structure of the terraces. A major output of the data presented in this Section is the direct evidence of the effect of surface structural order in the passivating monolayer on the corrosion resistance properties. For silver, it was observed that 2D pits could be initiated in the terraces in the less-resistant sites of the 2D anodic oxide, and could propagate in the terraces by dissolution at the newly created step edges but were blocked by the well-ordered regions of the 2D oxide ad-layer and by the 3D oxide nuclei. These results are confirmed for nickel. In addition, it is shown that not only the core of the disordered region but also the boundaries of the ordered regions are preferential sites of dissolution due to the detachment of the particles less strongly bonded to the passivating ad-layer. On Ag, it is observed that 3D pits could initiate and propagate in the second metal plane by the same mechanism. This mechanism observed on Ni for the topmost plane can be repeated in the metal planes newly exposed to the electrolyte after dissolution, thus allowing the 3D growth of the pits to occur after more prolonged polarization at the same potential.

## III.    STRUCTURE OF 3D PASSIVE FILMS ON METALS AND ALLOYS

Data on copper, nickel, iron, chromium and stainless steels are reviewed in this Section to illustrate the structure of 3D passive films and the effect of ageing on their crystallization.

### 1.   Crystalline Passive Films

### (*i*)  Copper

Figure 12 illustrates the influence of the oversaturation potential on the growth, crystallization and structure of the $Cu_2O$ oxide film formed in the potential range of Cu(I) oxidation.[36–38] At low oversaturation (see Fig. 12a obtained on Cu(111) at $E = -0.25$ $V_{SHE}$), poorly crystallized and one monolayer thick islands covering partially the substrate are formed after preferential nucleation at step edges. They are separated by islands of the ordered hydroxide ad-layer described above. At higher oversaturation (see Fig. 12b obtained on Cu(111) at $E = -0.2$ $V_{SHE}$), well crystallized and several monolayer thick films are formed. The equivalent thickness of the oxide layer can be deduced from subsequent measurements of the charge transfer during cathodic reduction scans. It was ~ 0.5 and 7 equivalent monolayers (ML, one ML corresponds to one (111)-oriented $O^{2-}$-$Cu^+$-$O^{2-}$ slab) after growth at –0.25 and –0.2 $V_{SHE}$, respectively. The observed lattice of the oxide layer (insert in Fig. 12b) is hexagonal with a parameter of ~ 0.3 nm, consistent with the Cu sub-lattice in the (111)-oriented cuprite. The oxide grows in parallel (or anti-parallel) epitaxy ($Cu_2O(111)$ $[1\,\bar{1}\,0] \parallel$ Cu(111) $[1\,\bar{1}\,0]$ or $[\bar{1}\,10]$).

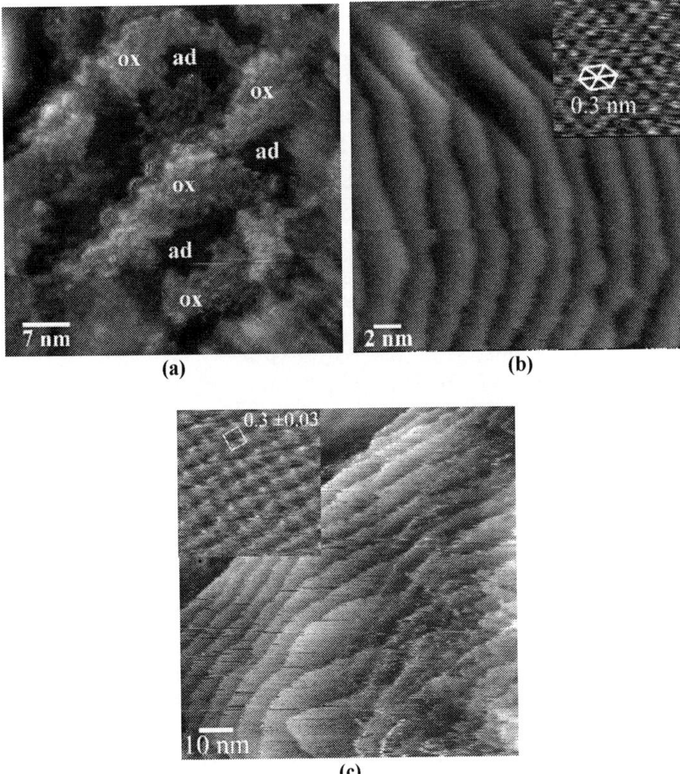

Figure 12. ECSTM images of the Cu(I) oxide grown on Cu(111) at –0.25 $V_{SHE}$ (a) and –0.20 $V_{SHE}$ (b), and on Cu(001) at –0.11 $V_{SHE}$ (c) in 0.1 M $NaOH_{(aq)}$. At low oversaturation (a), non-crystalline 2D oxide islands (ox.) separated by the adsorbed OH layer (ad) cover partially the substrate. At higher oversaturation (b,c), a 3D crystalline oxide layer fully covers the substrate. Its atomic lattice, shown in the inset, corresponds to that of $Cu_2O(111)$ (b) and $Cu_2O(001)$ (c) on Cu(111) and Cu(001), respectively. (a,b) Reprinted with permission from *J. Phys. Chem. B* **105** (2001) 4263, Copyright © 2001, The American Chemical Society. (c) Reprinted with permission from *J. Electroanal. Chem.* **554-555** (2003) 113, Copyright © 2003, Elsevier Science.

On Cu(001) (see Fig. 12c), the oxide layer, also a few ML thick, has a square symmetry and the same periodicity of 0.3 nm, consistent with the Cu-sub-lattice of (001)-oriented $Cu_2O$. The rotation of 45° of the close-packed $[1\bar{1}0]$ direction of the oxide lattice with respect to the close-packed direction of the Cu(001) lattice gives an epitaxial relationship noted as $Cu_2O(001)$ $[1\bar{1}0] \parallel Cu(001)[100]$. The 45° rotation is initiated by the formation of the adsorbed layer of OH groups described above, where the dimers of Cu atoms stabilized by bridging OH groups are aligned along the [100] and [010] directions.

On both substrates, the crystalline Cu(I) oxide layers have a nanostructured and faceted surface as shown in Fig. 12b and c. The surface faceting results from a tilt of a few degrees of the orientation of the oxide lattices with respect to the Cu lattice (see Fig. 13). The tilt is assigned to the relaxation of the epitaxial stress in the metal/oxide interface resulting from the large mismatch between the two lattices (17%). The height of the surface steps of the oxide layers corresponds to 1 ML of cuprite, indicating an identical chemical termination of the $Cu_2O(111)$ and $Cu_2O(001)$ oxide terraces. It is thought that the surface of the oxide layer is hydroxylated in the aqueous solution and that the measured lattice corresponds to OH and/or $OH^-$ groups forming a (1x1) layer on the $Cu^+$ planes of the (111)- and (001)-oriented cuprite layers.

Crystalline Cu(I)/Cu(II) duplex passive films having a total thickness of 4-5 nm are formed in the potential range of Cu(II) oxidation in 0.1 M $NaOH$.[39] On both Cu(111) and Cu(001), a terrace and step topography of the passivated surfaces is observed with terraces extending up to 20 nm in width and with a step height of $\sim 0.25$ nm corresponding to the thickness of one equivalent monolayer of CuO(001). Accordingly, the atomic lattices observed are consistent with a bulk-terminated CuO(001) surface which is characterized by a distorted hexagonal symmetry with in-plane nearest neighbor distances of $\sim 0.28$ nm along the closed packed directions (see Fig.14). The epitaxial relationships for the duplex layers are: $CuO(001)[\bar{1}10] \parallel Cu_2O(111)[1\bar{1}0] \parallel Cu(111)$ $[1\bar{1}0]$ or $[\bar{1}10]$ and $CuO(001)[\bar{1}10] \parallel Cu_2O(001)$ $[1\bar{1}0] \parallel Cu(001)[100]$, corresponding in both cases to the parallel alignment of the closed packed directions of the CuO and $Cu_2O$ lattices.

The common (001) orientation of the CuO outer layers of the duplex films on the two substrates is assigned to their surface hydroxylation at the passive film/electrolyte interface, necessary to stabilize the bulk-like termination of CuO(001) which is polar and

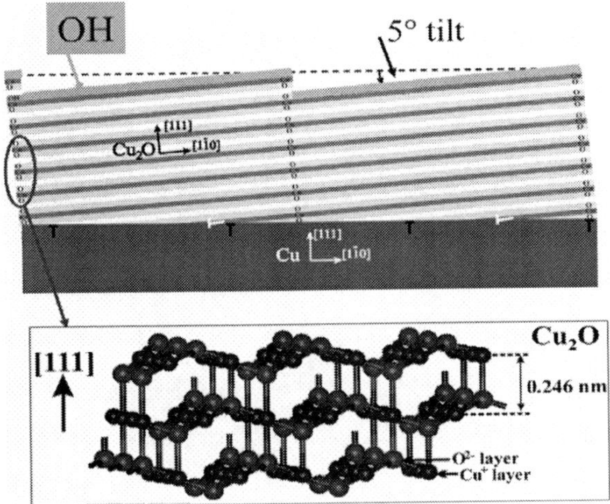


Figure 13. Model (section view) of the 7 ML thick Cu(I) oxide grown in tilted epitaxy on Cu(111). The stacking sequence of the $O^{2-}$ and $Cu^+$ planes in the oxide is illustrated by the gray lines. The oxide surface is terminated by a monolayer of hydroxyl/hydroxide groups. The faceted surface corresponds to a tilt of ~ 5° between the two lattices. Interfacial misfit dislocations and misorientation dislocations are illustrated by the T and ⊢— symbols, respectively. White and black symbols correspond respectively to dislocations of the oxide and metal lattices. Adapted with permission from *J. Phys. Chem. B* **105** (2001) 4263, Copyright © 2001, The American Chemical Society.

unstable when anhydrous. This assignment is supported by the step height measurements that indicate an identical chemical termination of all terraces. It is proposed that the surface is terminated by an $OH^-$ (or OH) layer in (1x1) registry on the topmost plane of the $Cu^{2+}$ sub-lattice, then $O^{2-}$ and $Cu^{2+}$ planes alternate towards the bulk of the oxide layer.

## (*ii*) Nickel

For Ni(111) in 1 mM NaOH at $E \geq -0.13$ $V_{SHE}$, i.e., beyond the top of the anodic peak corresponding to the growth of the passivating oxide, the 3D growth of the passive film is observed (see

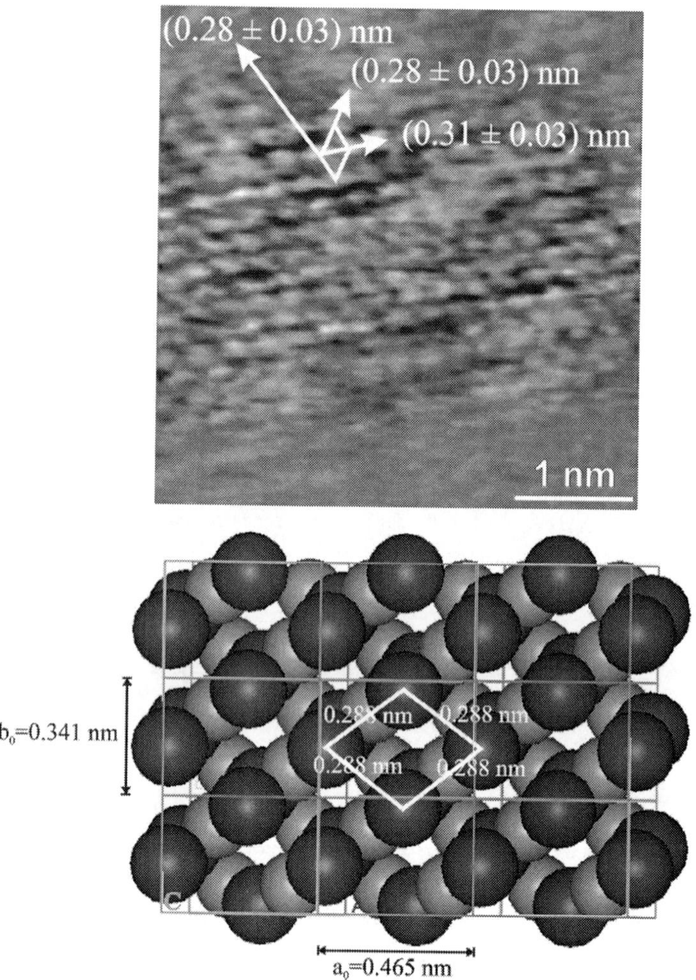


Figure 14. ECSTM image of the atomic lattice recorded on the terraces of the 3D Cu(II) anodic oxide formed on Cu(111) in 0.1 M NaOH(aq) at 0.83 $V_{SHE}$, and top view of the (001) face of CuO. The dark disks represent the $O^{2-}$, the bright disks $Cu^{2+}$. Reprinted with permission from *Corrosion Sci.* **46** (2004) 245, Copyright © 2004, Elsevier Science.

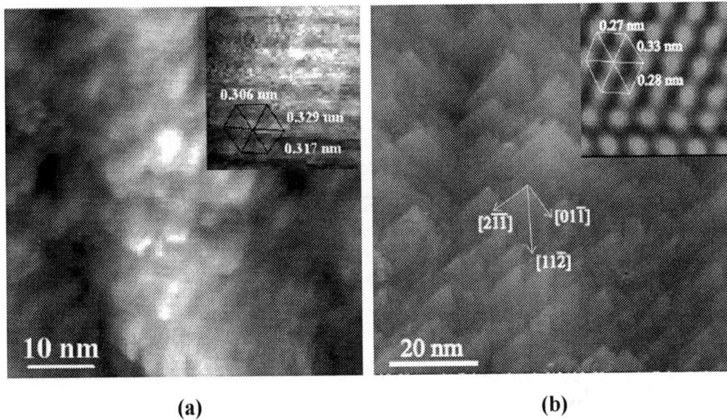


(a)                                                    (b)

Figure 15. ECSTM images of Ni(111) surfaces passivated (a) in 1 mM NaOH$_{(aq)}$ (pH ~ 11) at $E = -0.13$ V$_{SHE}$ and (b) in 0.05M H$_2$SO$_4$ + 0.095 M NaOH$_{(aq)}$ (pH ~ 2.9) at $E = +0.95$ V$_{SHE}$. The inserts show the high resolution images revealing the atomic lattices. (a) Reprinted from *J. Solid State Electrochem.* 9 (2005) 337, Copyright © 2005, Springer. (b) Reprinted with permission from *J. Electrochem. Soc.* 147 (2000) 1393, Copyright © 2000, The Electrochemical Society.

Fig. 15a). Its thickness can be estimated to increase from 1.5 to 4.5 nm with increasing potential based on data obtained in 0.5 M NaOH.[82] As on copper, it is characterized by the formation of a faceted topography indicative of the tilt between the oxide lattice of the passive film and the lattice of the substrate. The hexagonal lattice at –0.13 V and 0.22 V has a parameter of 0.32 ± 0.01 nm, slightly larger than that measured at –0.28 V when a 2D passivating oxide is formed. The parameter is in better agreement with the value of 0.317 nm expected for an unstrained 3D layer assigned to β-Ni(OH)$_2$(0001). This is consistent with a crystalline 3D outer hydroxide layer of the passive film formed in alkaline electrolytes, as opposed to the 2D outer hydroxide layer existing on the inner oxide layer in acid electrolytes.[80,82,84,92-103]

Figure 15b illustrates the typical topography of a Ni(111) single-crytal surface polarized in the middle of the passive domain in a sulfuric acid solution (pH ~ 2.9).[20] A thinner film, increasing up to 2 nm with increasing potential, is formed in this case.[20,22,48,83] The passivated surface is also faceted exhibiting terraces and steps. The presence of terraces at the surface of the crystallized passive film is indicative of a slightly tilted epitaxy between the NiO lattice forming the barrier oxide layer, and the Ni(111) substrate terraces. It has been proposed that this tilt partly relax the interfacial stress associated with the mismatch of 16% between the two lattices. This tilt has been confirmed by grazing X-ray measurements on Ni(111).[48] A similar surface faceting is observed for the Cu(I) oxide layer grown on copper (as described above) for which the lattice misfit between $Cu_2O$ and Cu lattices is similar.[36-38] The lattice measured on the terraces is hexagonal with a parameter of 0.3 ± 0.02 nm assigned to NiO(111). It must be pointed out that the (111) surface of NiO which has the NaCl structure is normally polar and unstable. It is however the surface which is obtained by passivation. The reason for this is that the surface is stabilized by adsorption of a monolayer of hydroxyl groups or by the presence of a monolayer of $\beta$-Ni(OH)$_2$ in parallel epitaxy with the NiO surface. These data show that the direction of growth of the oxide film is governed, at least in part, by the minimization of the oxide surface energy by the hydroxyl/hydroxide groups. The presence of water is thus a major factor for the structural aspects of the growth mechanism.

The lateral dimensions of the crystalline grains forming the passive films on nickel are relatively well documented. Values determined from the morphology observed by STM and AFM have been reported to range from ~ 2 nm in the initial stages of 2D growth to 30 to 230 nm for 3D films in stationary conditions of passivity.[17,20,22,24,25,104] A large dispersion could be found on the same sample, suggesting a varying degree of advancement of the coalescence of the oxide grains during the nucleation and growth mechanism. A lower average value of ~ 8 nm for the NiO(111) single-crystal domain size was obtained from GIXD data.[22,48] This difference may be assigned to the fact that STM and AFM measurements are unable to resolve the multiple twin or grain boundaries that can characterize the inner part of the passive film without markedly affecting the surface topography.


### (*iii*)   Iron


On iron, the crystallinity of the passive film formed in borate buffer solution (pH = 8.4) has also been evidenced by STM[29] and AFM.[30] The film thickness increases up to ~ 2 nm with increasing potential.[105] The *in situ* GIXD data have been fitted with a structure having the same sub-lattice of O anions as in $\gamma$-$Fe_2O_3$ and $Fe_3O_4$ but with different occupancies of the cation sites, thus forming the so-called LAMM phase.[49,50] The unit cell contains 32 oxide anions. 66 ± 10% of the 8 tetrahedral cation sites are occupied by divalent cations while 80 ± 10% of the 16 octahedral cation sites are occupied by trivalent cations. The occupancy of the octahedral interstitial by cations is 12 ± 4% and that of the tetrahedral interstitial is 0%.

Both the GIXD and STM data have shown that the film is nanocrystalline with numerous defects. The lateral grain size was determined to be 5 nm and 8 nm from the *in situ* GIXD data obtained on passivated Fe(110) and Fe(100) surfaces, respectively.[49,50] A value of 5 nm was obtained from *ex situ* STM data on passivated sputter-deposited pure iron films.[32] No beneficial effect of ageing under applied potential has been reported in these studies. However, investigation by *ex situ* STM as a function of the polarization potential for short polarization time ($t = 15$ min) revealed a coarsening of the crystalline terraces to values larger than 6 nm, suggesting a beneficial effect of increasing potential on the film crystallization for short polarization periods.[33] The lattice constant of the ordered terraces observed by STM was 0.32 ± 0.02 nm, in good agreement with the parameter of the (111)-oriented oxygen sub-lattice of the LAMM measured by GIXD (0.297 nm).

## 2.   Ageing Effects on Chromium-Rich Passive Films

### (*i*)   Chromium

The structure of chromium-rich passive films has also been investigated on chromium[26,27] and on ferritic and austenitic stainless steels.[42-46] A major finding for these systems is that potential and ageing under polarization are critical factors for the development of crystalline passive films and for the dehydroxylation of the passive film as evidenced by combined XPS and STM measurements.[26,44,46]

On chromium, the total thickness of the passive film formed in acid solution (0.5 M $H_2SO_{4(aq)}$) does not vary markedly with increasing potential or ageing (from 1.3 to 1.8 nm), as shown by XPS.[26] However, potential and ageing favor the development of the inner oxide layer. The film is mostly disordered at low potential where the oxide inner layer is not fully developed and where the passive film is highly hydrated and consists mainly of hydroxide. This supports the view that the passive film on chromium can consist of small $Cr_2O_3$ nanocrystals buried in a disordered chromium hydroxide matrix. At high potential where the inner part of the passive film is dehydrated and consists mostly of chromium oxide, larger crystals having a faceted topography and extending over several tens of nanometers have been observed by ECSTM.[27] The nanocrystals have a lattice consistent with the O sub-lattice in $\alpha$-$Cr_2O_3$(0001). The basal plane of the oxide is parallel to Cr(110). A special feature of the passive film on chromium is that the oxide nanocrystals can be cemented together by the chromium hydroxide disordered outer layer. It has been suggested[26] that the role of cement between grains played by chromium hydroxide, and of course the high stability of chromium oxide and chromium hydroxide, make this passive film extremely protective against corrosion.

### (*ii*)  Stainless Steels

For the passive layers formed on stainless steels in acid aqueous solution, the total thickness is the range 1.5–2 nm,[44,46] similar to that obtained on pure chromium. It has been observed that, for short polarization times ($\leq 2$ h), the crystallinity of the passive films decreases with increasing Cr content of the alloy.[42,43] Structural changes also occur during ageing under anodic polarization. The major modification is an increase of the crystallinity of the film and the coalescence of $Cr_2O_3$ nanocrystals in the inner oxide as observed on Fe-22Cr[44] and Fe-18Cr-13Ni alloys[46] studied over time periods of up to 65 h. This is illustrated by the images shown in Fig. 16.

The comparison of the rates of crystallization of Fe-22Cr(110) and Fe-18Cr-13Ni(100) revealed that the rate of crystallization is more rapid on the austenitic stainless steel than on the ferritic one. This was tentatively explained by a regulating effect of Ni on the supply of Cr on the alloy surface, a lower rate of Cr enrichment being in favor of a higher degree of crystallinity.[106]

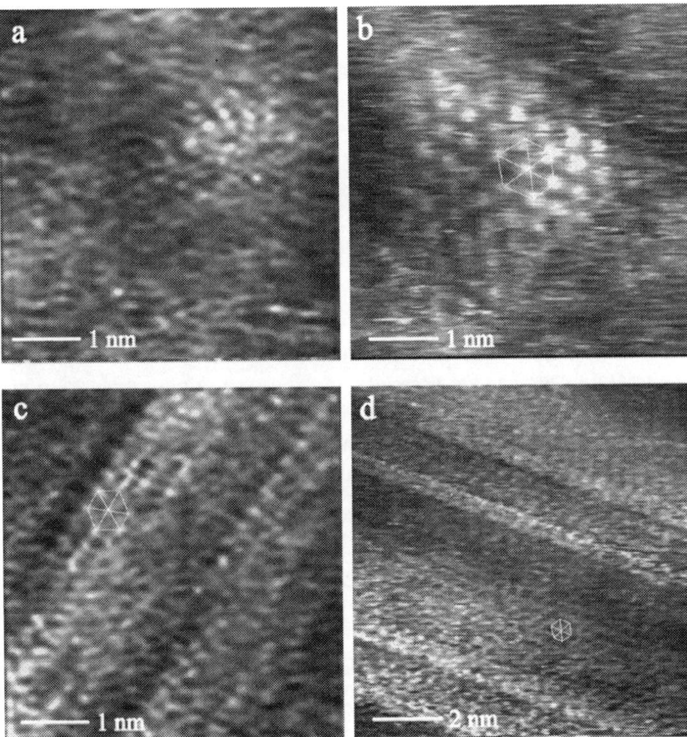

Figure 16. STM images of the Fe-22Cr(110) (left) and Fe-18Cr-13Ni(100) (right) surfaces recorded after passivation in 0.5 M $H_2SO_{4(aq)}$ at +0.5 $V_{SHE}$ for 2 h (a), (b) and for 22 h (c), (d). The nearly hexagonal lattice is marked. The effect of ageing under polarization is evidenced by the extension of the observed crystalline areas. (a,c) Reprinted with permission from *J. Electrochem. Soc.* **143** (1996) 1182, Copyright © 1996, The Electrochemical Society. (b,d) Reprinted with permission from *J. Electrochem. Soc.* **145** (1998) 909, Copyright © 1998, The Electrochemical Society.

## IV.　　DISSOLUTION IN THE PASSIVE STATE

This Section is focused on the effects of the passive film nanostructure on the dynamics of dissolution in the passive state.

The presence of crystalline defects (i.e., step edges) at the surface of the grains of the passive film plays a key role in the dissolution in the passive state as shown in Fig. 17. The data were obtained on Ni(111) surfaces passivated in a chloride-free sulfuric acid solution (pH = 2.9) to produce the characteristic surface described above.[109] Chlorides (0.05M NaCl) were subsequently introduced in the electrolyte without changing the pH. The sequence of three ECSTM images in Fig. 17a,b,c shows that the passive film dissolves at the edges of the facets resulting from the tilted epitaxy between the NiO(111) oxide and the Ni(111) substrate lattice. The observed dissolution is a 2D process leading to a progressively decreasing size of the dissolving facets by a step flow process. The process is similar to that of active dissolution of metal surfaces at moderate potential with no pit forming. The 2D step flow process is dependent on the step orientation: the step edges oriented along the closed-packed directions of the oxide lattice dissolve much less rapidly, due to the higher coordination of their surface atoms. This process leads to the stabilization of the facets with edges oriented along the close-packed directions of the oxide lattice, and produces steps that are oriented along the {100} planes, the most stable orientations of the NiO structure (see Fig. 17d).

The average dissolution rate of the facets, measured to be 0.44 $\pm$ 0.25 nm$^2$ s$^{-1}$ from such measurements, does not vary significantly with applied potential below or above the pitting potential value ($\sim$ 0.9 V$_{SHE}$).[104,109] This indicates that the potential drop at the passive film/electrolyte interface, associated with the surface reaction of dissolution of the oxide film, remains constant above the oxide grains. The average dissolution rate does not significantly decrease in the absence of chlorides in the electrolyte,[104,109] showing that the surface reaction of dissolution at the regular step edges at the surface of the nanograins of the well-passivated surface is not accelerated by the presence of chloride. The chlorides may however prevent the stabilization of the dissolving oxide film in the more disoriented steps forming the grain boundaries of the passive film where metastable pits are formed as shown below in Section V.2. Consistently with the absence of effect of chloride on the dissolution rate, the atomic lattice of the passive film formed in the absence of chloride is unmodified after the addition of chloride.[108]

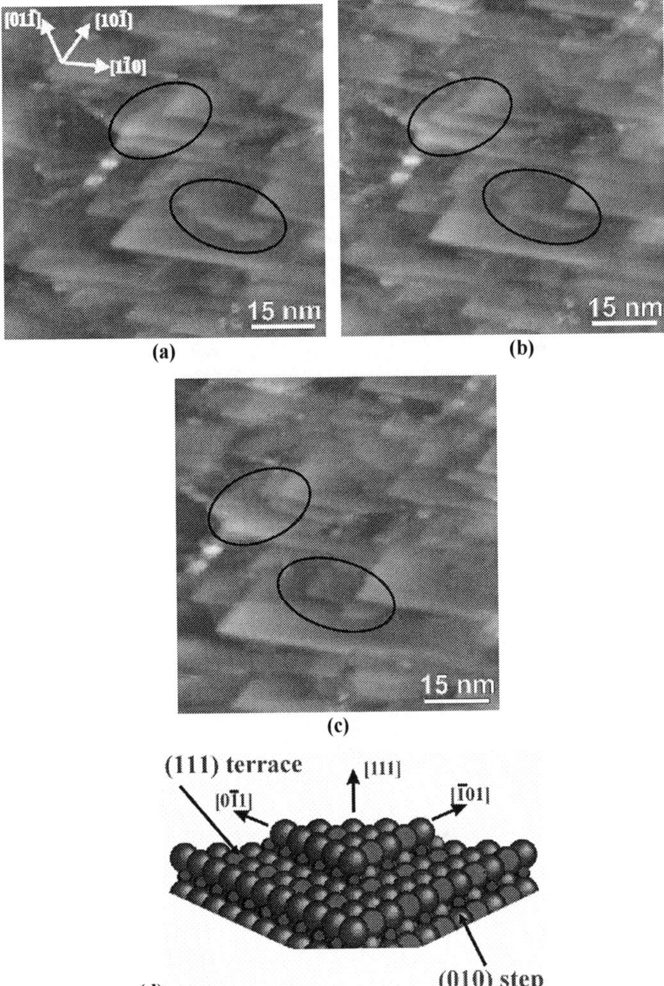


Figure 17. (a,b,c) Sequence of ECSTM images showing the localized dissolution of the passivated Ni(111) surface at +0.85 $V_{SHE}$ in 0.05M $H_2SO_4$ + 0.095M NaOH + 0.05M NaCl (pH = 2.9). The oxide crystallographic directions are indicated. The circles show the areas of localized dissolution. (d) Model of a (111) oriented facet delimited by edges oriented the close-packed directions of the NiO lattice. Oxygen (surface hydroxyls) and Ni are represented by large and small sheres, respectively. Reprinted with permission from *Surf. Interf. Anal.* **34** (2002) 139, Copyright © 2006, Wiley.

## V.  NANOSTRUCTURE OF PASSIVE FILM AND PASSIVITY BREAKDOWN

The understanding and control of localized corrosion (pitting) require to study at the (sub-)nanometer scale the mechanisms of initiation, including the role of chlorides on the growth and structure of passive films, and passivity breakdown.

In this Section, the relation between the nanostructure of 3D passive films and the properties of resistance to breakdown is addressed by reviewing recent STM and AFM data obtained on model nickel surfaces[25,104,107-109] free of defects related to the microstructure of the substrate (inclusions, second phase particles, grain boundaries). First, the modifications of the growth and structure induced by the presence of aggressive (i.e., chloride) ions in the aqueous environment are illustrated by data obtained on nickel in alkaline solutions. Second, the nanostructure modifications at inter-granular sites of passive films grown in the absence of chlorides and subsequently exposed to chlorides are discussed. A model of passivity breakdown emphasizing the role of inter-granular defects will be presented in Section VI.

### 1.  Effect of Chlorides on Coalescence and Crystallization of the Passive Film

Figure 18 shows topographic ECSTM images that evidence the effect of the increasing concentration of chlorides on the structure of the passive film on Ni(111).[25] In the absence of chlorides (Fig. 18a obtained in 0.1 M $NaOH_{(aq)}$ at $-0.175$ $V_{SHE}$, i.e., above the active-passive transition at $-0.44$ $V_{SHE}$) and in presence of a relatively low concentration of chlorides (Fig. 18b obtained in 0.1 M $NaOH_{(aq)}$ + 0.05 M NaCl at $-0.05$ V, i.e., also above the active-passive transition at $-0.39$ $V_{SHE}$), the images reveal a faceted topography characteristic of the crystallization of the passive film. Atomically resolved images revealing the crystalline lattice at the surface of the passive film could not be obtained in these elec-

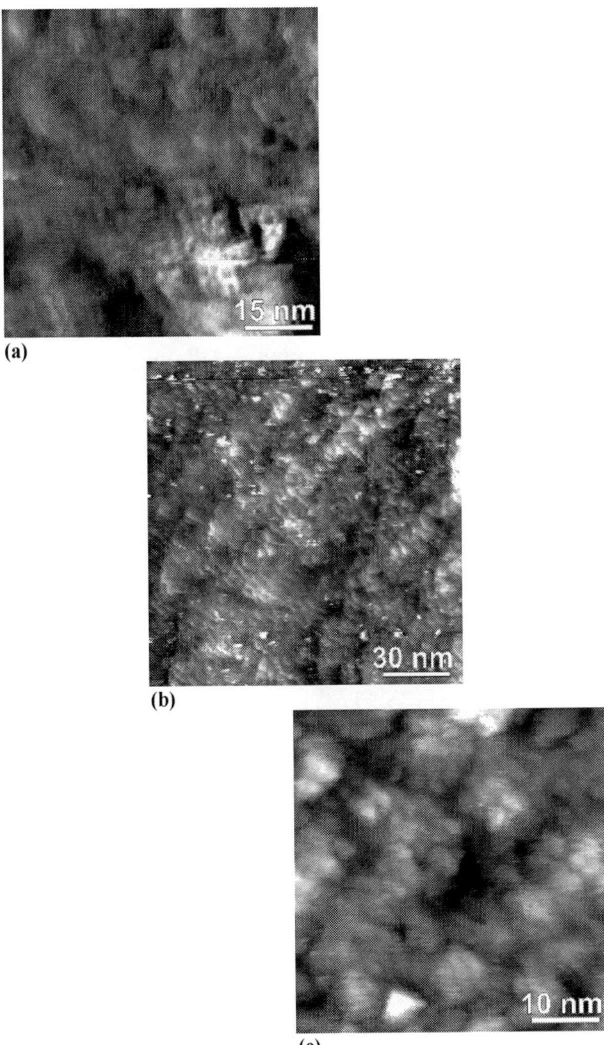

Figure 18. ECSTM images of the Ni(111) surface recorded (a) at −175 mV in 0.1 M NaOH$_{(aq)}$ ($z$ range = 0.8 nm), (b) at −50 mV in 0.1 M NaOH$_{(aq)}$ + 0.05 M NaCl ($z$ range = 0.6 nm), (c) at +200 mV in 0.1 M NaOH$_{(aq)}$ + 1 M NaCl ($z$ range = 2.4 nm). Reprinted with permission from *J. Electrochem. Soc.* **153** (2006) B453, Copyright © 2006, The Electrochemical Society.

trolytes, which was assigned to a relatively thick passive film (estimated to $\sim$ 3 nm)[83] decreasing the electron tunneling conductivity through the film and thus limiting the imaging resolution. The triangular morphology of some facets (better resolved in Fig. 18b but altered by tip artifacts in Fig. 18a) is consistent with an hexagonal symmetry of the crystalline structure similar to that of the surface lattices observed by ECSTM in 1 mM $NaOH_{(aq)}$ as described in Section III.1(*ii*) and by ECAFM in 10 mM $NaOH_{(aq)}$,[21] and assigned in both cases to the growth of a 3D passive film constituted of $\beta$-$Ni(OH)_2(0001)$.

The images shown in Fig. 18a and Fig. 18b evidence that there is no marked effect of the presence of chlorides on the crystallization of the 3D passive film for a concentration ratio $[Cl^-]/[OH^-] = 0.5$. However, the presence of chlorides modifies the time and potential required to observe the early stages of crystallization. In 1 mM $NaOH_{(aq)}$ in the absence of chlorides,[24] the early stages of crystallization are observed after 375 s polarization time at an overpotential of only 0.02 V with respect to the onset of the passivation peak, corresponding to a growth stage where the surface is covered by a 2D layer as described in Section III.1(*ii*). In 0.1 M $NaOH_{(aq)}$ + 0.05 M NaCl ($[Cl^-]/[OH^-] = 0.5$), a similar time of polarization (500 s) is required but at much larger overpotential (0.24 V) with respect to the onset of the active-passive transition. This indicates that the presence of chlorides delays the crystallization of the passive film in the transient stage of formation of a 2D layer. Higher overpotentials are required for crystallization but they induce the growth of the 3D film before its crystallization.[25]

Figure 18c shows a topographic ECSTM image of the Ni(111) surface obtained in 0.1 M $NaOH_{(aq)}$ + 1 M NaCl ($[Cl^-]/[OH^-] = 10$) at 0.2 $V_{SHE}$, i.e., far above the active-passive transition at $-$ 0.36 $V_{SHE}$. No faceted topography is observed in this case. The passive film has a nanogranular morphology. The lateral dimension of the individual grains ranges from 4 to 15 nm. The individual grains are aggregated and form clusters that extend up to 30 nm in their largest dimension. The shape of the grains does not show any preferential direction indicating an amorphous structure. The deepest boundaries between the clusters of grains have a measured depth of $\sim$ 3 nm consistent with the formation of a 3D layer.

Figure 19 shows a sequence of topographic EC-STM images of the Ni(111) surface obtained in 0.1 M $NaOH_{(aq)}$ + 0.1 M NaCl

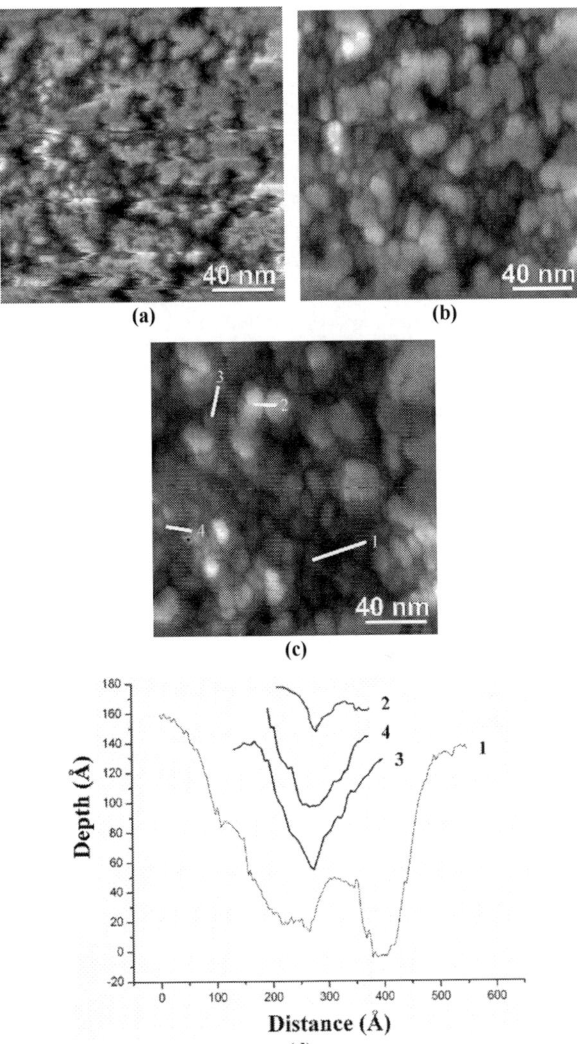

Figure 19. ECSTM images of the Ni(111) surface recorded in 0.1 M NaOH$_{(aq)}$ + 0.1 M NaCl ($z$ range = 10 nm (a), 17 nm (b), 18 nm (c)). (a), (b), and (c) were recorded 750 s, 870 s and 1230 s after a potential scan to -0.3 V$_{SHE}$. (d) line profiles along the segments in (c). Reprinted with permission from *J. Electrochem. Soc.* **153** (2006) B453, Copyright © 2006, The Electrochemical Society.

following a potential step to $-0.3$ V$_{SHE}$, depicting the growth of a 3D layer constituted of clusters of nanograins and the absence of crystallization for a concentration ratio $[Cl^-]/[OH^-] = 1$. In Fig. 19a, obtained after 750 s at $-300$ mV, the initial step topography of the substrate is fully substituted by a granular morphology, indicating the growth of a 3D layer. The lateral dimension of the grains ranges from 2 to 7 nm. The individual grains form clusters that extend up to 50 nm in the largest dimension. Their shape does not show any preferential direction indicating the absence of crystallization in the film. The deepest boundaries between the clusters of grains have a depth that ranges between 4 and 5 nm. This value shows the roughening of the surface produced during the growth of the 3D passive film.

In Fig. 19b, obtained after 870 s, the lateral dimensions of the individual grains increase and range between 6 and 15 nm, showing the coalescence of individual grains. There is still no sign of crystallization in the film. In Fig. 19c, obtained after 1230 s, the range of lateral dimensions of the individual grains is similar, indicating no further coalescence. The absence of crystallization of the passive film is confirmed. A major difference is the increase of the depth of boundaries separating the clusters of grains (see Fig. 19d). The widest boundary reaches a depth of $\sim$ 14 nm. This value is significantly larger than that measured in Fig. 19a showing that ageing in the presence of chloride ($[Cl^-]/[OH^-] = 1$) causes a marked roughening of the surface not observed at concentration ratios (($[Cl^-]/[OH^-] = 0.5$) or in the absence of chloride. The value of $\sim$ 14 nm is also much larger than the maximum total thickness of $\sim$ 4.5 nm reported for the Ni(II) passive film grown in 1 M NaOH$_{(aq)}$,[83] showing that 3D attacks of the nickel substrate, localized at the boundaries between grains, contribute to the observed surface roughening.

Thus, at higher chloride concentration ($[Cl^-]/[OH^-] \geq 1$), the crystallization of the passive film is observed to be fully blocked for overpotentials as large as 0.7 V with respect to the onset of the passivation peak. The passive film is amorphous with a nanogranular morphology. The lateral size of the individual grains is larger than that of the nuclei formed in the initial stages of growth, showing the coalescence of the nuclei. However, further coalescence to form larger grains of several tens of nanometers, the typical size of the grains constituting crystalline passive films as measured by

STM (see Section III.1 above) and AFM (see Section V.2 below), is blocked. This leads to the formation of clusters of nanograins having a size of several tens of nanometers. This blocking of the coalescence possibly results from the incorporation, during the growth, of chloride ions in the passive film. No data on the chemical composition of the passive film formed on Ni in chloride-containing alkaline solutions are available. However, the combined $Cl^-$ radiotracer and XPS measurements[36] for passive films grown in chloride-containing acid solutions show the incorporation of substantial amounts of chlorides both in the outer hydroxide and inner oxide layers of the passive film.[110] The preferential location of the chloride ions in the intergranular sites of the passive film formed in the chloride-containing alkaline solution might cause the observed blocking of the coalescence.

The intergranular sites corresponding to the boundaries between the clusters of nanograins of the passive film are preferential sites for the development of localized attacks of the substrate during the growth of the passive film. The higher overpotential required for nucleating the passive film in the presence of chlorides leads to the direct formation of 3D nuclei of the passive film. Dissolution of the metal substrate is expected to be active in the sites not yet passivated between the 3D nuclei. With time the 3D nuclei coalesce and grow up to the typical lateral size of ~ 10 nm, possibly limited by the accumulation of chloride in the intergranular sites. Dissolution is sustained in the intergranular sites where chlorides, possibly in higher concentration, still poison nucleation. The combined effects of chlorides lead to the growth of the clusters of nanograins separated by boundaries where sustained dissolution develops more opened intergranular structures. Eventually these boundaries are also passivated by the growth of nanograins, leading to a passive film which is inhomogeneous at the nanometer scale, with boundaries between the grain clusters being less protective sites against dissolution in the passive state.

## 2. Breakdown at Oxide Grain Boundaries

Figure 20 shows the modifications of the nanostructure of the passivated Ni(111) surface resulting from the increase of the potential in the passive range and the effect of the presence of chlorides.[104] After passivation at 0.55 $V_{SHE}$ in a chloride-free sulfuric acid solution (pH = 2.9) (Fig. 20a), the surface is completely covered with platelets assigned to the grains of the passive film. It is the faceted surface of these grains that is revealed by the typical higher magnification STM images shown in Section III.1(*ii*). The lateral size of the platelets, varying from 50 to 230 nm, suggests a varying degree of advancement of the coalescence of the oxide grains. Between the platelets, depressions with a depth varying from 0.4 to 1.4 nm are measured. Their formation is assigned to the dissolution occurring on the not yet passivated (or less protected) sites that are formed in the transient process of growth of the oxide film, prior to complete passivation of the surface.

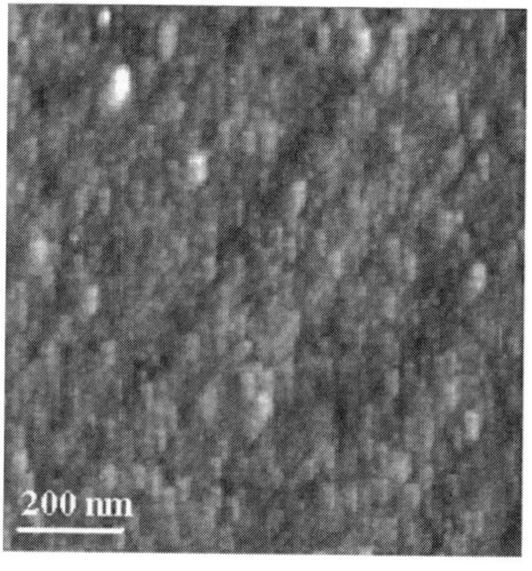

**(a)**

Figure 20. AFM images ($z$ range = 3 nm) of the Ni(111) surface after pre-passivation for 30 min in 0.05M $H_2SO_4$ + 0.095M NaOH (pH = 2.9) at +0.55 $V_{SHE}$ (a), and subsequent increase of the potential stepwise (steps of 0.1 V) every 30 min up to + 1.05 $V_{SHE}$ in the absence (b) or presence (c) of chloride (0.05 M NaCl). Reprinted with permission from *Local Probe Techniques for Corrosion Research*, EFC Publications N° 45, ISSN 1354-5116, Woodhead Publishing Ltd, Cambridge, England, 2007, p. 71, Copyright © 2007, Woodhead Publishing Ltd.

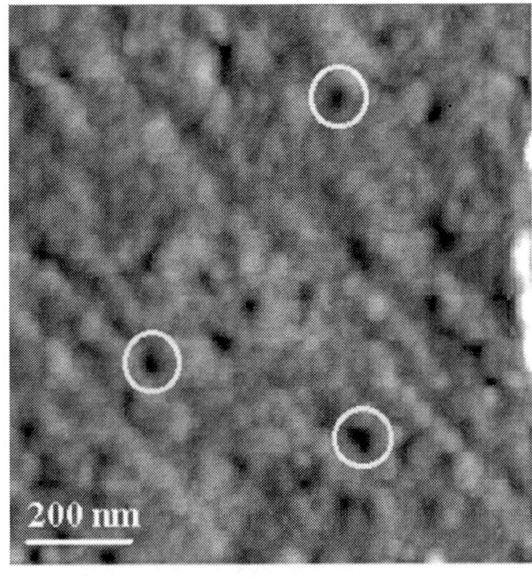

**(b)**

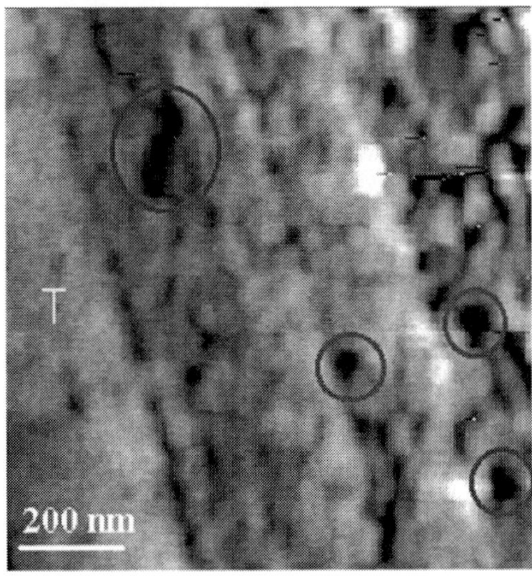

**(c)**

Figure 20. Continuation.

Increasing the potential in the passive range by successive potential steps up to 1.05 $V_{SHE}$ causes the roughening of the pre-passivated surface. In the absence of chlorides, the formation of local depressions of nanometer dimensions (20 to 30 nm at the surface) is observed with a density of $(3 \pm 2) \times 10^{10}$ cm$^{-2}$ (some are marked in Fig. 20b). The depth of these depressions (measured with the instrument carefully calibrated with the monoatomic step height of the Ni(111) substrate) ranges from 2.2 to 3.8 nm. This is larger than after pre-passivation at 0.55 $V_{SHE}$ and than the thickness of the passive film formed in these conditions ($< 2$ nm), indicating the locally enhanced corrosion of the substrate in these sites. This implies that the surface, pre-passivated at 0.55 $V_{SHE}$, has been locally depassivated, with a local enhancement of the corrosion of the substrate, and subsequently repassivated. It cannot be concluded however if depassivation results from local thinning or collapse mechanism of the passive film, as described below. Still, the major outcome of this experiment is that nanopits are formed in the passive state in the absence of chloride. The largest dimension of the nanopits measured by AFM corresponds to a charge transient of $\sim 8 \times 10^{-14}$ C per nanopit. Assuming a duration of 100 ms, the current transient would be $\sim 0.8$ pA. This is of the order of magnitude of $\sim 1$ pA that can be detected only with advanced transient measurement techniques, provided the electrode area is reduced to micrometer dimensions.[111]

In the presence of chlorides (Fig. 20c), the depressions observed between the grains have, for the most part, the same dimensions as those described above. However, significantly larger depressions are also observed (some are marked by circles in Fig. 20c). Their lateral dimension ranges between 40 and 50 nm at the surface and their depth between 5 to 6 nm. Their density is $(2 \pm 1) \times 10^{9}$ cm$^{-2}$.

Figure 21 shows nanoscale attacks of similar dimensions and localized at the grain boundaries of the passivated surface that were observed by STM after exposing to chlorides the Ni(111) surface pre-passivated at +0.9 $V_{SHE}$[107,108] These localized attacks were assigned to metastable pits on the basis of the absence of variations of the current characteristic of stable pitting during the corrosion test. The observed atomic structure inside these metastable pits (see insert in right panel) is crystalline with lattice parame-

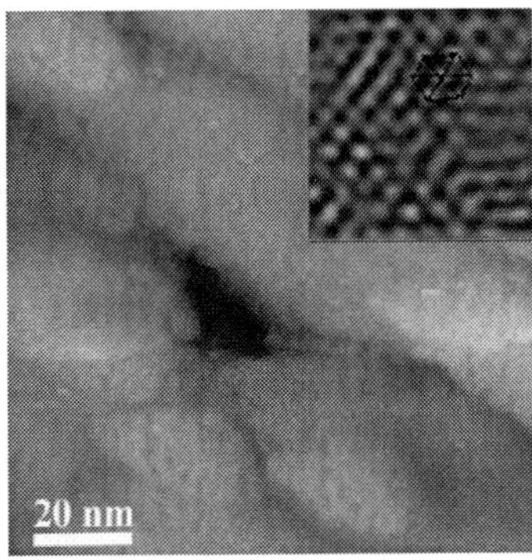



Figure 21. STM images of Ni(111) passivated in 0.05M $H_2SO_4$ + 0.095 M NaOH (pH ~ 2.9) at 0.9 $V_{SHE}$ for 30 min and subsequently exposed to 0.05 M NaCl for 90 min, showing metastable pits at two magnifications. The inset in the right images shows the crystalline lattice observed inside the metastable pit. Left: Reprinted with permission from *Critical Factors in Localized Corrosion III*, The Electrochemical Society Proceedings Series, PV 98-17, Pennington, NJ, 1999, p 552, Copyright © 1999, The Electrochemical Society. Right: Reprinted with permission from *Electrochem. Solid-State Lett.* **4** (2001) B1, Copyright © 2001, The Electrochemical Society.

ters similar to those measured on the passivated surface prior to exposure to chlorides, showing a similar structure of the surface repassivated in the presence of chlorides and confirming the metastable character of the pit.

Figure 20c also shows an area, marked T, corresponding to an extended terrace of the substrate that exhibits much less local attacks than the rest of the imaged surface where narrower terraces are observed due to a higher density of substrate step edges. This role of the defects of the substrate (monoatomic step edges) in passivity breakdown is confirmed in Fig. 21 which shows a preferential alignment of the metastable pits along the substrate step edges. The AFM and STM results reviewed above demonstrate that the breakdown events are located at the boundaries between the oxide crystals of the passive film, giving evidence that the grain boundaries of the passive film are preferential sites of passivity breakdown.

# VI. GRAIN BOUNDARY MODEL OF PASSIVITY
# BREAKDOWN AND PIT INITIATION

In the following we present and discuss a model for passivity breakdown and pit initiation based on the nanostructural data reviewed above.[112] Its originality is to take into account explicitly the role of the grain boundaries (or defective boundaries) of the barrier oxide layer of the passive film. It is considered that these defective sites are less resistive to ion transfer than the *defect-free* oxide lattice, which leads to a redistribution of the potential drops at the metal/oxide/electrolyte interfaces in the passive state. Possible passive film grain-boundary effects on ion diffusivities have been mentioned before they were actually observed,[113] and in a recent work on DFT calculation of ionic transport in iron oxide without grain boundary it was stated that oxide growth kinetics models could be reconciled with experimental data only if passive film grain-boundary effects were dominant.[114]

## 1.    Local Redistribution of the Potential Drop

Figure 22 shows a schematic representation of a metal surface covered by a barrier oxide layer with crystalline grains of nanometer dimensions. The grains are separated by boundaries that can be either ordered grain boundaries whose crystallographic orientation may vary locally (as expected for well-crystalline passive films formed on Cu, Ni and Fe for example), or slightly wider non-ordered boundary regions whose local atomic structure may also vary significantly from boundary to boundary and with time (as expected in the case of Cr and stainless steels for example). The model considers these boundaries as generic inter-granular defective sites separating  the different grains of the oxide lattice,  but a large population of these sites of varying ionic conductivity is expected due to their structure. No defects in the underlying substrate are considered.

The diagram in Fig. 22a shows schematically the potential at the metal surface and in the solution, and the potential drops at the metal/oxide (Me/Ox) interface, in the oxide film and at the oxide/electrolyte (Ox/El) interface for the *defect-free* oxide lattice. The potential at the metal surface is all over the same.  The poten-

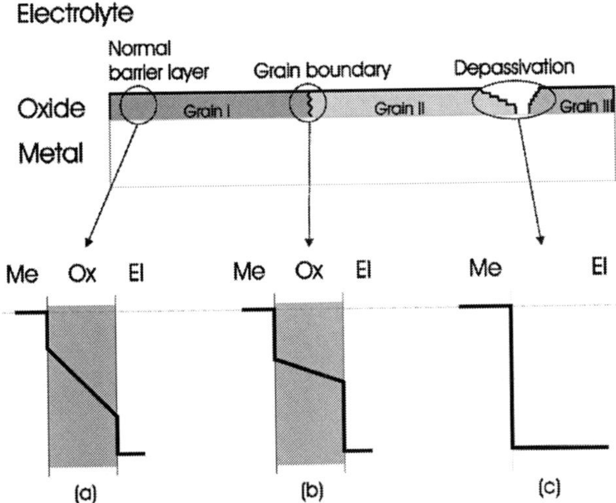


Figure 22. Model of a metal/oxide/electrolyte system with a surface oxide barrier layer consisting of nanograins separated by grain boundaries and schemes of the potential drop for (a) a normal defect-free barrier layer, (b) an inter-granular boundary, and (c) local depassivation (i.e., pit opening). Reprinted with permission from *Corrosion Sci.* **50** (2008) 2698, Copyright (2008), Elsevier Science.

tial within the solution close to the specimen's surface is the same if the electrolyte is sufficiently conductive and the local current density is low all over the surface (i.e., stationary conditions of passivity).

A locally different distribution of the potential drop is considered at an inter-granular defective site (Fig. 22b). The defect region is considered as less resistive to ion transfer since oxide grain boundaries can exhibit ionic conductivities many orders of magnitude larger than the bulk oxide.[115] As a consequence, the electric field strength should be smaller, i.e., the potential drop within the film is locally smaller. The missing part of the potential drop must be located at the Me/Ox and/or at the Ox/El interfaces, as shown in Fig. 22b. The redistribution of the potential drop is expected to vary locally depending on the local structure of the inter-granular defect.

The data on Cr and stainless steel have shown the growth of essentially amorphous non-ordered films in the initial stages of passivation. However, in all cases, increasing the passivation time and/or potential was observed to favor the film crystallization. Thus the passive film can be viewed as an evolving entity through which the residual ionic transport contributes to slowly heal out the numerous structural defects resulting from a rapid initial growth. As a result, the density of defective regions and their resistivity to ion transfer decreases and increases with time, respectively, leading to more protective films. Increasing the potential accelerates this ageing effect. This annealing effect is somewhat analogous to that resulting from the continuous breakdown and repair, suggested in the past when direct observation of nanoscale inter-granular defects in passive films was not yet available.[116-119] It is consistent with the observation of the decrease of the passive current with time still occurring after reaching a steady-state thickness of the barrier layer, and it is attested by the higher resistance to localized corrosion observed after *in situ* ageing.[120-122] Despite this beneficial ageing effect, a large population of sites (inter-granular boundaries) of faster ion transport than through the *defect-free* oxide lattice subsist after reaching the stationary conditions of passivity, and those having the highest ionic conductivity are considered as the sites the more susceptible to breakdown.

Several routes leading to passivity breakdown must be considered depending on the potential distribution in the defect region of the barrier layer.

## 2.  Local Thinning of the Passive Film

In the case of a larger potential drop at the Ox/El interface (Fig. 23), the metal-ion transfer from the oxide to the electrolyte should increase for cations forming the oxide lattice in the inter-granular boundary ($Me_L^{z+}$) and for interstitial cations more loosely bonded in the inter-granular boundary ($Me_{IG}^{z+}$). The increased transfer of the $Me_L^{z+}$ cations will lead to faster localized dissolution of the passive film. This enhanced dissolution at inter-granular boundaries is merely an increased localized dissolution in the passive state. However, statistical fluctuations of the interfacial potential drop are expected depending on the atomic structure of the inter-

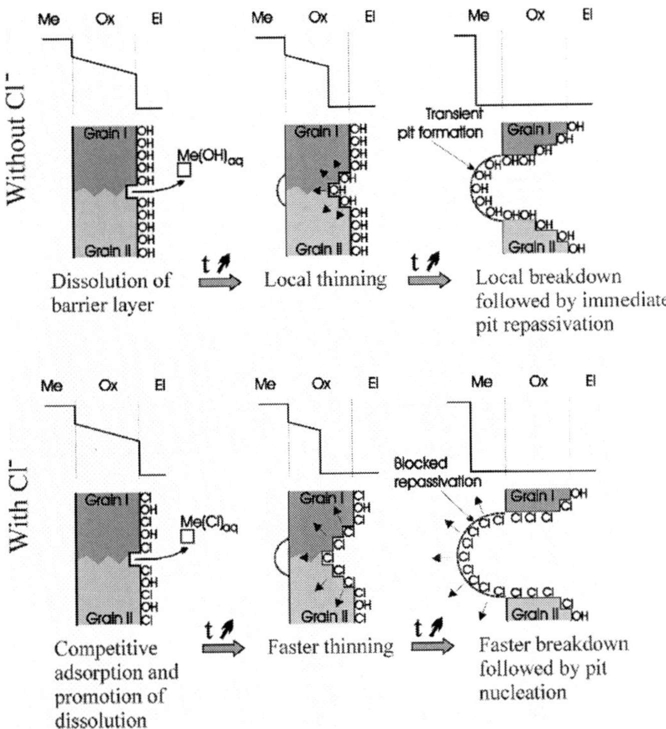


Figure 23. Mechanisms of local breakdown of passivity driven by the potential drop at the oxide/electrolyte interface of an inter-granular boundary of the barrier layer. The effect of chlorides is shown. Reprinted with permission from *Corrosion Sci.* **50** (2008) 2698, Copyright (2008), Elsevier Science.

granular boundaries of the barrier layer and on their evolution with time as discussed above. This will cause higher localized dissolution of the film until local depassivation (or pit trigger) takes place preferentially in the less resistive sites where faster local thinning is not compensated by film growth. At this stage, the much higher potential drop at the metal surface should lead to the immediate reformation of the passive layer. However, small pits will appear due to dissolution in the transient active state prior to repassivation. This scenario can explain the nanopit formation observed in the absence of chlorides as described in Section V.2.

If $Cl^-$ ions are present in the electrolyte, they compete with $OH^-$ for adsorption on surface sites. If $Cl^-$ are adsorbed on the oxide surface, the Me-$Cl^-$ (or MeO(H)-$Cl^-$) surface complexes are less strongly bonded to the oxide matrix and the activation energy for their transfer to the electrolyte is decreased. As a consequence the localized dissolution rate is increased and film growth is poisoned. This will lead to a faster thinning of the passive layer and localized depassivation in the less resistive sites (Fig. 23). This notion of local thinning of the passive film being a precursor to pitting has been previously proposed[123-124] and a mechanism of passivity breakdown based on local thinning has been discussed.[120] However, it was considered that passivity breakdown only occurs in the presence of $Cl^-$ due to their local adsorption and their catalytic effect on thinning. In the model reviewed here, the local thinning mechanism is related to the nanostructure of the barrier layer, and to the increase of the potential drop at the Ox/El interface of the inter-granular boundaries. In addition the high density of atomic defects present at the surface of the oxide in these sites is expected to lower the activation energy for the dissolution of metal cations forming the oxide lattice ($Me_L^{z+}$). This view implies that passivity breakdown events also occur in the absence of $Cl^-$ as proposed previously,[126] but are accelerated in their presence.

After local depassivation, $Cl^-$ also competes with $OH^-$ for adsorption on the metal surface. Repassivation will be poisoned by a competitive adsorption mechanism between chloride and hydroxide strongly affecting the nanostructure and crystallization of the surface in the metastable pits as shown by the results obtained on Ni (Section V.1). Blocking the crystallization can favor a higher local dissolution as shown for Ag and Ni covered by 2D passive films (Section II.2). Larger metastable pits will be formed in agreement with the AFM data presented in Section V.2. Repassivation will be prevented if the supply of $OH^-$ is not sufficient, which will happen if $Cl^-$ continuously replace $OH^-$. To block repassivation requires a high local concentration of $Cl^-$. This process will lead to selective pit nucleation at the less resistive sites of the large population of inter-granular boundaries of the barrier layer.

### 3.  Voiding at the Metal/Oxide Interface

A second possible mechanism leading to passivity breakdown is related to the transfer of the $Me_{IG}^{z+}$ cations in the electrolyte (Fig. 24a). These inter-granular cations are less strongly bonded to the oxide lattice and should dissolve more rapidly at the Ox/El interface than the $Me_L^{z+}$ cations. Local dissolution in this case is limited by the oxidation reaction, $Me \rightarrow Me^{z+} + V_{Me} + z\,e^-$, at the Me/Ox interface. A larger potential drop at this interface will yield a faster oxidation reaction locally. If ion transport in the barrier layer is dominated by cations, the faster transport of the new $Me_{IG}^{z+}$ cations in the inter-granular boundary of the barrier layer followed by their dissolution will cause the formation of localized depressed areas in the metal surface at the Me/Ox interface. This will initiate interfacial voiding if the injection of $V_{Me}$ vacancies is not compensated by their diffusion towards the bulk. The continuous growth of the metal voids at this interface will develop stress in the non-supported barrier layer and eventually lead to its rupture or collapse, i.e., local depassivation (Fig. 24a).

If the potential drop at the Met/Ox interface is not increased, the oxidation rate will not increase locally. However, the residual release of the $Me_{IG}^{z+}$ cations is expected to lead to local voiding in the oxide near the Me/Ox interface if an enlarged defective region in the inner part of the oxide layer coincides with a grain boundary in the outer part of the film. Stress-induced rupture of the non-supported outer part of the film will lead then to local depassivation.

Direct experimental evidence of microscopic voiding assigned to hydrogen pressurization at the metal/oxide interface has been reported for aluminum at the solid/liquid interface.[127] More recently, nanoscopic oxide voiding was observed for anodized aluminum below the stable pitting potential and independently of the microstructure of the metal.[128] The void origin was assigned to cation and anion vacancy saturation at the Met/Ox interface resulting from the ionic transport properties in the oxide, but it was concluded that the observed void attributes were not consistent with the vacancy condensates postulated by the point defect model.[2] Nanoscopic metal voiding has been observed at the buried

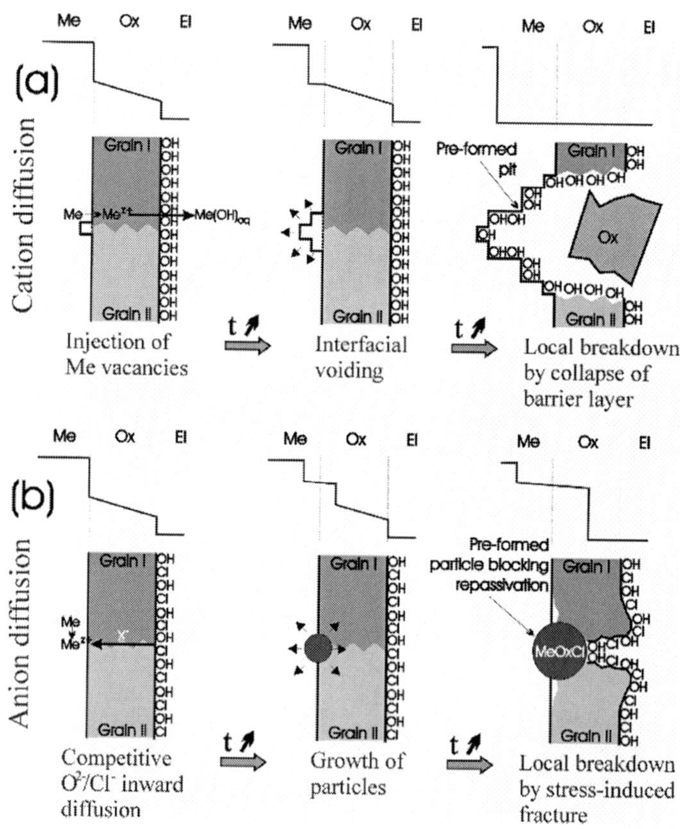



Figure 24. Mechanisms of local breakdown of passivity driven by the potential drop at the metal/oxide interface of an inter-granular boundary of the barrier layer. The effect of ion transport is shown: (a) predominant cation transport, (b) predominant anion tansport. Reprinted with permission from *Corrosion Sci.* **50** (2008) 2698, Copyright (2008), Elsevier Science.

TiAl/Al$_2$O$_3$ interface formed in gaseous environment,[129] and it was shown to result from the accumulation of the metal vacancies injected by the growth mechanism of the oxide, and dependent on the diffusion properties in the metallic substrate.

Both local thinning of the barrier film and voiding at the Met/Ox interface can take place simultaneously on different less resistive sites of the barrier layer or even combine on the same site to accelerate local depassivation. The pits formed at this stage may be different in size depending on the mechanism of local depassivation. Larger pits could form in the case of rupture of the passive film since it can be preceded by metal voiding underneath the film pre-forming local depressions in the metal surface. In the presence of Cl⁻, the mechanisms leading to pit repassivation or pit growth are expected to be independent of the mechanism of local depassivation.

Film breaking has already been discussed as a mechanism for passivity breakdown.[130,131] Mechanical stresses at weak sites were proposed as possible causes of rupture, but without clarifying the reason for their local occurrence. In the reviewed model it is proposed that voiding resulting from the faster transport through the inter-granular boundaries of the oxide barrier layer of the cations formed at the Me/Ox interface followed by their dissolution may be a major cause of stress-induced rupture of the non-supported passive film.

### 4.   Stress-Induced Fracture of the Passive Film

An alternative to metal voiding also leading to film breakdown must be considered depending on ion transport (Fig. 24b). If anions ($O^{2-}$ and Cl⁻) penetrate the barrier layer as inter-granular species and diffuse to the Me/Ox interface via these short-paths to react with the newly formed $Me_{IG}^{z+}$ cations, the formation of metal compounds at this buried interface will counteract voiding. However, the growth of the particles (containing Cl⁻ or not) will generate stress at the Met/Ox interface and in the barrier layer due to a Pilling-Bedworth ratio larger than 1. Continuous growth will fracture the barrier layer, leading to local breakdown. Repassivation will be blocked by particles containing a sufficient amount of Cl⁻, leading to immediate pit nucleation as previously proposed.[132]

Chloride penetration in passive films has been extensively discussed as a mechanism of passivity breakdown. It was first proposed that it occurs via highly defective barrier oxide layer[133] based on the postulate of an amorphous structure of the passive film, but without clarifying why breakdown occurs locally. Later, the point defect model[2,134,-136] proposed that Cl⁻ could penetrate and migrate via the oxygen vacancies of the oxide lattice, increasing the concentration gradient of interdependent cation vacancies and, if not annihilated by the oxidation reaction, their accumulation up to a critical concentration inside the film, thus initiating voiding and film breakdown. The model used here suggests that $Cl^-$ penetration and migration occurs predominantly via the intergranular sites of the oxide layer, and that this does not require an interdependent formation of cation vacancies due to the interfacial nature of these sites. As a consequence, $Cl^-$ will compete with $O^{2-}$ to combine with metal cations formed at the Me/Ox interface to form depassivating metal compounds.

The scenarios considered above do not take into account explicitly the presence of defects at the metal surface underneath the barrier layer. STM data on passivated nickel have shown that the nanopits resulting from the depassivation/repassivation events could be preferentially located along the atomic step edges of the metal at the Me/Ox interface. This is a direct evidence of the effect of atomic defects at the metal surface, lowering the activation energy for the $Me \rightarrow Me^{z+} + V_{Me} + ze^-$, interfacial reaction and thus promoting depassivation at the oxide grain boundaries coinciding with these defects by a mechanism involving the potential drop at the Me/Ox interface. Of course, local depassivation or pit trigger can also be promoted by the presence of microstructural defects of the substrate such as metal grain boundaries or interfaces between the matrix and inclusions or second phase particles for alloys,[137-139] also lowering the activation energy for the oxidation reaction at the Met/Ox interface. In all cases, the oxide formed above such interfaces of the substrate is expected to contain less resistive defective regions also sensitive to breakdown.


## VII.  CONCLUSION


In this chapter, the structural aspects of passivity and their impact on the corrosion properties of metal and alloy surfaces at the nanoscopic scale have been reviewed with emphasis on recent results obtained on metal surfaces by *in situ* nanoprobes (STM and AFM) implemented to electrochemistry. The data reviewed for Cu, Ag, Ni, Fe, Cr and stainless steels have demonstrated:

- The adsorption of hydroxide ions in the potential range below the one for oxide growth, and the role of structural precursors of the adsorbed layers in the growth of passive films, inducing surface reconstruction of the topmost metal plane to adopt the structural arrangement found in 3D passivating oxides.
- The preferential nucleation of passive films at pre-existing defects (i.e., step edges) of the metal surface.
- The local structure-sensitive corrosion resistance of metal surfaces covered by 2D passivating oxide/hydroxide layers in the transient stage of passive film growth. Dissolution is then promoted at the weakest non-ordered sites, initiating 2D nanopits on the substrate terraces and the growth of 3D nanopits by a repeated mechanism on the newly exposed terraces after prolonged polarization at low anodic overpotential.
- The crystallinity of passive films on numerous substrates and the effect of substrate structure on orientation of the oxide crystalline structure and its surface hydroxylation. Ageing under polarization is critical to the crystallization of chromium rich passive films.
- The surface faceting of the passivating oxide due to a few degree tilt of the oxide lattice with respect of the metal lattice.
- The role of preferential site played by step edges of the passive film surface, leading to a step flow mechanism of dissolution in the passive state.
- The presence of numerous grain boundaries in crystalline passive films, and the preferential role of these nanostructure-related defects in passivity breakdown and pit initiation.
- The effect of chlorides, inhibiting the coalescence and crystallization of passive films and thus promoting the densification of inter-granular boundaries and non-ordered regions both less resistant to corrosion.
- The occurrence of passivity breakdown at grain boundaries of the passive film both in the presence and in the absence of chlorides.

The reviewed data greatly contribute to the advances in the understanding and control of the passivation and localized corrosion of metal surfaces. Based on these data, a model for passivity breakdown and pit initiation taking into account explicitly the role of the grain boundaries (or defective boundaries) of the barrier oxide layer of the passive film has been reviewed. It is considered that these defective sites are less resistive to ion transfer than the *defect-free* oxide lattice, which leads to a redistribution of the potential drops at the metal/oxide/electrolyte interfaces in the passive state, leading to different breakdown routes depending on which interfacial potential drop predominates.

In the future, the impact of combined structural and chemical characterization of the nanostructural defects and heterogeneities of surfaces protected by ultra-thin layers should allow a major breakthrough in the control and design of new materials, corrosion resistant at the nanoscopic scale.

## REFERENCES


[1]G.S. Frankel, *J. Electrochem. Soc.* **145** (1998) 2186.

[2]D.D. Macdonald, *Pure Appl. Chem.* **71** (1999) 951.

[3]M. Schütze, Editor, *Corrosion and Environmental Degradation* (Wiley-VCH, Weinheim, 2000).

[4]J.W. Schultze, M.M. Lohrengel, *Electrochim. Acta* **45** (2000) 2499.

[5]P. Marcus, Editor, *Corrosion Mechanisms in Theory and Practice*, 2nd edition (Marcel Dekker Inc., New York, 2002).

[6]H.-H. Strehblow, Passivity of Metals, in *Advances in Electrochemical Science and Engineering*, R.C. Alkire, D.M. Kolb, Editors, Wiley-VCH, Weinheim, 2003, pp. 271-374.

[7]K. Itaya, *Prog. Surf. Sci.* **58** (1998) 121.

[8]O.M. Magnussen, *Chem. Rev.* **102** (2002) 679.

[9]I.C. Oppenheim, D.J. Trevor, C.E.D. Chidsey, P.L. Trevor, K. Sieradzki, *Science* **254** (1991) 687.

[10]T.P. Moffat, F.-R. Fan, A.J. Bard, *J. Electrochem. Soc.* **138** (1991) 3224.

[11]S.J. Chen, F. Sanz, D.F. Ogletree, V.M. Hallmark, T.M. Devine, M. Salmeron, *Surf. Sci.* **292** (1993) 289.

[12]M. Stratmann, M. Rohwerder, *Nature* **410** (2001) 420.

[13]H. Martín, P.Carro, A. Hernández Creus, J. Morales, G. Fernández, P. Esparza, S. González, R.C. Salvarezza, A.J. Arvia, *J. Phys. Chem. B* **104** (2002) 8239.

[14]F.U. Renner, A. Stierle, H. Dosch, D.M. Kolb, T.L. Lee, J. Zegenhagen, *Nature* **439** (2006) 707.

[15]F.U. Renner, A. Stierle, H. Dosch, D.M. Kolb, J. Zegenhagen, *Electrochem. Comm.* **9** (2007) 1639.

[16]V. Maurice, H. Talah, P. Marcus, *Surf. Sci.* **284** (1993) L431.

[17]V. Maurice, H. Talah, P. Marcus, *Surf. Sci.* **304** (1994) 98.

[18]S.-L. Yau, F.-R. Fan, T.P. Moffat. A.J. Bard. *J. Phys. Chem.* **98** (1994) 5493.

[19]T. Suzuki, T. Yamada, K. Itaya, *J. Phys. Chem.* **100** (1996) 8954.

[20]D. Zuili, V. Maurice, P. Marcus, *J. Electrochem. Soc.* **147** (2000) 1393.

[21]N. Hirai, H. Okada, S. Hara, *Transaction JIM* **44** (2003) 727.

[22]J. Scherer, B.M. Ocko, O.M. Magnussen, *Electrochim. Acta* **48** (2003) 1169.

[23]M. Nakamura, N. Ikemiya, A. Iwasaki, Y. Suzuki, M. Ito, *J. Electroanal. Chem.* **566** (2004) 385.

[24]A. Seyeux, V. Maurice, L.H. Klein, P. Marcus, *J. Solid State Electrochem.* **9** (2005) 337.

[25]A. Seyeux, V. Maurice, L.H. Klein, P. Marcus. *J. Electrochem. Soc.* **153** (2006) B453.

[26]V. Maurice, W. Yang, P. Marcus. *J. Electrochem. Soc.* **141** (1994) 3016.

[27]D. Zuili, V. Maurice, P. Marcus, *J. Phys. Chem. B* **103** (1999) 7896.

[28]R.C. Bhardwaj, A Gonzalez-Martin, J.O'M. Bockris, *J. Electroanal. Chem.* **307** (1991) 195.

[29]M.P. Ryan, R.C. Newman, G.E. Thompson. *J. Electrochem. Soc.* **142** (1995) L177.

[30]J Li, D.J. Meier, *J. Electroanal. Chem.* **454** (1998).

[31]I. Diez-Pérez, P. Gorostiza, F. Sanz, C. Müller, *J. Electrochem. Soc.* **148** (2001) 307.

[32]E.E. Rees, M.P. Ryan, D.S. MacPhail, *Electrochem. Solid-State Lett.* **5** (2002) B21.

[33]H. Deng , H. Nanjo, P. Quian, A. Santosa, I. Ishikawa, Y. Kurata, *Electrochim. Acta* **52** (2007) 4272.

[34]N. Ikemiya, T. Kubo, S. Hara, *Surf. Sci.* **323** (1995) 81.

[35]V. Maurice, H.-H. Strehblow, P. Marcus, *Surf. Sci.* **458** (2000) 185.

[36]J. Kunze, V. Maurice, L.H. Klein, H.-H. Strehblow, P. Marcus, *J. Phys. Chem. B* **105** (2001) 4263.

[37]J. Kunze, V. Maurice, L.H. Klein, H.-H. Strehblow, P. Marcus, *Electrochim. Acta* **48** (2003) 1157.

[38]J. Kunze, V. Maurice, L.H. Klein, H.-H. Strehblow, P. Marcus, *J. Electroanal. Chem.* **554-555** (2003) 113.

[39]J. Kunze, V. Maurice, L.H. Klein, H.-H. Strehblow, P. Marcus, *Corrosion Sci.* **46** (2004) 245.

[40]S. Ando, T. Suzuki, K. Itaya, *J. Electroanal. Chem.* **431** (1997) 277.

[41]A. Foelske, J. Kunze, H.-H. Strehblow. *Surf. Sci.* **554** (2004) 10.

[42]M.P. Ryan, R.C. Newman, G.E. Thompson. *Philosophical Mag. B* **70** (1994) 241.

[43]M.P. Ryan, R.C. Newman, G.E. Thompson. *J. Electrochem. Soc.* **141** (1994) L164.

[44]V. Maurice, W. Yang, P. Marcus. *J. Electrochem. Soc.* **143** (1996) 1182.

[45]H. Nanjo, R.C. Newman, N. Sanada, *Appl. Surf. Sci.* **121** (1997) 253.

[46]V. Maurice, W. Yang, P. Marcus, *J. Electrochem. Soc.* **145** (1998) 909.

[47]A. Machet, A. Galtayries, S. Zanna, L. H. Klein, V. Maurice, P. Jolivet, M. Foucault, P. Combrade, P. Scott, P. Marcus, *Electrochim. Acta* **49** (2004) 3957.

[48]O.M. Magnussen, J. Scherer, B.M. Ocko, R.J. Behm, *J. Phys. Chem. B* **104** (2000) 1222.

[49]M.F. Toney, A.J. Davenport, L.J. Oblonsky, M.P. Ryan, C.M. Vitus, *Phys. Rev. Lett.* **79** (1997) 4282.

[50]A.J. Davenport, L.J. Oblonsky, M.P. Ryan, M.F. Toney, *J. Electrochem. Soc.* **147** (2000) 2162.

[51] H.-H. Strehblow, B. Titze, *Electrochim. Acta* **25** (1980) 839.

[52] W. Kautek, J.G. Gordon, *J. Electrochem. Soc.* **137** (1990) 2672.

[53] B. Millet, C. Fiaud, C. Hinnen, E.M.M. Sutter, *Corrosion Sci.* **37** (1995) 1903.

[54] Y. Feng, K.-S. Siow, W.-K. Teo, K.-L. Tan, A.-K. Hsieh, *Corrosion* **53** (1997) 389.

[55] U. Sander, H.-H. Strehblow, J.K. Dohrmann, *J. Phys. Chem.* **85** (1981) 447.

[56] M.R.G. De Chiavo, R.C. Salvarezza, A. Arvia, *J. Appl. Electrochem.* **14** (1984) 165.

[57] S.T. Mayer, R.H. Muller, *J. Electrochem. Soc.* **139** , (1992) 426.

[58] H.Y.H. Chan, C.G. Takoudis, M.J. Weaver, *J. Phys. Chem. B* **103** (1999) 357.

[59] C.A. Melendres, G.A. Bowmaker, J.-M. Leger, B.J. Beden, *J. Electroanal. Chem.* **449** (1998) 215.

[60] S. Härtinger, B. Pettinger, K. Doblhofer, *J. Electroanal. Chem.* **397** (1995) 335.

[61] J.M.M. Droog, P.T. Alderliesten, G.A. Bootsma, *J. Electroanal. Chem.* **99** (1979) 173.

[62] J.M.M. Droog, F. Huisman, *J. Electroanal. Chem.* **115** (1980) 211.

[63] J.M.M. Droog, *J. Electroanal. Chem.* **115** (1980) 225.

[64] N. Iwasaki, Y. Sasaki, Y. Nishina, *Surf. Sci.* **198** (1988) 524.

[65] D. Hecht, P. Borten, H.-H. Strehblow, *Surf. Sci.* **365** (1996) 263.

[66] D. Hecht, R. Frahm, H.-H. Strehblow, *J. Phys. Chem.* **100** (1996) 10831.

[67] D. Hecht, H.-H. Strehblow, *J. Electroanal. Chem.* **440** (1997) 211.

[68] E.R. Savinova, P. Kraft, B. Pettinger, K. Doblhofer, *J. Electroanal. Chem.* **430** (1997) 47.

[69] D. Lützenkirchen-Hecht, H.-H. Strehblow, *Electrochim. Acta* **43** (1998) 2957.

[70] D. Lützenkirchen-Hecht, H.-H. Strehblow, *Ber. Bunsenges. Phys. Chem.* **102** (1998) 826.

[71] D. Lützenkirchen-Hecht, C.U. Waligura, H.-H. Strehblow, *Corrosion Sci.* **40** (1998) 1037.

[72] D.Y. Zemlyanov, E.R. Savinova, A. Scheybal, K. Doblhofer, R. Schlögl, *Surf. Sci.* **418** (1998) 441.

[73] K.J. Stevenson, X. Gao, D.W. Hatchett, H.S. White, *J. Electroanal. Chem.* **447** (1998) 43.

[74] B.M. Jovic, V.D. Jovic, G.R. Stafford. *Electrochem. Comm.* **1** (1999) 247.

[75] E.R. Savinova, D. Zemlyanov, A. Scheybal, T. Schedelniedrig, K. Doblhofer, R. Schlögl, *Langmuir* **15** (1999) 6546.

[76] E.R. Savinova, D. Zemlyanov, B. Pettinger, A. Scheybal, R. Schlögl, K. Doblhofer, *Electrochim. Acta* **46** (2000) 175.

[77] J. Kunze, H.-H. Strehblow, G. Staikov, *Electrochem. Comm.* **6** (2004) 132.

[78] V. Maurice L.H. Klein, H.-H. Strehblow, P. Marcus, *J. Phys. Chem. C* **111** (2007) 16351.

[79] B. Beden, D. Floner, J.-M. Leger, C. Lamy, *Surf. Sci.* **162** (1985) 822.

[80] F. Hahn, B. Beden, M.J. Croissant, C. Lamy, *Electrochim. Acta* **31** (1986) 335.

[81] R. Simpraga, B.E. Conway, *J. Electroanal. Chem.* **280** (1990) 341.

[82] W.K. Paik, Z. Szklarska-Smialowska, *Surf. Sci.* **96** (1980) 401.

[83] H.-W. Hoppe, H.-H. Strehblow, *Surf. Interf. Anal.* **14** (1989) 121.

[84] D. Floner, C. Lamy, J.M. Leger, *Surf. Sci.* **234** (1990) 87.

[85] M. Grden, K. Klimek, A. Czerwinski, *J. Solid State Electrochem.* **8** (2004) 390.

[86] M. Grden, K. Klimek, *J. Electroanal. Chem.* **581** (2005) 122.

[87] H. Yang, J.L. Whitten, *Surf. Sci.* **370** (1997) 136.

[88] S. Ando, T. Suzuki, K. Itaya, *J. Electroanal. Chem.* **412** (1996) 139.

[89]M. Dietterle, T. Will, D.M. Kolb, *Surf. Sci.* 327 (1995) L495.

[90]T. Teshima, K.Ogaki, K. Itaya, *J. Phys. Chem. B* **101 (1997)** 2046.

[91]A. Seyeux, V. Maurice, L.H. Klein, P. Marcus, *Electrochim. Acta* (2008) in press.

[92]T. Dickinson, A.F. Povey, P.M.A. Sherwood, *J. Chem. Soc. Faraday Trans. 1* **73** (1977) 327.

[93]P. Marcus, J. Oudar, I. Olefjord, *J. Microsc. Spectrosc. Electron.* **4** (1979) 63.

[94]H.-W. Hoppe, H.-H. Strehblow, *Corrosion Sci.* **20** (1980) 167.

[95]B.P. Lochel, H.-H. Strehblow, *J. Electrochem. Soc.* **131** (1984) 713.

[96]D.F. Mitchell, G.I. Sproule, M.J. Graham, *Appl. Surf. Sci.* **21** (1985) 199.

[97]P. Delichere, A. Hugot-Le Goff, N. Yu, *J. Electrochem. Soc.* **133** (1986) 2106.

[98]R.E. Hummel, R.J. Smith, E.D. Verink, *Corrosion Sci.* **27** (1987) 803.

[99]H.-W. Hoppe, H.-H. Strehblow, *Surf. Interf. Anal.* **14** (1989) 121.

[100]B.M. Biwer, M.J. Pellin, M.W. Schauer, D.M. Gruen, *Surf. Interf. Anal.* **14** (1989) 635.

[101]F.T. Wagner, T.E. Moylan, *J. Electrochem. Soc.* **136** (1989) 2498.

[102]H.-W. Hoppe, H.-H. Strehblow, *Corrosion Sci.* **31** (1990) 167.

[103]P. Marcus, in *Electrochemistry at Well-Defined Surfaces*, J. Oudar, P. Marcus, J. Clavilier, Editors, Special Volume of J. Chimie Physique 88 (1991) 1697.

[104]V. Maurice, T. Nakamura, L. Klein, P. Marcus, Initial stages of localised corrosion by pitting of passivated nickel surfaces studied by STM and AFM, in *Local Probe Techniques for Corrosion Research*, EFC Publications N° 45, R. Oltra, V. Maurice, R. Akid, P. Marcus, Editors, Woodhead Publishing Ltd, Cambridge, England, 2007, p. 71.

[105]Z. Lu, D.D. Macdonald, *Electrochim. Acta* (2008) DOI: 10.1016/j.electacta.2008.05.021.

[106]P. Marcus, V. Maurice, in *Passivity and Its Breakdown*, P. Natishan, H.S. Isaacs, M. Janik-Czachor, V.A. Macagno, P. Marcus, M. Seo, Editors, The Electrochemical Society Proceedings Series, PV 97-26, Pennington, NJ, 1998, p. 254.

[107]V. Maurice, V. Inard, P. Marcus, in *Critical Factors in Localized Corrosion III*, P.M. Natishan, R.G. Kelly, G.S. Frankel, R.C. Newman, Editors, The Electrochemical Society Proceedings Series, PV 98-17, Pennington, NJ, 1999, p 552.

[108]V. Maurice, L.H. Klein, P. Marcus, *Electrochem. Solid-State Lett.* **4** (2001) B1.

[109]V. Maurice, L.H. Klein, P. Marcus, *Surf. Interf. Anal.* **34** (2002) 139.

[110]P. Marcus, J.-M. Herbelin, *Corrosion Sci.* **140** (1993) 1571.

[111]G.T. Burstein, C. Liu, *Corrosion Sci.* **50** (2008) 2.

[112]P. Marcus, V. Maurice, H.-H. Strehblow, *Corrosion Sci.* **50** (2008) 2698.

[113]D.D. Macdonald, *J. Electrochem. Soc.* **139** (1992) 3434.

[114]S.C. Hendy, N.J. Laycock, M.P. Ryan, *J. Electrochem. Soc.* **152** (2005) B271.

[115]J. H. Harding, K.J.W. Atkinson, R.W. Grimes, *J. Am. Ceram. Soc.* **86** (2003) 554.

[116]B. MacDougall, *J. Electrochem. Soc.* **124** (1977) 1185.

[117]B. MacDougall, *J. Electrochem. Soc.* **126** (1979) 919.

[18]B. MacDougall, *J. Electrochem. Soc.* **127** (1980) 789.

[119]B. MacDougall, D.F. Mitchell, M.J. Graham, *J. Electrochem. Soc.* **132** (1985) 2895.

[120]H.-H. Strehblow, Mechanisms of Pitting Corrosion, in *Corrosion Mechanisms in Theory and Practice*, P. Marcus, Editor, 2nd edition, Marcel Dekker Inc., New York, 2002, pp. 243-285.

[121]W.P. Yang, D. Costa, P. Marcus, *J. Electrochem. Soc.* **141** (1994) 111.

[122]W.P. Yang, D. Costa, P. Marcus, *J. Electrochem. Soc.* **141** (1994) 2669.

[123]H.-H. Strehblow, *Werkst. Korros.* **27** (1976) 792.

[124]B. MacDougall, *J. Electrochem. Soc.* **124** (1977) 1185.

[125]C.J. Boxley, H.S. White, *J. Electrochem. Soc.* **151** (2004) B256.

[126]D.E. Williams, R.C. Newman, Q. Song, R.G Kelly, *Nature* **350** (1991) 216.

[127]C.B. Bargeron, R.B. Givens, *Corrosion (Houston)* **36** (1980) 618.

[128]K. Zavadil, J. Ohlhausen, G. Kotula, *J. Electrochem. Soc.* **153** (2006) B296.

[129]V. Maurice, G. Despert, S. Zanna, M.-P. Bacos, P. Marcus, *Nature Mater.* **3** (2004) 687.

[130]J.A. Richardson, G.C. Wood, *Corrosion Sci.* **10** (1970) 313.

[131]N. Sato, *Electrochim. Acta* **16** (1971) 1683.

[132]S.P. Mattin, G.T. Burnstein, *Philos. Mag. Lett.* **76** (1997) 341.

[133]T.P. Hoar, *Corrosion Sci.* **5** (1965) 279.

[134]L.F. Lin, C.Y. Chao, D.D. Macdonald, *J. Electrochem. Soc.* **128** (1981) 1194.

[135]M. Urquidi, D.D. Macdonald, *J. Electrochem. Soc.* **132** (1985) 555.

[136]D.D. Macdonald, *Pure Appl. Chem.* **71** (1999) 951.

[137]D.E. Williams, T.F. Mohiuddin, Y.Y. Zhu, *J. Electrochem. Soc.* **145** (1998) 2664.

[138]M.P. Ryan, D.E. Williams, R.J. Chater, B.M. Hutton, D.S. McPhail, *Nature* **415** (2002) 770.

[139]Q. Meng, G.S. Frankel, H.O. Colijn, S.H. Goss, *Nature* **424** (2004) 389.

# 2

# Anodic Oxide Films on Aluminum: Their Significance for Corrosion Protection and Micro- and Nano-Technologies

Hideaki Takahashi,* Masatoshi Sakairi,**
and Tatsuya Kikuchi**

*Asahikawa National College of Technology, Shunkoh-dai 2-2, Asahikawa,
071-8142, Japan
**Graduate School of Engineering, Hokkaido University, N13, W8, Kita-Ku,
060-8628, Sapporo, Japan

## I. APPLICATIONS OF ANODIC OXIDE FILMS

It was only 120 years ago that humans became able to obtain aluminum metal industrially by applying electricity to reduce bauxite ore. Hence, aluminum is much newer than other metals such as copper, iron, and gold, which have been used since pre-historical times. This is surprising since aluminum comprises 7.56 % of all elements near the surface of the earth, and is found in abundant amounts, next to only oxygen and silicon. The reason why aluminum metal only became available fairly recently is that aluminum has a strong chemical affinity to oxygen, and this prevents reduction of aluminum oxide by chemical reaction with carbon at high temperatures, unlike iron- and copper-oxides. Reduction of alumi-

*Modern Aspects of Electrochemistry,* Number 46, edited by S. –I. Pyun and J. -W. Lee, Springer, New York, 2009.

S.-I. Pyun and J.-W. Lee (eds.), *Modern Aspects of Electrochemistry 46:*     59
*Progress in Corrosion Science and Engineering I,*
DOI 10.1007/978-0-387-92263-8_2, © Springer Science+Business Media, LLC 2009

     H. Takahashi, M. Sakairi, and T. Kikuchi

## Table 1
### Applications for Anodic Films on Aluminum.

| Property | Function | Application |
| --- | --- | --- |
| Physico-mechanical | Hardness<br>Wear-resistance | Screw threads, gears, bolts/nuts, break discs, clutches, fans, nozzles, valves, cam tracks, fuel pumps, injectors |
| | Lubrication | Pistons/piton rings, rollers/bearings |
| | Membrane | Gas separators, molecule separators, super grids, templates for plasma etching, molds for nano-rods and -tubes |
| Chemical | Corrosion resistance | Aircraft, vehicles, boats, trains, architecture, household goods, sport goods, office supplies |
| | Wettability | Al litho (PS plate) |
| | Adhesion | Organic coating, gravure printing rolls |
| | Catalysis | Catalysts, catalysis supports |
| Electro-magnetic | Insulating | Printed circuit boards, IC boards, electric cables, ozonbe generators |
| | Dielectric | Electrolytic capacitors, humidity sensors, gas sensors |
| | Magnetic | Magnetic recording devices, rotary encoders |
| Optical | Luminescence | Electro-luminescence displays |
| | Absorption | Sun-light energy absorbing plates |
| Thermal | Emittance | Far infra red emitters |

num oxide was first realized by H. Davy in 1807, using Voltaic piles, which had been invented in 1800 by the Italian scientist, A. Volta.

Annually about 38 million tons of aluminum metal are newly produced worldwide from bauxite ore and most aluminum products are subjected to surface treatments such as anodizing, chemical conversion coating, painting, metal plating, and other treatments. Anodizing is one of the most common surface treatments of aluminum, and is performed for corrosion protection as well as for a wide variety of other purposes.[1-3] Table 1 shows examples of applications of aluminum with anodic oxide films.

Anodizing is generally performed for the corrosion protection of aircraft, automobiles, boats, trains, buildings, household articles, sport goods, office supplies, electronics, and other products. Hard anodizing is performed to improve the hardness and wear-resistance of machinery parts, for instance screw threads, gears, bolts, and brake discs. To improve the lubricating performance of pistons/piston rings and rollers/bearings, $MoS_2$ is filled into the pores of porous type anodic oxide films (PAOF). As membranes for the separation of gases and other chemical compounds, PAOF are also used after removal from the substrate where they were formed. Adhesion with organic coatings and wettability of anodic oxide films play an important role in pre-sensitized plates for offset printing. The high inner surface area of porous anodic oxide films makes them useful as catalysts and catalyst supports.

Barrier type anodic oxide films (BAOF) with high insulating and dielectric properties are used in printed circuit boards, IC-supporting plates, aluminum cables, electrolytic capacitors, humidity sensors, and other applications. Deposition of ferromagnetic substances, Fe, Ni, and Co, into the pores of PAOF causes anisotropy of magnetic properties, allowing usage as memory devices. Electro-luminescence of anodic oxide films on aluminum is applied to the fabrication of luminescent display boards, and black coloring with metal deposition is used for sun energy absorbing boards. Far infrared emitters can also be made using aluminum alloys covered with anodic oxide films.

The authors have been working on basic researches focusing on the structure and properties of anodic oxide films on aluminum, and have also performed research on applications developing novel technologies based on anodizing. This chapter will describe the structure and properties of anodic oxide films on aluminum and their significance for corrosion protection and micro- and nano-technologies.

## II.  STRUCTURE AND PROPERTIES OF OXIDE AND HYDROXIDE FILMS

Oxide and hydroxide films formed on aluminum may be classified into five categories: pretreated-surface film, thermal oxide film, hydrothermal oxide film, barrier type anodic oxide film (BAOF), and porous type anodic oxide film (PAOF) (Fig. 1).

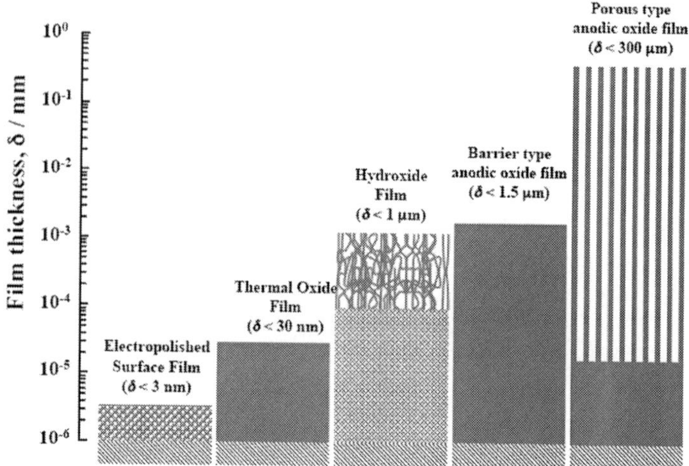


Figure 1. Structure of oxide and hydroxide films on aluminum.

## 1.    Pretreated-Surface Film

Pretreatments like degreasing, etching, polishing, and others are carried out before other surface treatments to control the uniformity of the chemical and physical properties of the aluminum surface. Phosphoric acid/chromic acid solution immersion (PC-treatment) is also used to remove oxide films from the aluminum substrate without damage to the substrate. Table 2 shows solution compositions and conditions of typical pretreatments of aluminum and aluminum alloys: mechanical polishing, electropolishing, chemical polishing, alkaline etching, and PC-treatment.

A suitable pretreatment or combination of pretreatments can be selected, depending on the structure and alloy composition of specimens and the subsequent process to be carried out after the pretreatment. For pure aluminum, electropolishing gives the smoothest surface in all pretreatments, but mechanical polishing is more significant for aluminum alloys, since preferential dissolution of alloying elements or the aluminum matrix makes surfaces rough with electropolishing. Mechanically polished surfaces are contaminated by small particles of diamond/alumina included   in

## Table 2
## Chemical Composition and Conditions for Typical Pre-Treatments of Aluminum.

| Pretreatment | Chemical composition | Condition |
|---|---|---|
| Mechanical polishing | Emery paper polishing with water flow<br>Buff polishing with diamond/alumina paste | |
| Electropolishing | 220-ml 60% $HClO_4$/780-ml $CH_3COOH$ | 283–285 K, 28–30 V, 3–5 min |
| Chemical polishing | 100 ml $HNO_3$/600ml $H_3PO_4$ | 368 K, 3–5 min |
| Alkaline etching | 0.5 M NaOH<br>50% $HNO_3$ | 293–298 K, 15 min<br>343 K, 1 min |
| PC treatment | 0.2-M $CrO_3$/0.5-M $H_3PO_4$ | 358–363 K, 5–20 min |

polishing paste, and the contaminants need to be removed by light etching. Chemical polishing doesn't require electricity, but requires relatively high temperatures. Alkaline etching is accompanied with a post-treatment in nitric acid solutions, and results in a less smooth surface. A thin oxide film is present on specimens after all the pretreatments. Table 3 shows the thicknesses and chemical compositions of films formed by electropolishing, chemical polishing, and PC-treatment.[4]

Surface films after the pretreatments have thicknesses of 3–5 nm and are contaminated with electrolyte anions from the solutions used. It is noteworthy that the surface film formed by electropolishing includes small amounts of $Cl^-$ ions, which have been incorporated due to the reduction of $ClO_4^-$ ions during electropolishing. Chemical polishing results in the formation of oxide film

## Table 3
## Thickness and Chemical Compositions of Surface Films Formed by Electropolishing, Chemical Polishing, and PC-Treatment. Reprinted with permission from *J. Metal Finishing Soc., Jpn.*, 36, (1985) 96. Copyright (1985), Metal Finishing Society Japan.

| Pretreatment | Thickness<br>nm | Chemical composition |
|---|---|---|
| Electropolishing | 306 | $AlO_x(OH)_{2.91-2x}Cl_{0.09}\cdot(x-0.62)H_2O$ |
| Chemical-polishing | 4.7 | $AlO_x(OH)_{2.45-2x}(PO_4)_{0.17}(NO_2)_{0.04}\cdot(x-0.73)H_2O$ |
| PC-treatment | 4.3 | $AlCr(III)_{0.43}O_x(OH)_{3.30-2x}(PO_4)_{0.33}\cdot(x-0.07)H_2O$ |

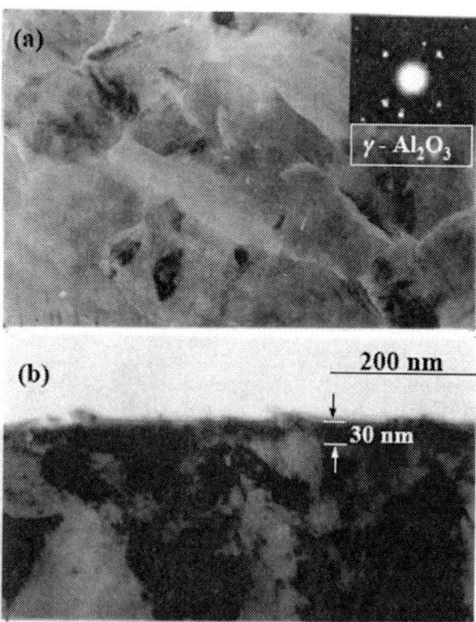



Figure 2. TEM images of a) removed thermal oxide film and b) vertical cross section of the film. Heating was carried out for 3h at 823 K in air. Reprinted with permission from *J. Electron Microsc.*, **40** (1991) 101. Copyright (1991), Japanese Society of Electron Microscopy, Oxford University Press.

containing phosphate and nitrite ($NO_2^-$) ions, while PC-treatment leads to the formation of oxide films containing Cr (III) and phosphate ions. The $Cl^-$ ions included in the surface film formed by electropolishing cause the formation of BAOF with relatively many imperfections, and PC-treatment after electropolishing improves the soundness of the BAOF.[5]

## 2.   Thermal Oxide Film

When aluminum is heated in air at 523–873 K, thermal oxide films composed of $\gamma$-$Al_2O_3$ are formed (Fig. 2):[6]

$$2\,Al + 3\,O_2 \rightarrow 2\,Al_2O_3 \tag{1}$$

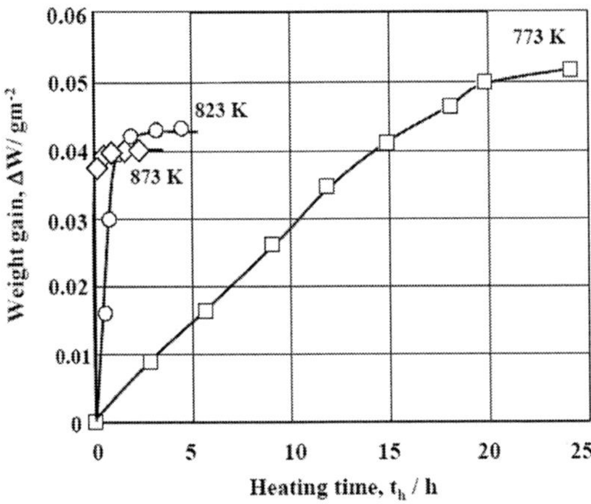



Figure 3. Weight gains at different heating temperatures.

The growth of thermal oxide films follows a parabolic rate law,[7] resulting in 30–50 nm film thicknesses after a few to a thousand hours of heating. The growth rate of the thermal oxide films strongly depends on temperature (Fig. 3),[6] and the kinetic energy for the parabolic growth rate is about 240 kJ/mol.[7] Thermal oxide film consists of an outer amorphous oxide layer and an inner crystalline oxide layer (Fig. 4), and crystalline oxide islands grow laterally by oxygen transport via channels in the amorphous oxide layer.[8]

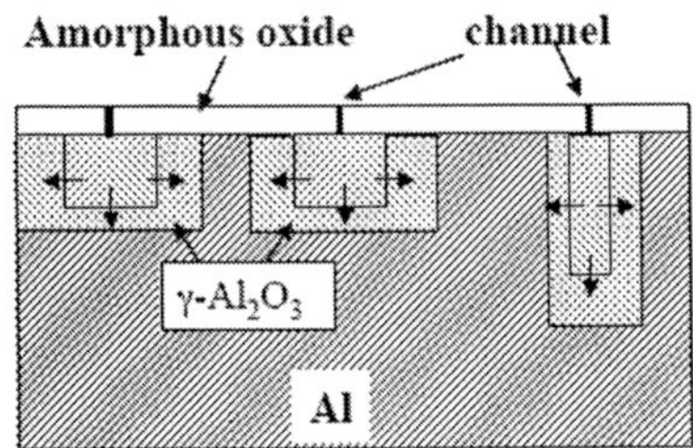



Figure 4. Schematic model of growth of thermal oxide films.

**Table 4**

**Chemical Compositions of Hydroxide Films Formed on Aluminum by Immersing in Boiling Distilled Water and Neutral Phosphate Solutions at Different Concentrations. Reprinted with permission from *J. Mineralogical Soc. Jpn.*, *"Koubutsu-gaku Zashi"*, 19 (1990) 387. Copyright (1990), Mineralogical Society of Japan.**

| Concentration of phosphate kmol m$^{-3}$ | Chemical composition of hydroxide films |
|---|---|
| 0 (distilled water) | $Al_2O_3 \cdot 2.4H_2O$ |
| $10^{-4}$ | $Al_2O_{2.6}(PO_4)_{0.28} \cdot 1.5H_2O$ |
| $10^{-3}$ | $Al_2O_{2.5}(PO_4)_{0.36} \cdot 1.6H_2O$ |
| $10^{-2}$ | $Al_2O_{2.2}(P_2O_7)_{0.4} \cdot 3.0H_2O$ |
| $10^{-1}$ | $Al_2O_{1.7}(P_2O_7)_{0.67} \cdot 3.3H_2O$ |

Coalescence of the islands results in the cease of film growth. The resulting film thickness strongly depends on the heating temperature,[6] and crystal orientation of the substrate.[8] In aluminum alloys containing Mg, oxidation takes place by a rapid nucleation of $MgAl_2O_4$, and MgO.[9] With PC-treated specimens there is slower oxidation than with electropolished or alkaline-etched specimens, and this suggests that chromic oxide and phosphate ions included in the surface film after PC-treatment suppress the transport of oxygen during thermal treatment.[6]

### 3.   Hydroxide Film

Immersion of aluminum in boiling pure water results in the formation of hydroxide films, which are composed of pseudo-boehmite.[10] The chemical composition of this hydroxide film has been determined to be $Al_2O_3 \cdot 2.4H_2O$ by X-ray photoelectron spectroscopy (XPS, Table 4). The crystallinity of the hydroxide film can be measured by Fourier transform infrared spectrometry (FTIR), which shows a sharp peak at 1100 cm$^{-1}$ and two broad peaks at 3300 and 600 cm$^{-1}$ (Fig. 5) for the boiled specimen, and the spectrum with light perpendicularly polarized to the surface gives more pronounced peaks, suggesting anisotropic structure (Fig. 6).[10]

The hydroxide films consist of two layers: a dense inner hydroxide layer and a porous outer hydroxide layer, and are less than

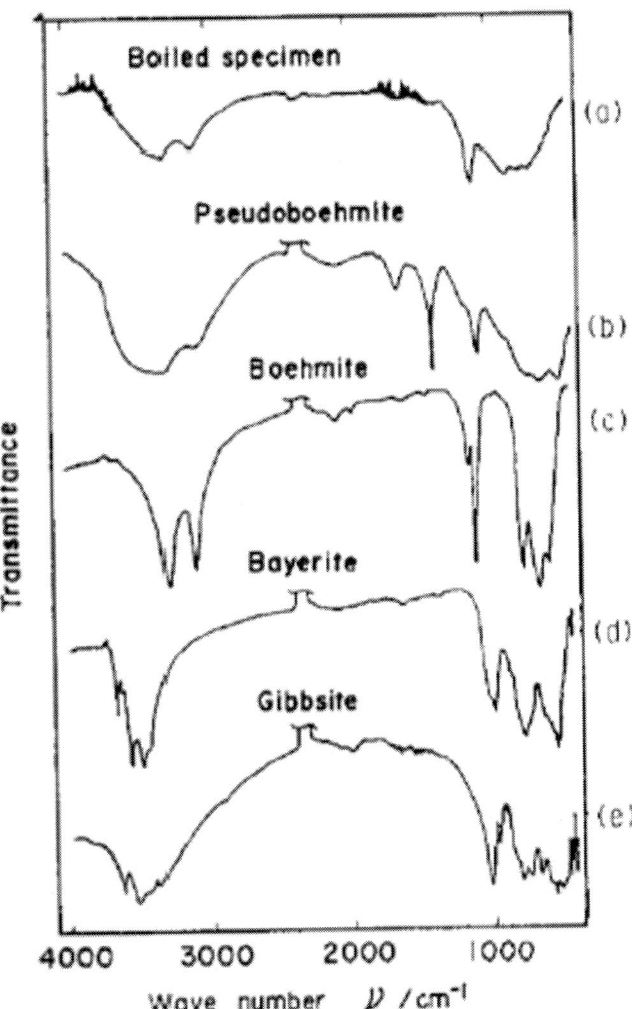



Figure 5. FTIR spectrum of (a) aluminum immersed in boiling distilled water for 60 min, and spectra of standard particles of (b) psuedo-boehmite, (c) boehmite, (d) bayerlite, and (e) gibbsite. Reprinted with permission from *J. Surface Sci. Soc. Jpn.*, **8**, (1987) 279. Copyright (1987), Surface Science Society of Japan.

1 μm thick (Fig. 7). Immersion in pure water at temperatures below 313 K results in films composed of bayerite.[11]

The structure and thickness of hydrous oxide films are highly sensitive to the chemical composition of solutions for hydrothermal treatment. In neutral solutions containing phosphate ions, the formation of hydroxide films is inhibited, and thinner hydroxide films are formed at higher concentrations between $10^{-4}-10^{-2}$ M (Fig. 8).[12] In $10^{-1}$ M phosphate solutions, relatively thick hydroxide films are formed on less smooth surfaces, because local dissolution of the substrate is enhanced. Analysis by XPS suggests

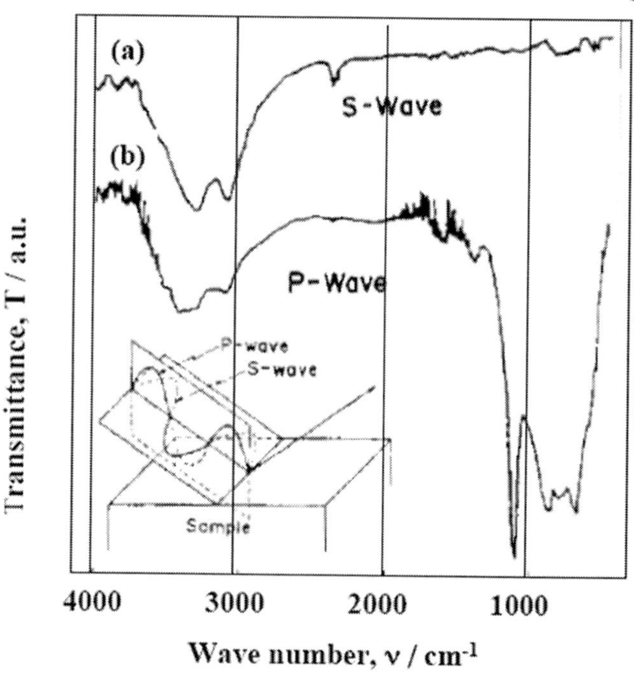


Figure 6. FTIR spectra of aluminum after immersion in boiling pure water, obtained with polarized light: (a) P-wave, (b) S-wave. Reprinted with permission from *J. Surface Sci. Soc. Jpn.*, **8**, (1987) 279. Copyright (1987), Surface Science Society of Japan.

that the chemical composition of hydroxide films strongly depends on the phosphate concentration (Table 4). With increasing phosphate concentration, the amount of phosphate incorporated in the hydroxide  droxide films increases and the incorporated phosphates change from ortho- to pyro-phosphates.

Boiling in amine solutions enhances the formation of hydroxide films, since a higher pH of the solution promotes the dissolution of the substrate.[13] The pH of pure water increases gradually with time during immersion of aluminum at 373 K, while the pH of solutions containing ethylamine decreases slightly with time (Fig. 9). The following reactions are considered to take place in pure water:

$$Al + 3\ H_2O \rightarrow x[Al(OH)_2]^+ + (1-x)\ Al(OH)_3 + x\ OH^- + 3\ H_2 \tag{2}$$

and in amine solutions,

Figure 7. TEM image of a vertical cross section of specimen immersed in boiling doubly distilled water for 1h. Reprinted with permission from *J. Metal Finishing Soc. Jpn.*,  38, (1987) 67. Copyright (1987), Metal Finishing Society of Japan.

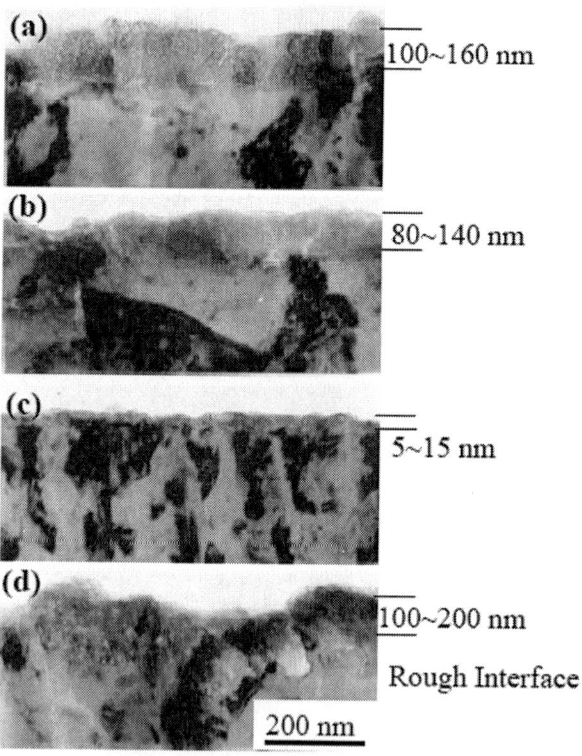


Figure 8. TEM images of the vertical cross sections of aluminum immersed in neutral phosphate solutions (pH = 7) at 373 K for 1 h: (a) $10^{-4}$, (b) $10^{-3}$, (c) $10^{-2}$, and (d) $10^{-1}$ M. Reprinted with permission from *J. Mineralogical Soc. Jpn., "Koubutsugaku Zashi"*, **19** (1990) 387. Copyright (1990), Mineralogical Society of Japan.

$$Al + 3 H_2O + x\, OH^- = x\, [Al(OH)_4]^- + (1 - x)\, Al(OH)_3 + 3 H_2$$

$$(3)$$

The mass of hydroxide films formed in $10^{-1}$ M ethylamine solution is three times those formed in pure water (Fig. 10). The XPS analysis suggests that no nitrogen-bearing species are included in the hydroxides formed in ethylamine solutions. Silicate

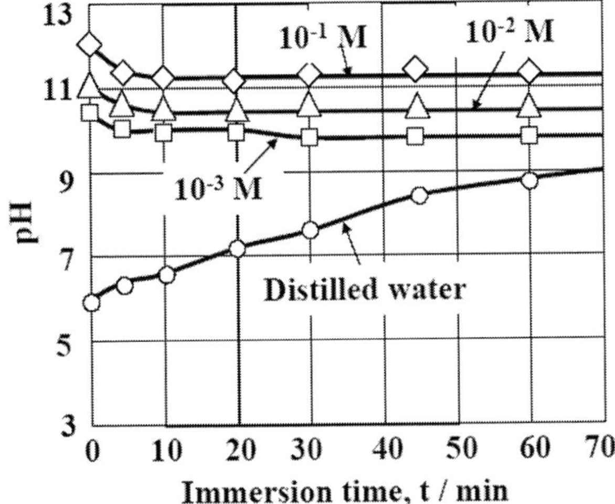



Figure 9. Change in the pH of distilled water and ethylamine solutions of different concentrations during aluminum immersion. Solution volume: 30 cm³. Specimen surface area : 16 cm².

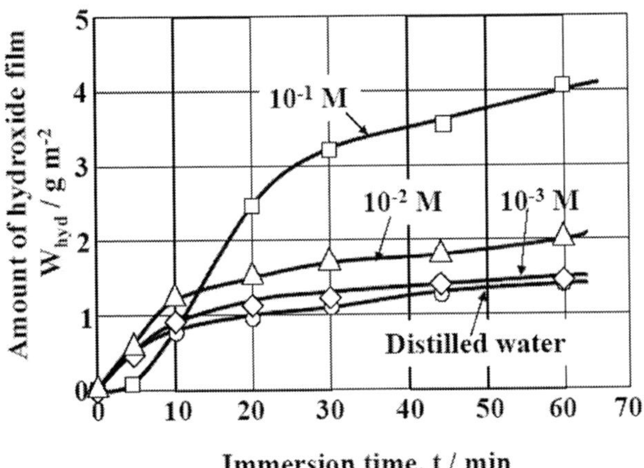



Figure 10. Changes in the amount of hydroxide film formed with immersion time in distilled water and ethylamine solutions of different concentrations.

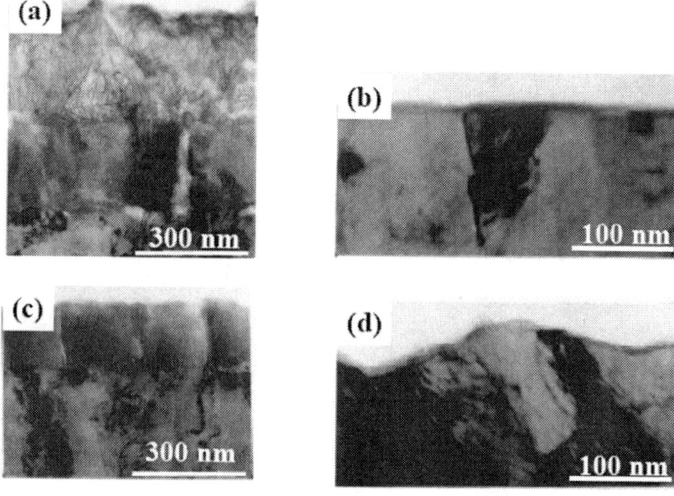

Figure 11. TEM images of the vertical cross sections of aluminum immersed for 1h at 373 K in neutral silicate solutions at (a) $10^{-4}$ and (b) $10^{-2}$ M and in neutral citrate solutions at (c) $10^{-4}$ and (d) $10^{-2}$ M. Reprinted from *Corr. Sci.*, **31**, (1990) 243. Copyright (1990), with permission from Elsevier.

ions in boiling neutral solutions do not affect the formation of hydroxide films at concentrations as low as $10^{-4}$ M (Fig. 11a), while, at $10^{-2}$ M, the hydroxide formation is suppressed vigorously resulting in 5–8 nm thick films (Fig. 11b).[14] Citrate ions suppress hydroxide formation at $10^{-4}$ M by forming complexes with aluminum ions (Fig. 11c), while citrate ions enhance the dissolution of the substrate at $10^{-2}$ M, so that thinner hydroxide films are formed on rougher surfaces (Fig. 11d).[14]

## 4.  Barrier-Type Anodic Oxide Film (BAOF)

When aluminum and aluminum alloys are polarized anodically in electrolytic solutions, oxide films are formed on the substrates. This procedure is termed *anodizing* and the oxide films formed by this procedure are termed *anodic oxide film*. Anodic oxide films can be classified into two groups: barrier and porous type anodic oxide films.

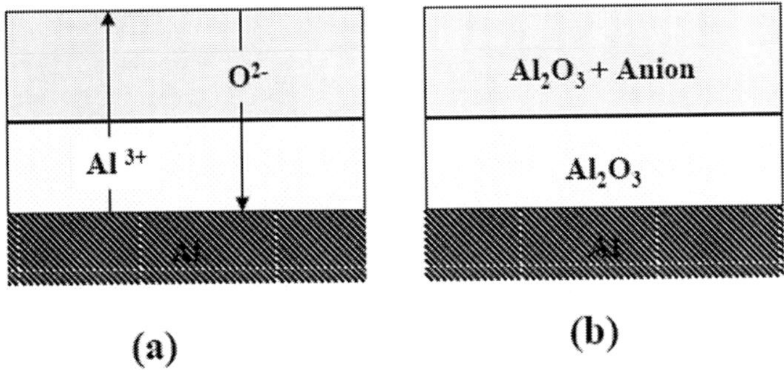


Figure 12. Schematic model of a) ion transport during BAOF growth and b) chemical composition of outer and inner layers.

Barrier type anodic oxide film (BAOF) is obtained in neutral solutions containing borate, phosphate, adipate, and other solutions, and consists of a thin compact amorphous oxide layer.[15] Electrochemical reactions   on the anode and cathode are   expressed by

$$2\ Al + 3\ H_2O \rightarrow Al_2O_3 + 6\ H^+ + 6\ e^- \quad \text{(anode)} \tag{4}$$

$$6\ H^+ + 6\ e^- \rightarrow 3\ H_2 \qquad\qquad \text{(cathode)} \tag{5}$$

Oxide formation in BAOF occurs at the interfaces between solution and oxide, and between oxide and substrate by the transport of $Al^{3+}$ ions and $O^{2-}$ ions across the oxide film (Fig. 12a). The oxide formation at the oxide/solution interface is accompanied by incorporation of electrolyte anions, so that BAOF consists of an outer anion-incorporated oxide layer and an inner pure-alumina layer (Fig. 12b). The ionic current, $i_i$, during BAOF formation can be expressed by the following equation:[16]

$$i_i = A\ \exp(BE) \tag{6}$$

where $A$ and $B$ are proportionality constants and $E$ is the average electric field across the BAOF. Equation (6) can be substituted with the following equation:

$$i_i = (nvq)\ \exp[-(W - qaE)/kT] \tag{7}$$

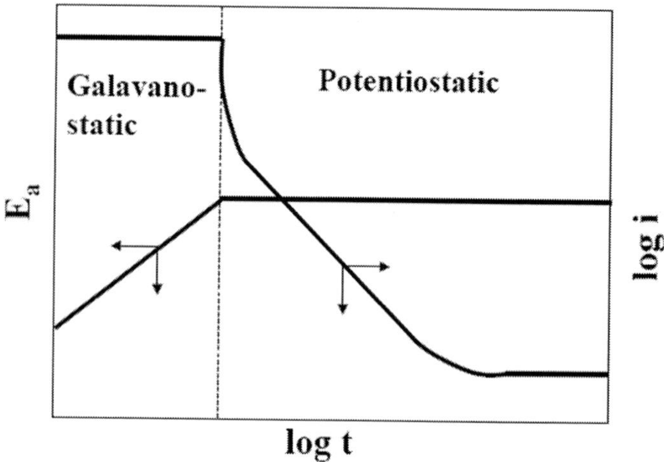


Figure 13. Changes in anode potential, $E_a$, and anodic current, $i_a$, under galvanostatic and potentiostatic anodizing conditions.

where $n$ is the number of mobile ions, $\nu$ the vibration frequency of the mobile ions, $q$ the charge of mobile ions, $W$ the activation energy for transport through the film, $a$ the half distance between interstitial positions in the oxide lattice, $k$ the Boltzman constant, and $T$ the absolute temperature. Equation (7) was first derived by Cabrera and Mott[17] and later modified by Verway.[18]

At the start of anodizing under galvanostatic conditions, the anode potential increases linearly with time, and subsequently a rapid switch to potentiostatic anodizing results in a current decay reaching a small steady value (Fig. 13).[19] The current decay shows a slope of −1 in log. $i$ vs. log $t_a$ plots, suggesting the anti-logarithmic growth law for oxide formation. The steady current observed after long anodizing is termed a *leakage current* and corresponds to an electronic current through defects of oxide as well as an ionic one that is responsible for the renewal of oxide by dissolution and formation. The leakage current is higher at higher temperatures, due to a more significant oxide renewal, and this results in the growth of a porous layer over the compact barrier layer (Fig. 14).[15]

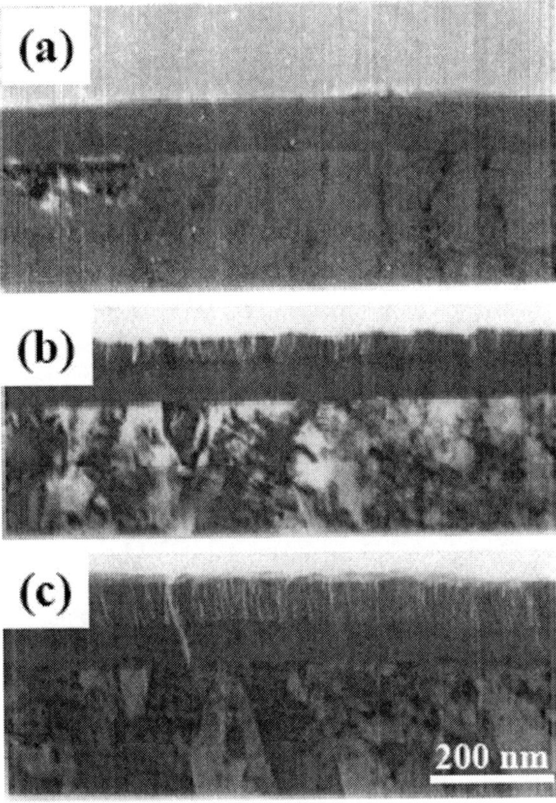

Figure 14. TEM images of vertical cross sections of aluminum anodized potentiostatically at 50 V in a neutral borate solution at 313 K for (a) 5 , (b) 60, and (c) 120 min. Reprinted from *Electrochim. Acta,* **25**, (1980) 279. Copyright (1990), with permission from Elsevier.

The film thickness, $\delta$, is proportional to the potential, $E_a$, applied during anodizing, and can be expressed by the following equation:[15,20–21]

$$\delta = K \times E_a \qquad (8)$$

where $K$ is a proportionality constant, generally 1.3–1.6 nm/V. The outer layer of BAOF includes electrolyte anions as shown in Fig. 12b, and the depth profile of anion concentrations has a steady value throughout the outer layer.[20-22] The outer layer can easily be distinguished from the inner layer by an electron-beam-induced crystallization method, in which thin slices of vertical cross sections of oxide films are irradiated with electron beams to crystallize only the inner layer.[23] The ratio of the thickness of the outer layer to that of the inner layer is mainly determined by the transport numbers of $Al^{3+}$ ions and $O^{2-}$ ions. However, the thickness ratio depends on the electrolytic solution used even under the same anodizing conditions, since anions incorporated into the oxide are transported either outwards or inwards during anodizing (Fig. 15).[24] This can be ascertained by comparing the positions of markers and the interface between the inner/outer layers. In borate solutions, the marker locates at the interface between the inner and outer layers, and this suggests an absence of transport of borate during anodizing (Fig. 15a). The inner/outer layer interface in the oxide film formed in phosphate solutions locates underneath  the marker position, suggesting an inward transport of phosphate ions (Fig. 15b). Anodizing in chromate solutions leads to the incorporation of chromic ions, and the inner/outer layer interface is below the marker position, suggesting outward transport of chromic ions (Fig. 15c).

Galvanostatic anodizing in neutral solutions gives rise to a linear increase in $\delta$ with time, and eventually the oxide formation

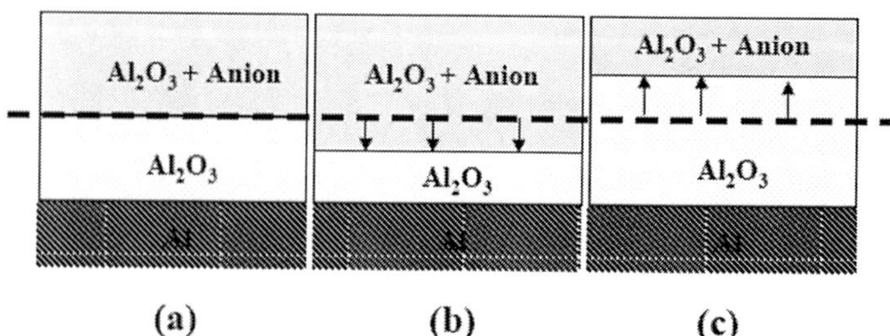


Figure 15. Schematic model of the transport of anions during anodizing: (a) no transport, (b) inward transport, and (c) outward transport.

**Table 5**
**Chemical Compositions, pH, and Specific Resistivities of Borate Solutions. Reprinted with permission from *J. Electrochem. Soc.*,144, (1997) 866. Copyright (1997), The Electrochemical Society.**

| No. | Chemical composition | pH | Specific resistivity $\Omega$cm |
|-----|----------------------|-----|----------------|
| I | 0.5-M $H_3BO_3$ | 3.9 | 34 500 |
| II | 0.5-M $H_3BO_3$/0.005-M $Na_2B_4O_7$ | 6.0 | 1 470 |
| III | 0.5-M $H_3BO_3$/0.05-M $Na_2B_4O_7$ | 7.4 | 180 |

stops abruptly, at a *film breakdown* potential, $E_{bd}$, which strongly depends on the electrolyte and the concentration of anodizing solution (Table 5, Fig. 16).[25] In boric acid / borate solutions, the film

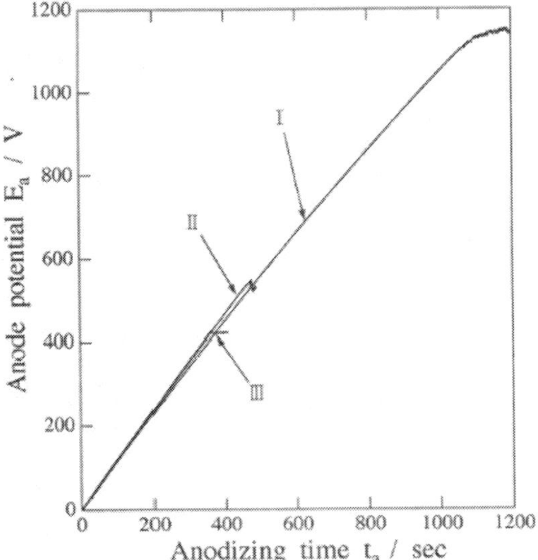


Figure 16. Changes in anode potential, $E_a$, with time, $t_a$, during anodizing in neutral borate solutions at different concentrations at 298 K and 25 A m$^{-2}$. Reprinted with permission from *J. Electrochem. Soc.*, **144**, (1997) 866. Copyright (1997), The Electrocheical Society.

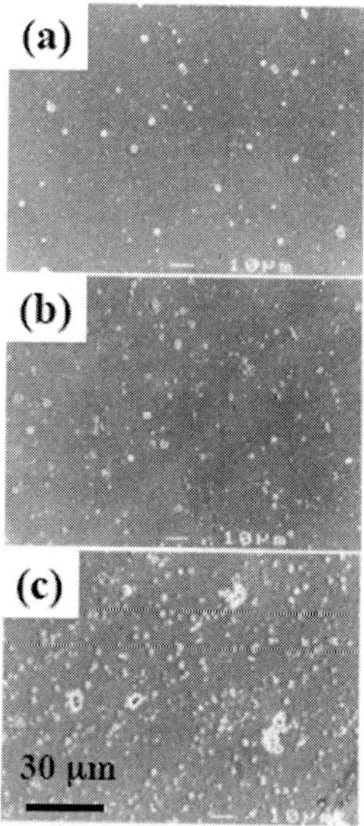


Figure 17. SEM images of the surface of anodic oxide films formed by anodizing in 0.5 M $H_3BO_3$ up to (a) 200 , (b) 700, and (c) 1180 V. Reprinted with permission from *J. Electrochem. Soc.*, **144**, (1997) 866. Copyright (1997), The Electrochemical Society.

breakdown potential increases with decreasing borate concentration, or increasing solution specific resistivity (Table 5). The 0.5 M boric acid solution, which has a 34,500 $\Omega$cm specific resistivity, shows $E_{bd}$ values as high as 1,000 V, resulting in film thicknesses of 1.3–1.6 μm, with numerous imperfections included in the oxide films (Fig. 17). The number of imperfections increases only slightly with increasing $E_a$, but the size of the imperfections becomes much larger (Fig. 18).

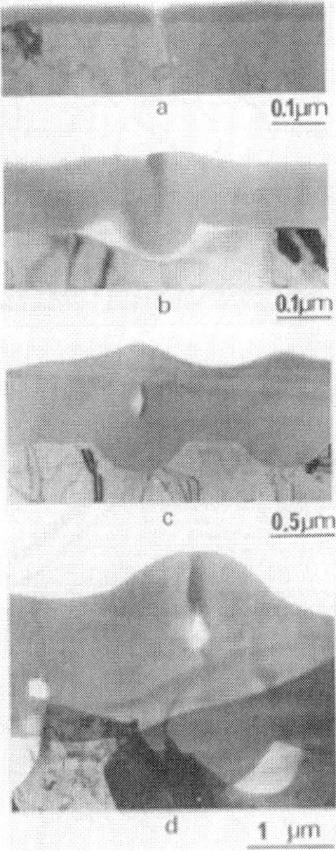

Figure 18. TEM images of the vertical cross sections of aluminum anodized in 0.5 M $H_3BO_3$ up to (a) 20, (b) 100, and (c) 4500 V, and (d)1180 V. Reprinted with permission from *J. Electrochem. Soc.*, **144**, (1997) 866. Copyright (1997), The Electrochemical Society.

Anodizing in $5 \times 10^{-5}$ M KOH solution, which has 166,700 $\Omega$ cm of specific resistivity, gives a value of $E_{bd}$ as high as 1600 V, and forms BAOF with many imperfections.[26]

The film breakdown during anodizing in neutral solutions can be explained in terms of electron avalanches,[27] electrostrictive forces,[28] or mechanical stresses.[29]

## 5.    Porous Type Anodic Oxide Film (PAOF)

### (*i*)   *Structure*

Porous type anodic oxide film (PAOF) is obtained by anodizing in acid solutions like $H_2SO_4$, oxalic acid, $H_3PO_4$, and $H_2CrO_4$, and possesses the *hexagonal cell model structure* morphology, shown in Fig. 19.[30]

Anodizing conditions in different solutions are shown in Table 6. The oxide film consists of numerous fine hexagonal cells perpendicular to the metal substrate, and each cell has a pore at the center. The pores are separated from the metal substrate by a thin hemispherical *barrier layer*, existing at the interface between the oxide film and the metal substrate.

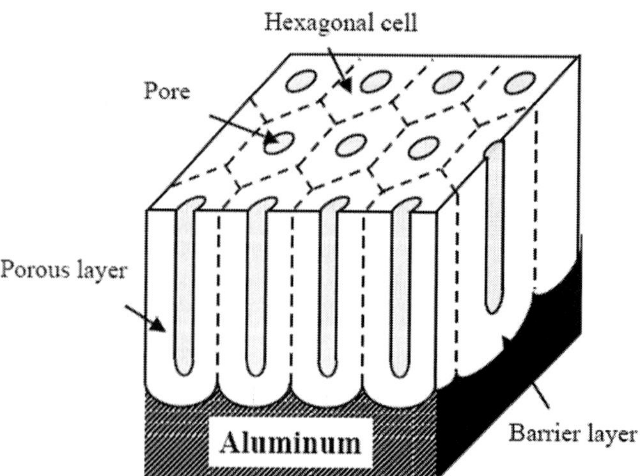


Figure 19. Schematic model of the structure of porous anodic oxide film formed on aluminum.

**Table 6**
**Conditions for Anodizing in Different Solutions.**

| Procedure | Chemical composition | T ($^0$C) | Voltage (V) | c.d. (A dm$^{-2}$) | Appearance, etc. |
|---|---|---|---|---|---|
| Sulfuric acid | Sulfuric acid 5–25 wt% <br> Aluminum sulfate 0.1–5 wt% | 15–25 | 15–20 (DC) | 0.8–3 | Transparent |
| Oxalic acid | Oxalic acid 3–5 wt% | 20–30 | 25 (DC) | 1–1.5 | Yellowish |
| Chromic acid | Chromic acid 10 wt% | 45–55 | 40–50 (DC) | 0.3–1 | Grayish |
| Phosphoric acid | Phosphoric acid 3–10 wt% | 20–30 | 40–100 (DC) | 0.5–2 | Light blue |
| Hard anodizing | Sulfuric acid 10–20 wt% <br> Aluminum sulfate 0.1–10 wt% <br> Oxalic acid 1 wt% | 0–5 | 25–60 (DC) | 2–4 | Transparent, $H_v > 600$ |
| Kalcolor | Sulfuric acid 5 wt% <br> Sulfo-silicilic acid 10 wt% | 22–25 | 25–70 (DC) | 2–3.2 | Bronze-black |
| Alkaline anodizing | Sodium hydroxide 8–12 wt% <br> Hydrogen peroxide 2–3 wt% <br> Sodium phosphate 0.1–0.5 wt% | 10–20 | 30–70 (DC) | 1–4 | Resistant in basic solution |
| A.C. anodizing | Sulfuric acid 15–30 % | 0–40 | 15–30 (AC) | 3–12 | Soft, flexible, sulfide included |
| Molten salt anodizing | Potassium bisulfate 66 mol% <br> Sodium bisulfate 33 mol% | 180 | > 160 (DC) | 1 | α-alumina |

Galvanostatic anodizing gives rise a steady potential value, $E_s$, after an initial transient period, and allows a steady increase in the film thickness, maintaining the number of cells, $N_c$, the size of cells, $D_c$, and pores, $D_p$, and the barrier layer thickness, $\delta_b$, during anodizing. The thickness of PAOF, $\delta_{pf}$, is proportional to the anodizing time, $t_a$, and current density, $i_a$.

$$\delta_{pf} = k_t \times i_a \times t_a \tag{9}$$

where $k_t$ (= 3.6 to 4.5 x $10^{-5}$ $\mu$m C$^{-1}$ m$^2$) is the proportionality constant, which is higher at　lower temperatures and lower acid concentrations. Figures 20 and 21 show the relationship between the current density, $i_a$, plotted logarithmically, and steady values of the anode potential, $E_s$, obtained in $H_2SO_4$ and oxalic acid solutions.[2]

The $E_s$ vs. log $i_a$ relationship is an *S-shaped curve* in both solutions, and it shifts to higher current density regions at higher temperatures and higher acid concentrations. Anodizing in $H_2SO_4$ solutions, which requires lower $E_s$ than in oxalic solutions, has the advantage of lower electric power consumption. Anodizing should be performed under the conditions at the mid regions of the $E_s$ vs. log $i_a$ relationship since anodizing at the lower regions easily leads to *powdering* of the oxide and in the higher regions to *burning*. Powdering is the phenomenon where the outer parts of PAOF become powdery by thinning of the pore walls, due to chemical dissolution, while burning is the phenomenon where the oxide film grows non-uniformly by local heating, due to a non-uniform current distribution.

The values of $N_c$, $D_c$, $D_p$, and $\delta_b$ are functions of $E_s$, as shown in Figs. 22-25, and only slightly dependent on temperature, kind of acid solution, and acid concentration[2]. With increasing $E_s$, the values of $N_c$ decrease, while $D_c$, $D_p$, and $\delta_b$ increase. The curves in Figs. 22-25 lead to the following equations:

$$N_c(\text{m}^{-2}) = 14 \times 10^{15} \times E_s^{-1.27}(\text{V}) \tag{10}$$

$$D_p(\text{nm}) = 14 + 0.21 \times E_s(\text{V}) \ [Es < 15 \text{ V}] \tag{11}$$

$$= 4.2 + 0.84 \times E_s(\text{V}) \ [E_s > 15 \text{ V}] \tag{12}$$

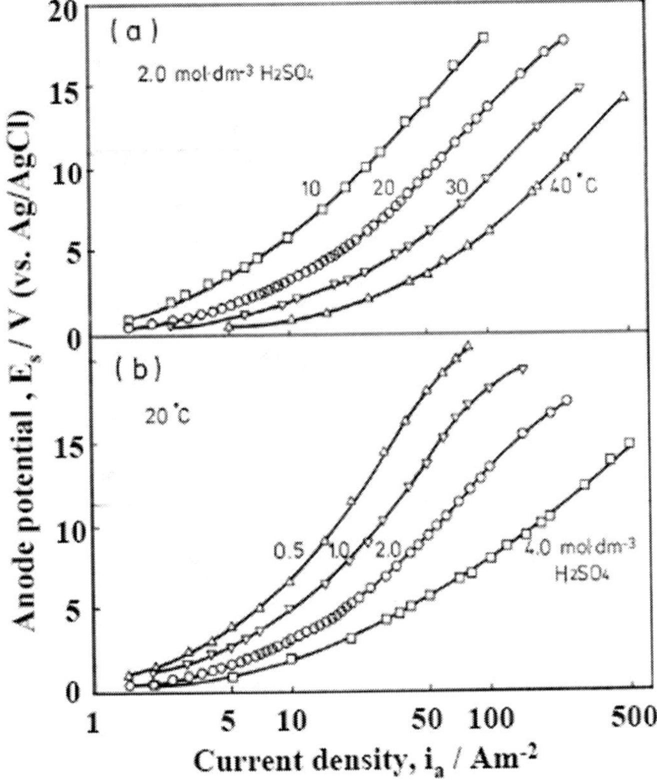



Figure 20. Relationship between steady anode potentials, $E_s$, and current density, $i_a$, in $H_2SO_4$ solutions. Reprinted with permission from *J. Metal Finishing Soc. Jpn*, **33**, (1982) 156. Copyright (1982), Surf. Finishing So. Jpn.

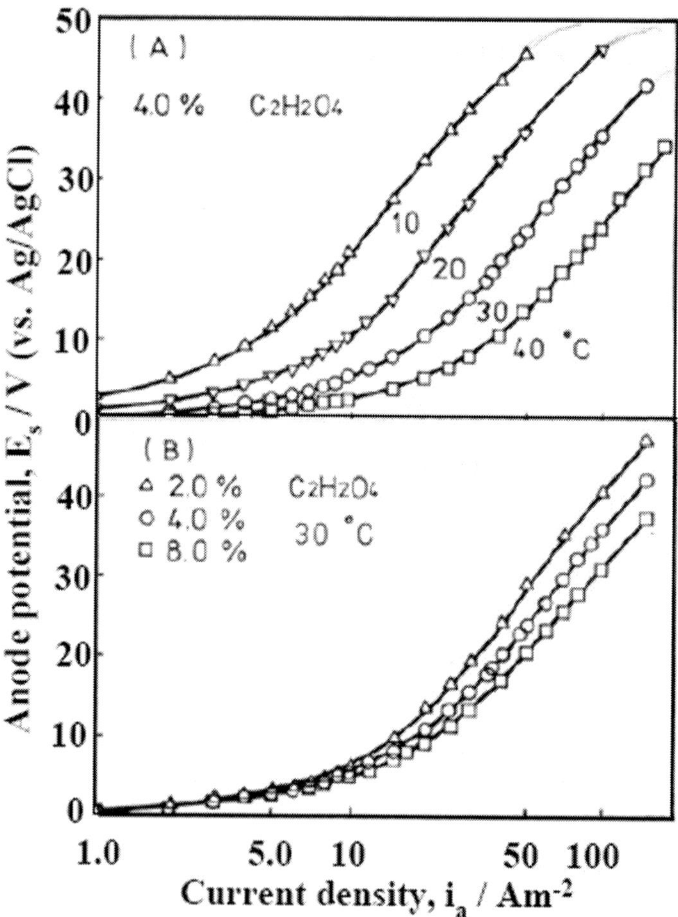


Figure 21. Relationship between steady anode potentials, $E_s$, and current density, $i_a$, in $H_2C_2O_4$ solutions. Reprinted with permission from *J. Metal Finishing Soc. Jpn, **34**, (1983) 548.* Copyright (1983), Surf. Finishing So. Jpn.

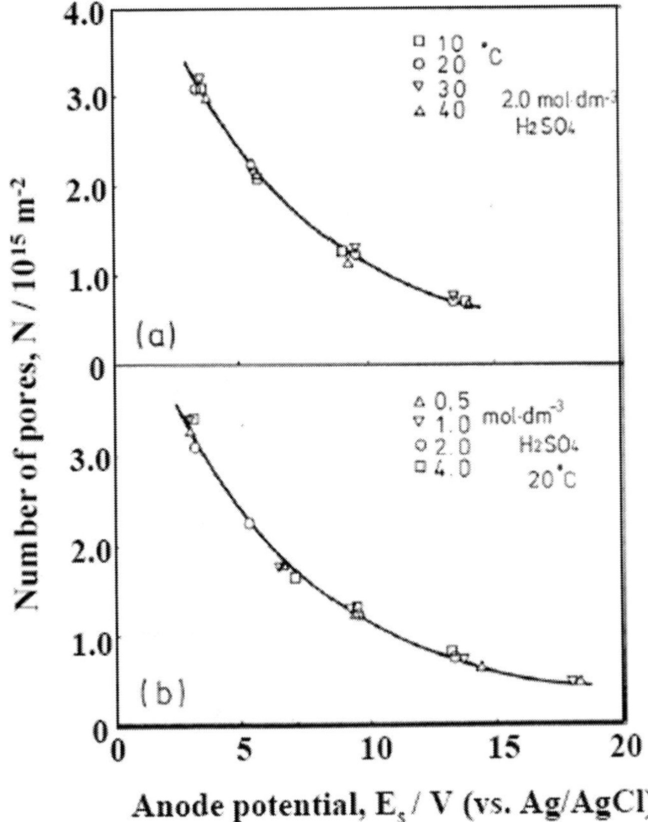


Figure 22. Number of pores vs. voltage curves obtained for different (a) temperatures and (b) electrolyte concentrations. Reprinted with permission from *J. Metal Finishing Soc. Jpn*, **33**, (1982) 156. Copyright (1982), Surf. Finishing So. Jpn.

$$D_c(\text{nm}) = 6.5 + E_s\,(\text{V}) \qquad\qquad (13)$$

$$\delta_b(\text{nm}) = 2.1 \times E_s^{0.83}\,(\text{V}) \qquad\qquad (14)$$

Hence, the porosity, $\alpha = N\pi D_p^2/4$,[19] of the oxide film can be expressed in the following equation:

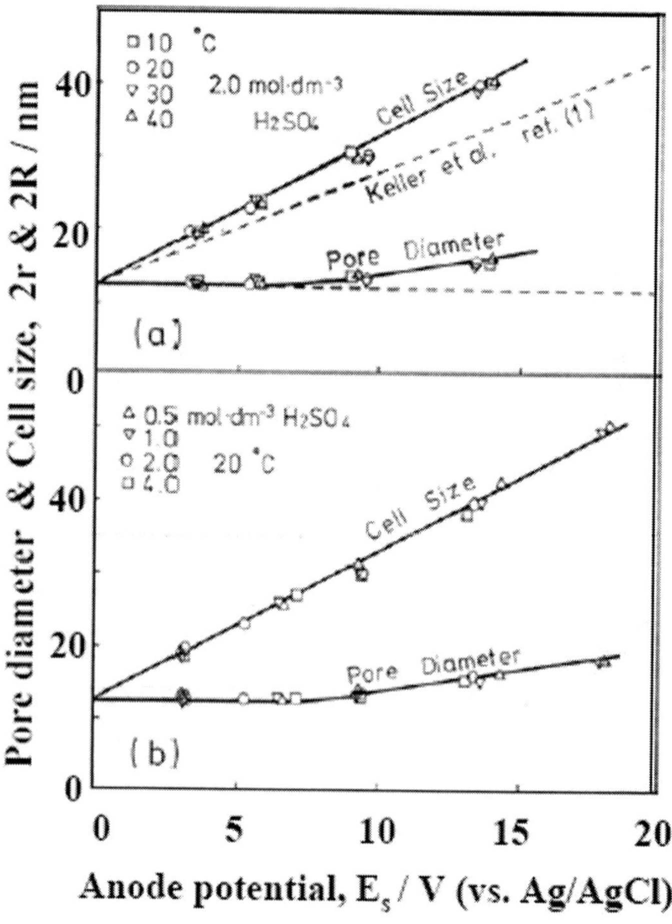



Figure 23. Changes in cell size, 2R, and pore-diameter, 2r, as functions of (a) temperature and (b) electrolyte concentration. Data reported by Keller et al. are indicated as broken lines in (a). Reprinted with permission from *J. Metal Finishing Soc. Jpn,* **34**, (1983) 548. Copyright (1983), Surf. Finishing So. Jpn.

$$\alpha = 0.8 \times E_a^{-0.59} \ (\text{V}) \tag{15}$$

The $\alpha$ value decreases sharply with increasing $E_s$ (Fig. 25).

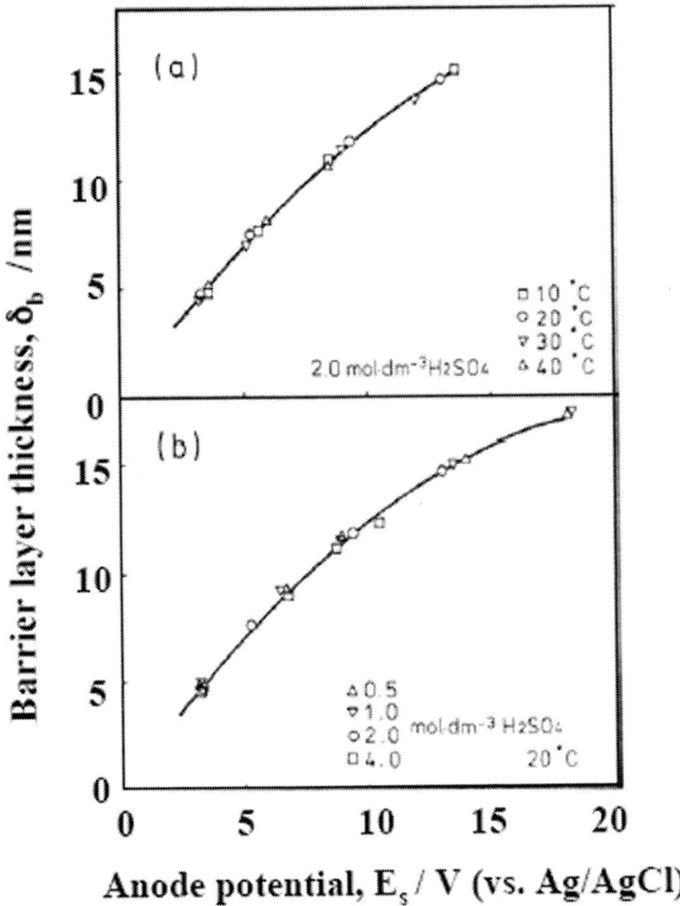


Figure 24. Barrier layer thickness vs. voltage curves obtained for different (a) temperatures and (b) electrolyte concentrations. Reprinted with permission from *J. Metal Finishing Soc. Jpn*, **33**, (1982) 156. Copyright (1982), Surf. Finishing So. Jpn.

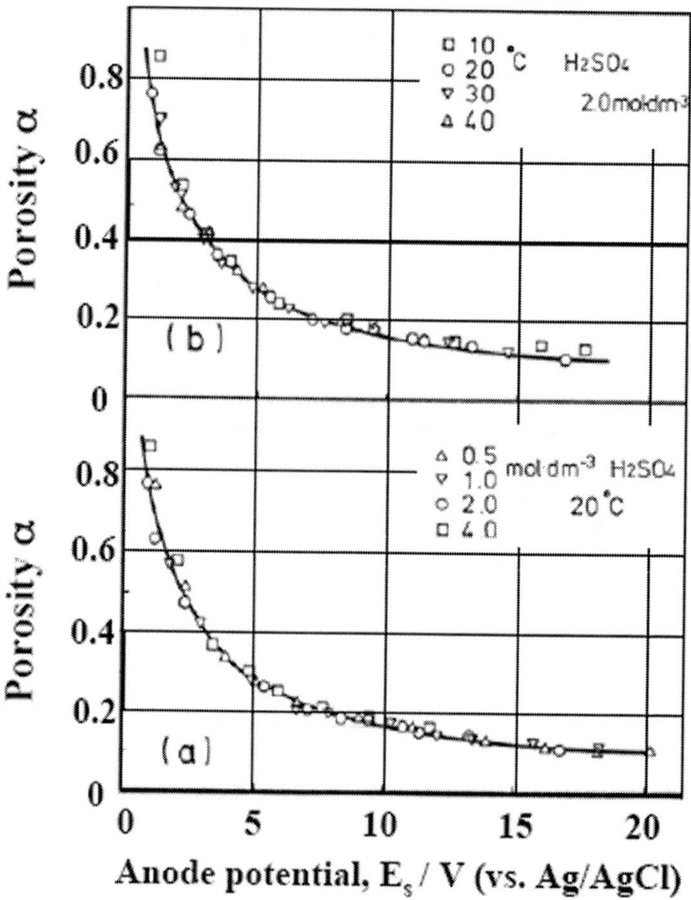



Figure 25. Changes in porosity, a, with steady anode potential, $E_s$, obtained in $H_2SO_4$ solutions. Reprinted with permission from *J. Metal Finishing Soc. Jpn,* **33**, (1982) 156. Copyright (1982), Surf. Finishing So. Jpn.

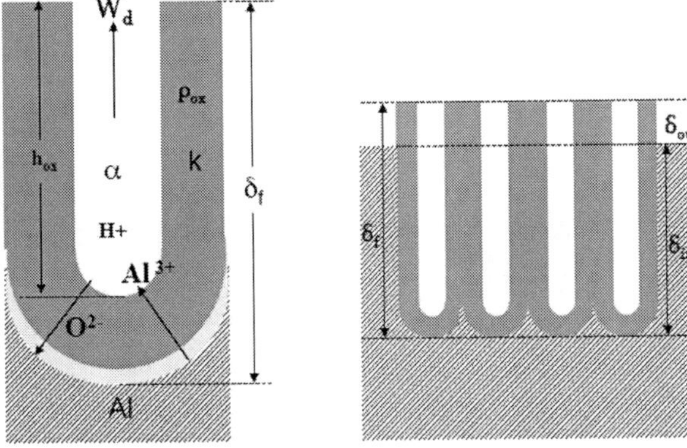


Figure 26. Schematic model of (a) ion transport and (b) volume expansion during PAOF growth by anodizing.

## (ii) Growth under Steady Conditions

The mechanism for PAOF growth during anodizing under steady current conditions in acid solutions are illustrated in Fig. 26a. A high electric field is sustained by the scallop-shaped barrier layer, and this gives rise to the transport of $Al^{3+}$ and $O^{2-}$ across the barrier layer. New oxide is formed at the interface between the barrier layer and the metal substrate, while $Al^{3+}$ ions are ejected directly into the solution after reaching the interface between the solution and oxide (pore bottom), i.e. dissolving into the solution without forming oxide (electrochemical dissolution). The direct ejection of $Al^{3+}$ ions is due to a rapid chemical dissolution of oxide at the   pore bottom at low pH.   Thus, in the steady condition, the thickness of the barrier layer, $\delta_b$, remains constant, and pores become deeper at a steady rate. The number of pores, $N_p$, pore diameter, $2r$, and cell size, $2R$, are also constant. The total current, $i_a$, is the sum of the current for electrochemical dissolution, $i_d$, and the oxide formation current at the oxide/substrate interface, $i_{ox}$, assuming that the electronic current is negligibly small:

$$i_a = i_d + i_{ox} \tag{16}$$

The amount of aluminum oxidized during anodizing, $W_{T(Al)}$, can be expressed by

$$W_{T(Al)} = (\, i_a t_a / nF)M_{Al} \tag{17}$$

where $t_a$ is anodizing time, $n$ the number of electrons for Al oxidation, $F$ the Faraday constant, and $M_{Al}$ the atomic mass of Al. The total amounts of dissolved $Al^{3+}$ ions during anodizing, $W_{d(Al)}$ is the sum of the Al amounts dissolved electrochemically and chemically, and can be expressed by

$$W_{d(Al)} = W_{T(Al)} T_{Al} + \alpha h_{ox}\rho_{ox}k \tag{18}$$

where $T_{Al}$ is the transport number of $Al^{3+}$ ions across the barrier layer, $\alpha$ is the porosity of pores, $h_{ox}$, the depth of pore, $\rho_{ox}$, the density of the oxide, and $k$ the mass per cent of $Al^{3+}$ ions in the oxide. The amount of $Al^{3+}$ ions incorporated in the oxide film, $W_{ox(Al)}$, is the difference between $W_{T(Al)}$ and $W_{d(Al)}$:

$$W_{ox(Al)} = W_{T(Al)} - W_{d(Al)} \tag{19}$$

The volume of aluminum oxidized, $V_{T(Al)}$, is correlated with $W_{T(Al)}$ by the following equation:

$$V_{T(Al)} = W_{T(Al)} / \rho_{Al} \tag{20}$$

where $\rho_{Al}$ is the density of Al. The apparent volume of oxide, $V_{T(ox)}$, is the sum of the actual volume of oxide, $V_{ox}$ and the pore volume, $V_p$, and is expressed by

$$V_{T(ox)} = V_{ox} + V_p = W_{T(ox)} / \rho_{ox} \tag{21}$$

$$= W_{T(Al)} (1 - T_{Al}) / (k\rho_{ox}) \tag{22}$$

where $W_{T(ox)}$ is the sum of the amount of oxide, $W_{ox}$, and $\alpha h_{ox}\rho_{ox}$. The Pilling-Bedworth ratio, $R_{P.B.}$, which is the ratio of the apparent oxide volume to the oxidized Al volume, can be derived by dividing Eq. (20) by Eq. (22), giving

$$R_{P.B.} = V_{T(ox)} / V_{T(Al)} = (1 - T_{Al}) / (k\rho_{ox}/\rho_{Al}) \tag{23}$$

Assuming $T_{Al} = 0.3$,[31] $k = 0.46$,[32] $\rho_{ox} = 2.8 \times 10^{-3}$ kg / mol,[33] and $\rho_{Al} = 2.7 \times 10^{-3}$ kg / mol, the value of $R_{P.B.}$ can be estimated to be 1.47. This strongly suggests that aluminum specimens undergo volume expansions by the formation of PAOF. The volume expansion ratio, $R_{exp}$, is defined by

$$R_{exp} = \delta_{out} / \delta_{in} = \delta_{out} / (\delta_f - \delta_{out}) \tag{24}$$

where $\delta_{out}$ is the film thickness over the initial surface of the specimen and $\delta_{in}$ is the film thickness inside the specimen (Fig. 26b). The $R_{exp}$ is correlated with $R_{P.B.}$ by the following equation:

$$R_{exp} = R_{P.B.} - 1 = (1 - T_{Al}) / (k\rho_{ox}/\rho_{Al}) - 1 \tag{25}$$

Figure 27 shows a SEM image of a vertical cross section of an aluminum specimen anodized in 0.22 M oxalic acid at T = 293 K with 150 Am$^{-2}$ for 4h. It is clear that the specimen expands by anodizing, and the $R_{exp}$ value is 0.4 under this anodizing condition.

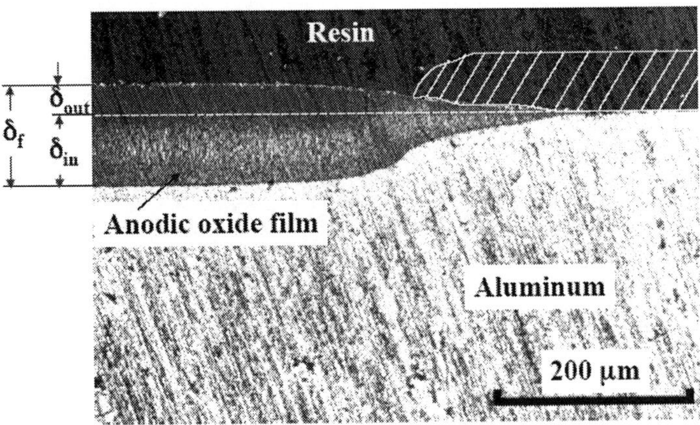


Figure 27. SEM image of the vertical cross section of an aluminum specimen anodized in 0.22 kmol m$^{-3}$ oxalic acid at T = 293 K with 150 Am$^{-2}$ for 4h. The hatched area shows an area where insulating film was attached.

As shown in Fig. 26, $O^{2-}$ ions transport under a high electric field across the barrier layer during anodizing. Electrolytic anions also transport inwards across the barrier layer at a rate slower than $O^{2-}$ ions. As a result, PAOF contains small amounts of anions of the electrolyte used. These anions distribute across the pore wall and barrier layer as shown in Fig. 28.[34] Longer anodizing causes the chemical dissolution of oxide, leading to the formation of cone-shaped pores (Fig. 29).[35] Eventually, the pore walls become substantially thinner, and are destroyed. This is the phenomenon of *powdering* of the oxide film, and the onset of powdering terminates the growth of  PAOF.  Lower temperatures and lower acid concentrations enable the formation of thicker films without powdering, and maximum thicknesses of several hundred μm are possible under optimum conditions.

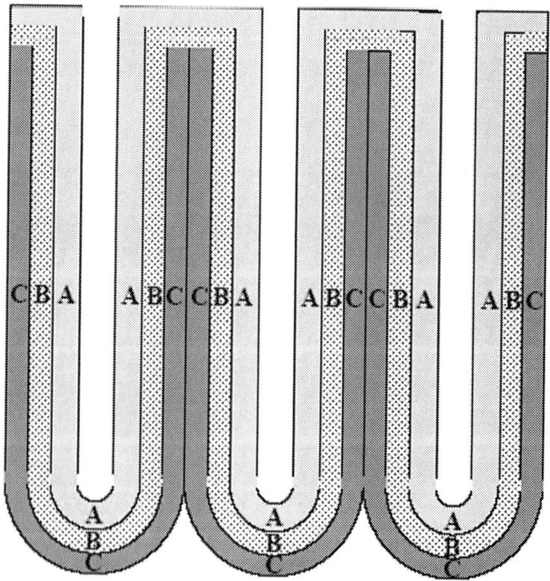


Figure 28. Schematic model showing the chemical structure of porous anodic oxide films. A.  $Al_{2-x}H_{3x}O_{3-y}$ $(Anion^{p-})_{2y/p}$, B. $Al_2O_{3-y}$ $(Anion^{p-})_{2y/p}$, C. $Al_2O_3$

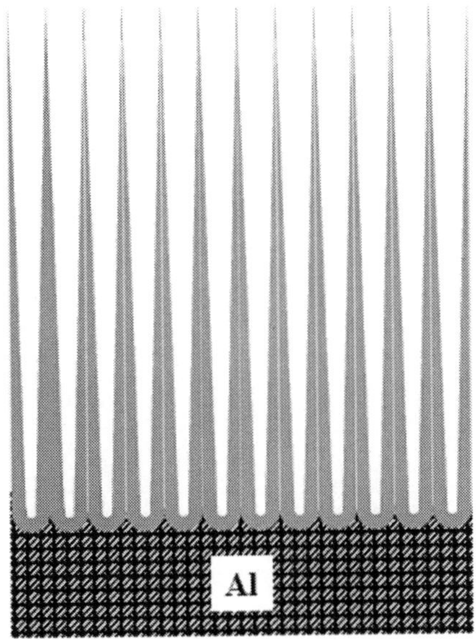



Figure 29. Cone-shaped pores formed by chemical dissolution during long anodizing.

Anodizing of aluminum alloys gives rise to the formation of PAOF with a modified structure by preferential dissolution or incorporation of alloying elements. Aluminum alloys with high copper contents show lower film growth rates and film structure with many imperfections, due to a preferential dissolution of copper / aluminum intermetallic compound phases,[36] while aluminum alloys with high silicon contents incorporate silicon particles in the oxide film and grow at low rates, due to $O_2$ evolution on the silicon incorporated in the oxide film.[37,38] An aluminum die-casting alloy, ADC12, which contains high concentrations of copper and silicon, and has been solidified rapidly in dies at pressures as high as 0.5 M Pa, is the one of the most difficult material to anodize. It has non-uniform distribution of Si- and Cu-containing second-phases and a high concentration of both Si and Cu dissolved in the $\alpha$–phase of aluminum and this results in the formation of a PAOF with uneven thickness at low growth rates.[39]

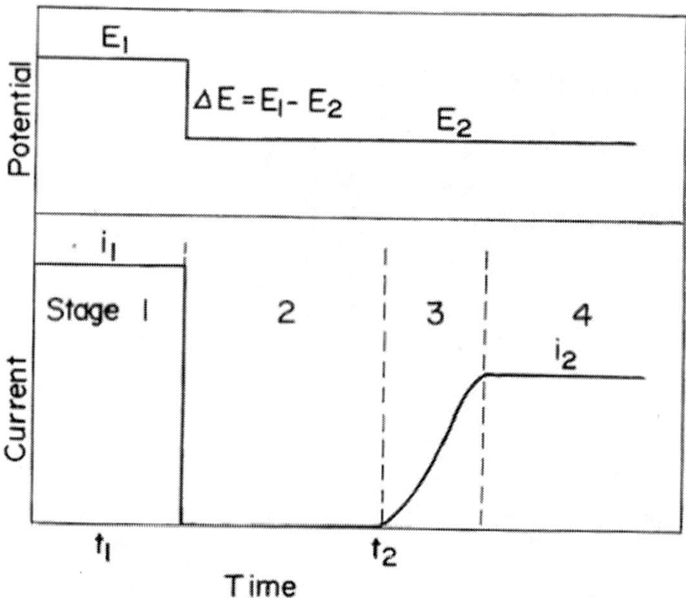


Figure 30. Characteristic changes in current during anodizing at $E_2$ after a decrease from $E_1$. Reprinted with permission from *J. Electron Microscopy*, **22**, (1973) 149. Copyright (1973), Jpn. Electron Microscope Society, Oxford University Press

### (iii)  *Growth under Transient Conditions*

As described in the previous Section, anodizing of aluminum in acid solutions at a constant potential, $E_1$, results in a steady current, $i_1$, allowing the steady growth rate of porous anodic oxide films shown in Fig. 26. When the potential is abruptly decreased to $E_2$ during anodizing, the current becomes zero immediately after the potential change, and then gradually increases with time before reaching a steady value, $i_2$ (Fig. 30). This phenomenon is termed the *current recovery effect* and was first reported by Murphy and Michelson.[40] A schematic model of the change in the film structure during the current recovery period is shown in Fig. 31.[41]

In Stage 1, the porous layer becomes thicker at a steady rate, keeping the   barrier layer thickness, $d_1$, constant   (see Fig. 31a).

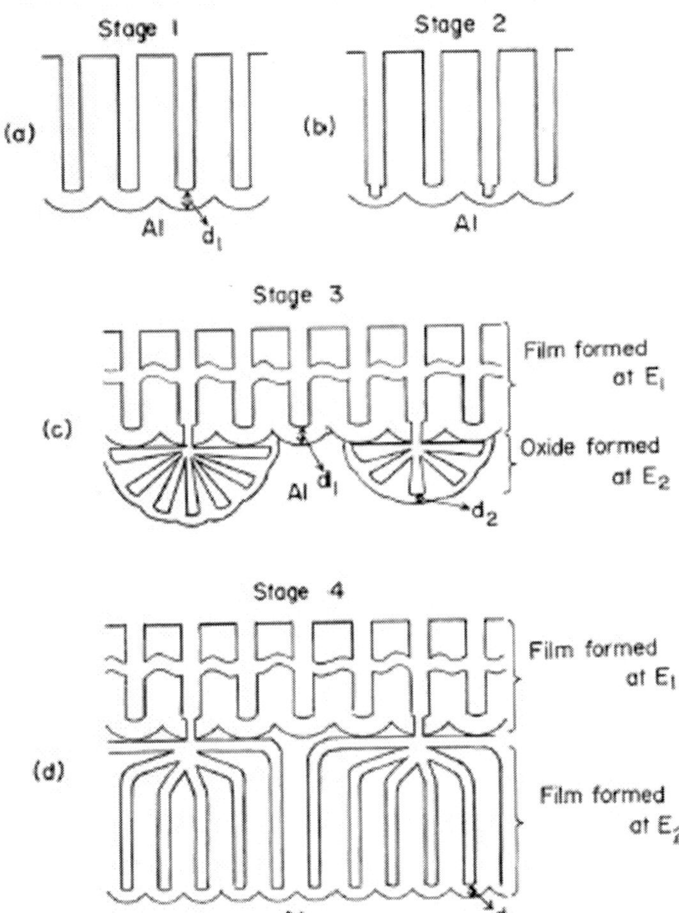



Figure 31. Mode of change in film structure during the current recovery period. Reprinted with permission from *J. Electron Microscopy*, **22**, 149 (1973) 157. Copyright (1973), Jpn. Electron Microscope Society, Oxford University Press.

When the potential is lowered to $E_2$, the growth of the film almost ceases because of the considerable decrease in the electric field across the barrier layer, and then the thickness of the barrier layer starts to decrease   due to the slow dissolution of the barrier layer

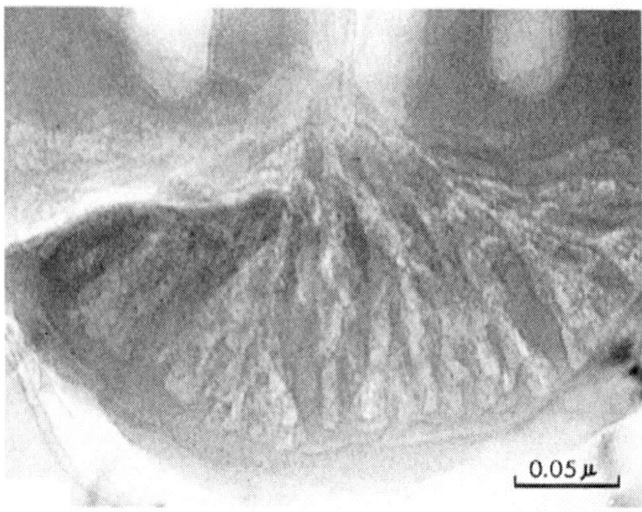

Figure 32. TEM image of the vertical section of anodic oxide film at Stage 3 of current recovery period. Reprinted with permission from *J. Electron Microscopy*, **22**, 149 (1973) 157. Copyright (1973), Japan Electron Microscope Society, Oxford University Press.

into the electrolyte (Stage 2, Fig. 31b). The current at Stage 2 shows a very small value, and increases slightly with time. This is because the electric field across the barrier layer becomes progressively higher due to the dissolution.

In Stage 3, the anodizing current starts to increase steeply as the result of the considerable thinning of the barrier layer or the steep increase of the field strength, the rate of dissolution also speeds up. At this time, the formation of the new semispherical oxides, in which many smaller cells develop in the radial directions, takes place (Fig. 31c). New oxide semi-spheres initiate at the bottom of the pores that have been produced during Stage 1. With further growth, the oxide semi-spheres come into contact, and the pores in the semi-spheres tend to grow in parallel and extend perpendicular to the substrate metal. The thickness of the barrier layer becomes the thickness of the barrier layer of the oxide semi-spheres, $d_2$ in Fig. 31. Figure 32 shows TEM images of a vertical cross section of an anodic oxide film at Stage 3.[41]

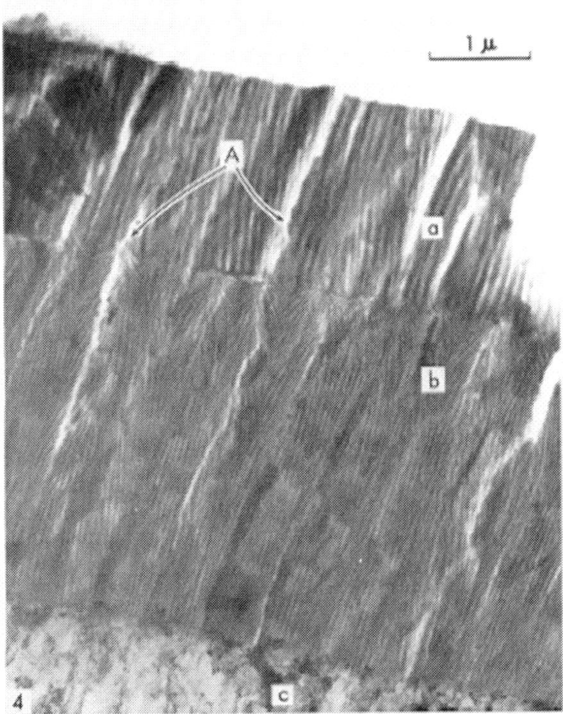


Figure 33. TEM image of a vertical cross section of anodic oxide film at Stage 4 - a: oxide layer formed at $E_1$, b: oxide layer formed at $E_2$, c: the metal substrate. Cracks indicated by the A were formed by the preparation of the specimen. Reprinted with permission from *J. Electron Microscopy*, **22**, 149 (1973) 157. Copyright (1973), Japan. Electron Microscope Society, Oxford University Press.

In Stage 4, the formation of oxide proceeds at a steady rate, and the current, $i_2$, remains constant. The cellular structure of the inner part of the film now corresponds to $E_2$ (Fig. 31d). The potential distribution across the barrier layer is the same as in Stage 3. Figure 33 is a TEM image clearly showing the two layer structure with $E_1$ and $E_2$ as in Fig. 31.

# III.    CHANGES IN FILM STRUCTURE BY COMBINATION OF TREATMENTS

As described in Section II, there are four typical surface treatments of aluminum: thermal treatment, hydrothermal treatment, anodizing in acid solutions, and anodizing in neutral solutions, and these treatments result in the formation of thermal oxide film, hydroxide film, PAOF, or BAOF. An understanding of the change in the structure of these oxide films by combining each treatment with other treatments is useful for developing technologies to create new functions on the aluminum surface (Fig. 34).

## 1.    Change of PAOF Structure by Heating, Boiling, and Anodizing in a Neutral Solution

When PAOF-covered aluminum is heated in air, there is dehydration and crystallization of the oxide film. In addition, cracks are formed in the oxide film because of the higher thermal    expansion

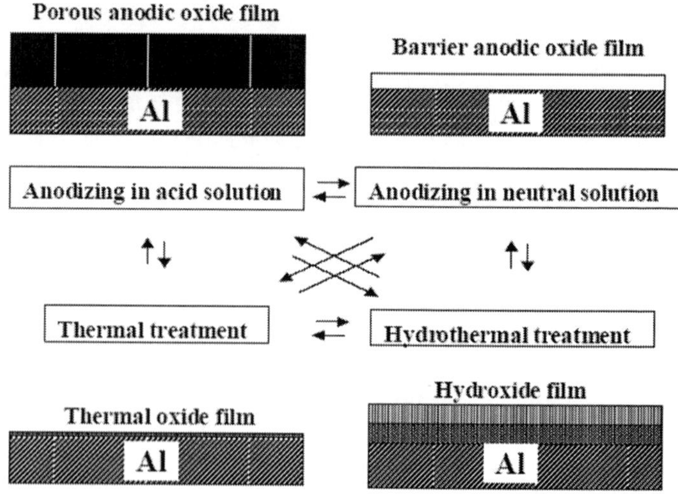


Figure 34. Schematic outline of potential combinations of the four typical treatments with other treatments.

coefficient of the metal substrate than that of the oxide film (Fig. 35). The crack formation strongly depends on heating temperature, film thickness, and post-treatment of *pore-sealing* (detailed below). Below 673 K, cracks more easily form in PAOF at higher heating temperatures on thicker oxide films and are enhanced by pore sealing (Fig. 35a-b). Above 773 K, cracks don't form, but continuous channels form on the substrate along the grain boundary, probably due to a softening of the substrate during the enlargement of grains at the higher temperatures.[42]

Immersion of aluminum covered with POAF in boiling pure water results in *pore-sealing* with hydroxide formed by the following equation (Fig. 36),

$$Al_2O_3 + x\,H_2O \rightarrow Al_2O_3 \cdot x\,H_2O \qquad (x = 1.5\text{--}2) \qquad (26)$$

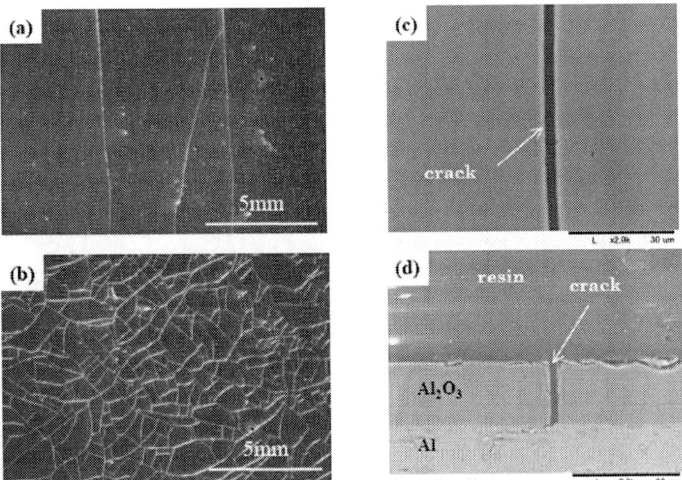


Figure 35. SCLM images (a, b) and SEM images (c, d) of aluminum covered with 24 µm thick PAOF after heating at $T_h$ = 573 K for 180 min: (a-c) surface, (d) vertical cross section, (a, c, d) without pore-sealing, (b) with pore-sealing.

The volume expansion by the formation of psuedo-boehmite causes the pores to become sealed, as shown in Fig. 36.[43] A ten-minute  sealing treatment is  sufficient to seal the pores completely in boiling water, and with longer sealing treatments water slowly penetrates through the hydroxide to allow further hydration. Sealing treatment also results in the outermost part of the POAF to become highly crystallized. Figure 37 shows SEM images of a PAOF before and after pore-sealing for 30 min, illustrating the formation of a highly crystallized hydroxide layer with about 1 μm thickness on the outermost part of PAOF, as well as the complete sealing of pores by hydroxide.[44]

The sealing may be achieved by pressurized steam sealing which enables quicker sealing, but the equipment needed is elaborate and difficult to operate. Cold sealing, which is performed in solutions containing $Ni^{2+}$ and $F^-$ ions at ambient temperatures, has gained wide-spread use because of low energy consumption.[45] The reaction during cold sealing is considered to be of the following form:

$$Al_2O_3 + x\ Ni^{2+} + y\ F^- + 3\ H_2O \rightarrow$$
$$Al_2Ni_x\ F_y(OH)_z + (6 - z)\ OH^- \qquad (27)$$

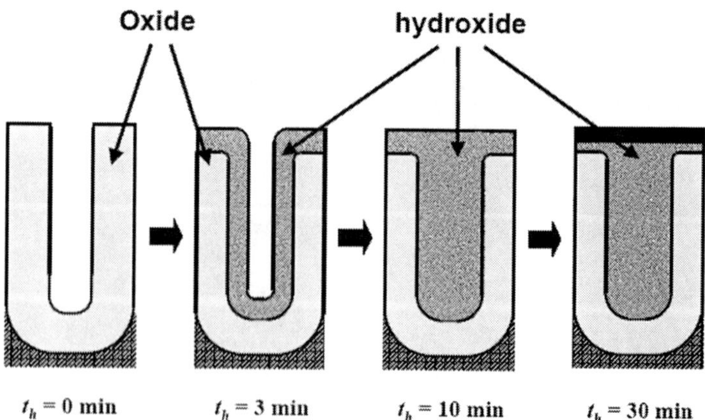


Figure 36. Process of pore sealing with hydroxide during dipping in boiling pure water. Reprinted with permission from *J. Metal Finishing Soc. Jpn*, **33**, (1982) 242. Copyright (1982), Surf. Finishing So. Jpn.

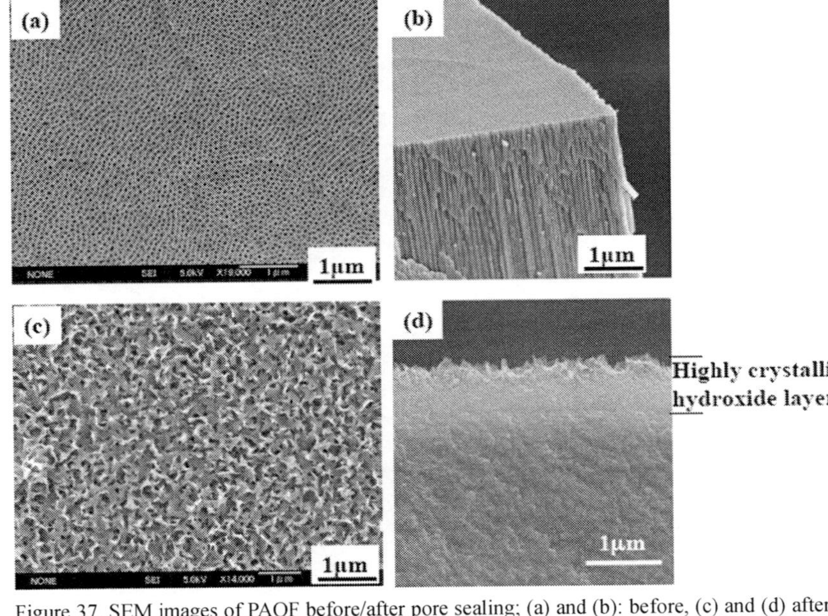


Figure 37. SEM images of PAOF before/after pore sealing; (a) and (b): before, (c) and (d) after, (a) and (c) surface, (b) and (d) vertical cross section. Reprinted from *Electrochim. Acta*, **52**, (2007) 4724. Copyright (2007), with permission from Elsevier.

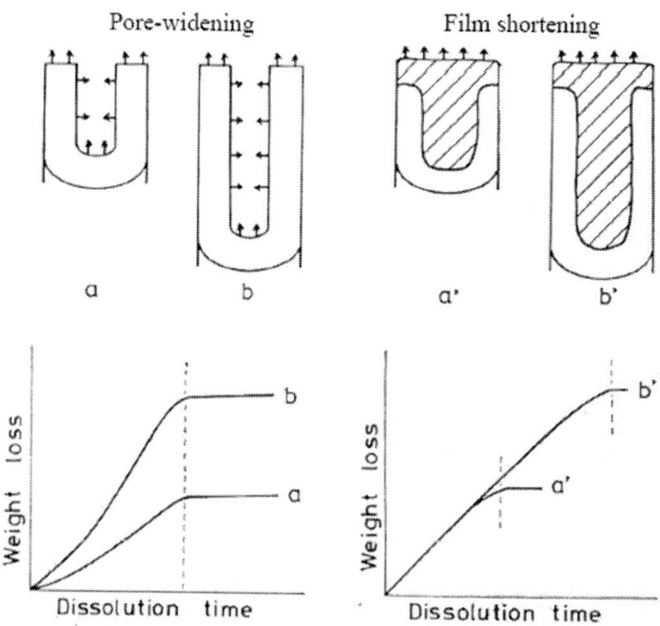


Figure 38. Model explaining the dissolution characteristics of the porous anodic oxide films before and after hot water treatment. Reprinted with permission from *J. Metal Finishing Soc. Jpn*, **34**, (1983) 44. Copyright (1983), Surf. Finishing So. Jpn.

Aluminum-nickel-fluoro-complexes are deposited in the pores and the result is a sealing of the pores.

Dissolution characteristics of PAOF in acid solutions change considerably by hydrothermal treatment (Fig. 38).[43] The dissolution of POAF proceeds via *pore widening* before pore sealing,[35] and *film shortening* occurs after pore sealing. In the pore widening, the time for the complete dissolution of POAF is independent of the film thickness, while in film shortening the time is proportional to the film thickness. The dissolution rate decreases with increasing hydrothermal treatment time, and there is an induction time before dissolution starts and it becomes longer after 30 min of hydrothermal treatment, due to crystallization of the outermost layer (Fig. 39).

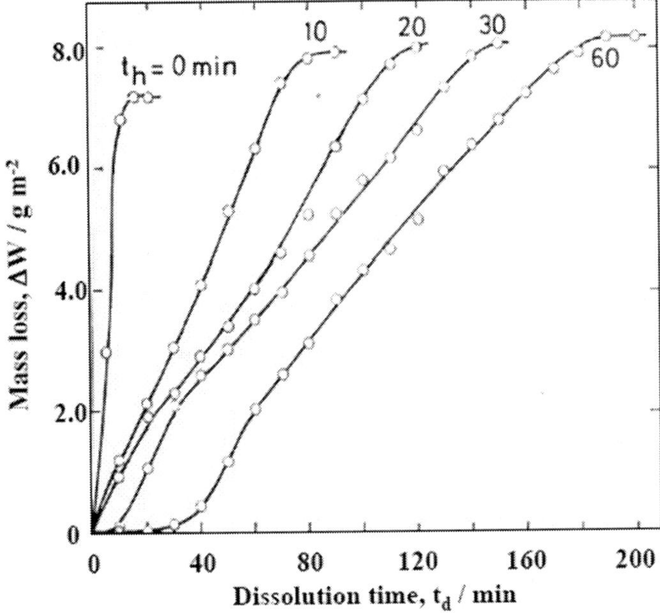


Figure 39. Effect of pore sealing time on the mass loss vs. dissolution time relationship. Reprinted with permission from *J. Metal Finishing Soc. Jpn*, **34**, (1983) 44. Copyright (1983), Surf. Finishing So. Jpn.

Anodizing of PAOF-covered aluminum in neutral solutions gives rise to the *pore-filling* phenomenon. Here the pores are filled with new oxide from the pore bottom, and a thin oxide layer is also formed at the interface between oxide film and metal substrate (Fig. 40a).[46,47] During a galvanostatic re-anodizing, anode potential, $E_a$, shows a jump at the very initial stage, and then increases linearly with a slope of $m_1$ before a knee point at $t_p$. The slope of the curve after $t_p$, $m_2$, is the same as that obtained for electropolished aluminum (broken curve in Fig. 40b), since pore filling is completed at $t_p$.

The porosity of PAOF, $\alpha$, is correlated with $m_1$ and $m_2$ by the following equation,[47]

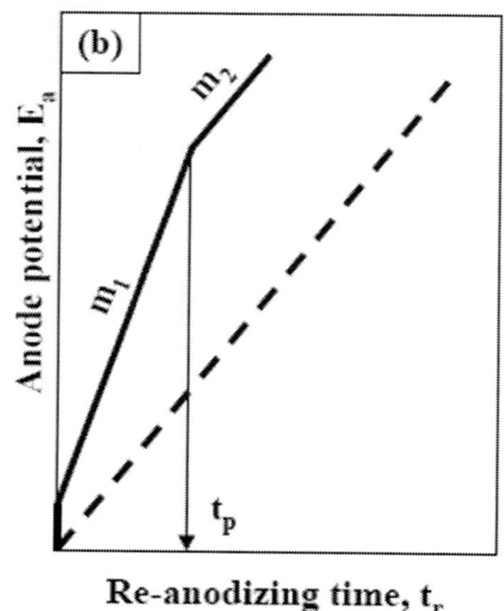


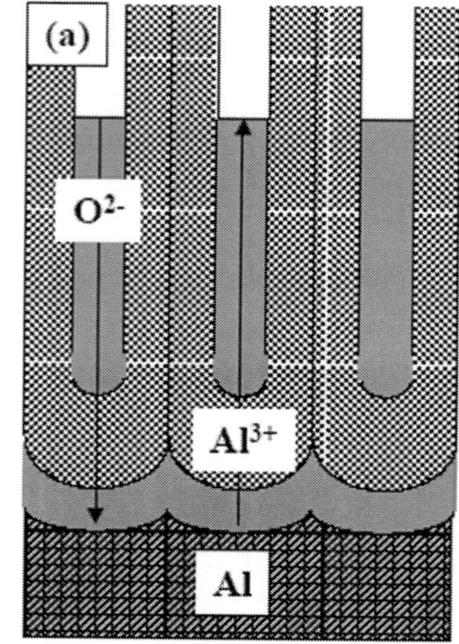


Figure 40. Mode of pore-filling during re-anodizing in a neutral solution of aluminum covered with PAOF.

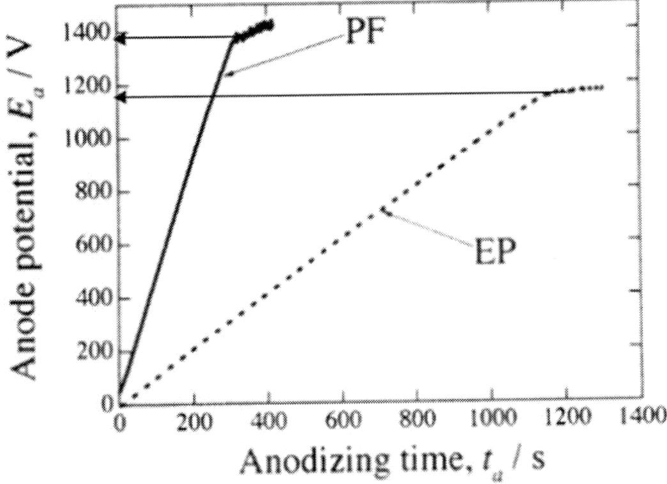


Figure 41. Change in anode potential with time during a galvanostatic anodizing in 0.5 M-H₃BO₃ on electropolished specimen (EP) and PAOF-covered specimen (PF). Reprinted with permission from *J. Surf. Finishing Soc. Jpn*, **53**, (2002) 142. Copyright (2002), Surf. Finishing Soc. Jpn.

$$\alpha = T_{Al}(m_2/m_1) / \{1 - (1 - T_{Al})(m_2/m_1)\} \tag{28}$$

where $T_{Al}$ is the transport number of $Al^{3+}$. Equation (28) shows that $m_2/m_1$ becomes higher as $\alpha$ decreases. Hence, monitoring $E_a$ vs. $t_r$ curves during galvanostatic pore filling can be used to estimate the porosity of PAOF.

The pore-filling process is utilized in capacitor manufacturing to produce capacitors for inverter systems. The oxide films formed by pore filling are stable and can be subjected to frequent changes in loads, making them ideally suited for this kind of application. This process is also suitable to increase the film breakdown potential, $E_{bd}$. Figure 41 shows the change in anode potential, $E_a$, with time during galvanostatic anodizing in 0.5 M H₃BO₃ on an electropolished specimen (EP) and a PAOF-covered specimen (PF).

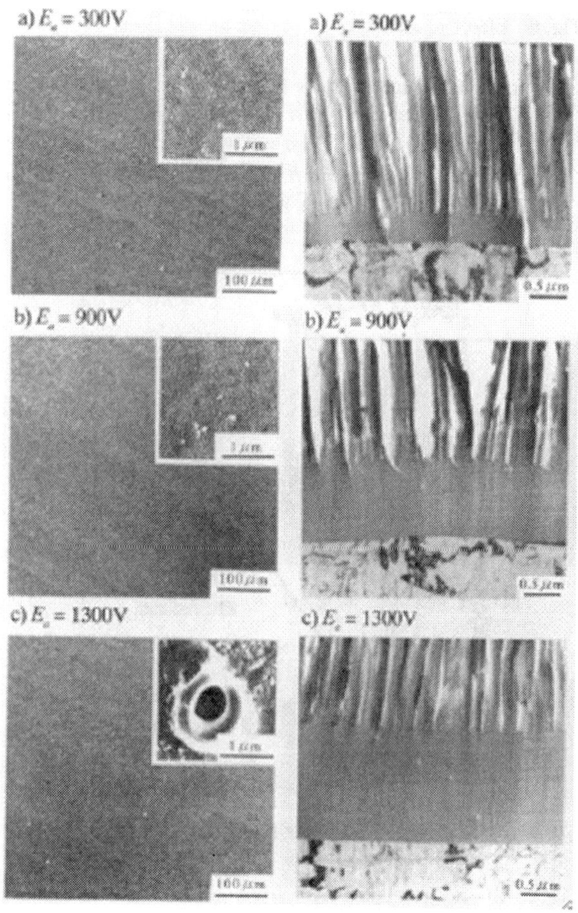



Figure 42. Change in the structure of PAOF with time during re-anodizing of PAOF-covered specimen. Reprinted with permission from *J. Surf. Finishing Soc. Jpn,* **53**, (2002) 142. Copyright (2002), Surf. Finishing Soc. Jpn.

The breakdown potential of the PF-specimen is more than 200 V higher than that of the EP-specimen. The advantage of pore filling method that imperfections are hardly formed in the oxide film before $E_{bd}$, unlike anodizing of electropolished specimens (Fig. 42).

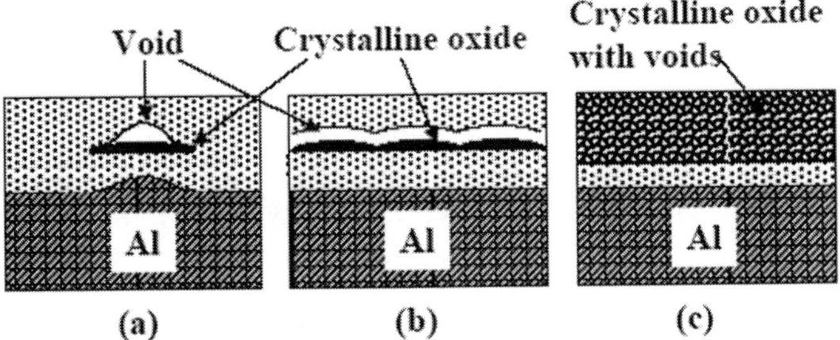



Figure 43. Schematic model of different types of anodic oxide film formed on aluminum after a variety of thermal treatments.

## 2.  Change in BAOF Structure by Heating, Boiling, and Anodizing in an Acid Solution

Heating of BAOF-covered aluminum gives rise to dehydration of the oxide. A TGA analysis up to 873 K shows that about 45 µg/1 mg of the original oxide mass is lost up, suggesting a composition of $Al_2O_3\cdot0.27H_2O$ for water content in a BAOF.[15] Hydrothermal treatment of BAOF-covered aluminum leads to the hydration of the oxide film as well as to the formation of hydroxide by reaction with the aluminum substrate. Anodizing of BAOF-covered aluminum in acid solutions causes the formation of PAOF underneath the BAOF (Fig. 43c). The PAOF has a pore-branched structure, since PAOF formation is initiated locally through imperfections in BAOF.[48]

## 3.  Change of the Structure of Thermal Oxide Films by Boiling, and Anodizing in a Neutral solution and in an Acid Solution

The changes in the structure of thermal oxide films caused by boiling in water and anodizing in acid solution are similar to the changes in barrier type oxide films undergoing the same treatment. Anodizing of thermal oxide film-covered aluminum in neutral solutions results in the formation of a crystalline oxide film, because the thermal oxide film, composed of $\gamma\text{-}Al_2O_3$, acts as seeds for the crystalline oxide, enhancing the crystallization of amorphous oxide under a high electric field. As a result, the structure of the anodic oxide films formed strongly depends on the pretreatments, heating temperature, heating time and anodizing conditions.

Kobayashi and Shimizu found that platelet patches of $\gamma$-$Al_2O_3$ form at the middle part of anodic oxide films formed by anodizing of electropolished aluminum after heating at 823 K for 15 min and that there are voids over each crystalline oxide patch (Fig. 43a).[49] Anodizing was carried out in 0.1 M ammonium pentaborate solution at 293 K with a constant c. d. of 50 A m$^{-2}$ up to 100 V. Crevecoeur and de Wit found a continuous layer of $\gamma$-$Al_2O_3$, accompanied with voids on it, at the middle part of anodic oxide films after heating 823 K for 15 min (Fig. 43b).[50] Here samples were anodized at 292 K in a pentaborate ethylene glycol/water solution by a potential scanning at 0.5 V/s up to 400 V. Alwitt et al. heated alkali-etched aluminum at 498 K for 20 min and anodized in ammonium citrate solution at 10 A m$^{-2}$ at 343 K, to obtain two-layer anodic oxide films, consisting of an outer crystalline oxide layer and an inner amorphous oxide layer, as shown in Fig. 43c.[51]

Figure 44 shows the change in anode potential, $E_a$, with time, $t_a$, during anodizing in 0.5 M $H_3BO_3$/0.05 $MNa_2B_4O_7$ solution at 353 K, for electropolished aluminum with/without heating at 823 K for 3 h.[6] The specimen without heating shows a linear increase in $E_a$ from zero and the $E_a$ vs. $t_a$ curve has a somewhat flat slope, while the specimen with heating shows a jump at the very initial stage, and displays a steeper curve. The anodizing of unheated specimens at relatively high temperatures causes an electrochemical dissolution of the oxide films, leading to the formation of an amorphous anodic oxide that has a two layer structure: an outer porous layer and an inner dense layer (Fig. 45a).

The heated specimens, however, have a one-layer crystalline oxide film structure (Fig. 45b). Thermal oxide films act as seeds for crystalline oxide that with time transforms the whole of the anodic oxide film into crystalline oxide, resulting in a suppression of the electrochemical dissolution of oxide during anodizing. A comparison of Figs. 44 and 45 enables an estimate of the film thickness per 1 V of anode potential, showing numerical values of

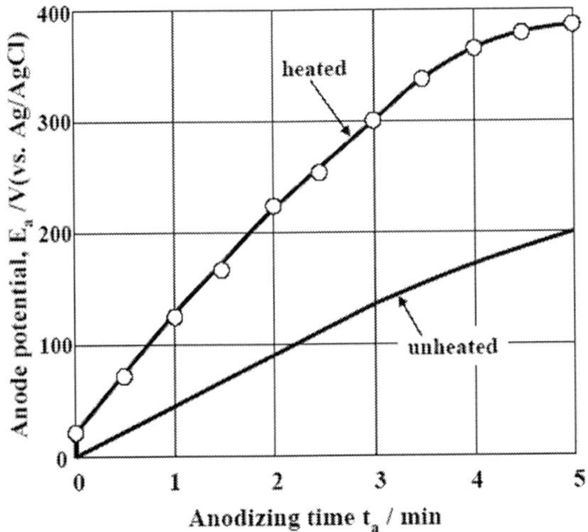

Figure 44. Change in anode potential with time during anodizing of aluminum with / without heating.

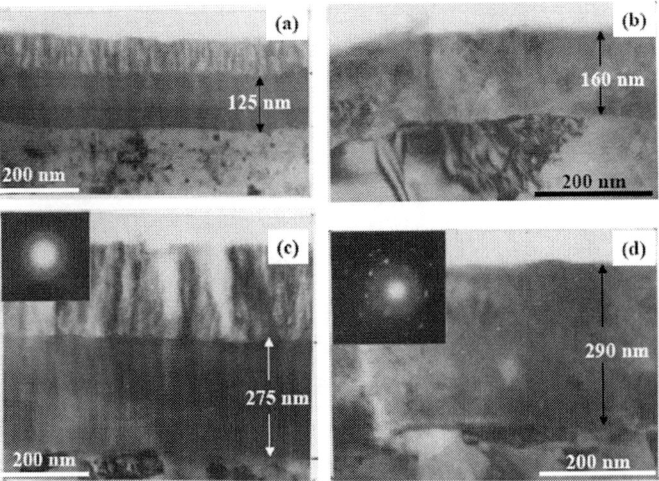

Figure 45. TEM images of vertical cross section of specimens anodized for 2 (a and b) and 5 min (c and d) under conditions in Fig. 44. Photos-(a), and –(c) were obtained for the specimen with heating, and photos-(b) and –(d) were without heating. Reprinted with permission from *J. Electron Microscopy*, **40**, (1991) 101. Copyright (1991), Jpn. Electron Microscope Society, Oxford University Press

1.4 nm/V for unheated specimens and 0.75 nm/V for heated specimens at both 2 and 5 min. Hence, it is clear that the anodic oxide films formed after heat treatment can sustain much higher electric fields than those formed on unheated specimens. This dielectric property is very useful in the development of new types of capacitors with higher capacitance.

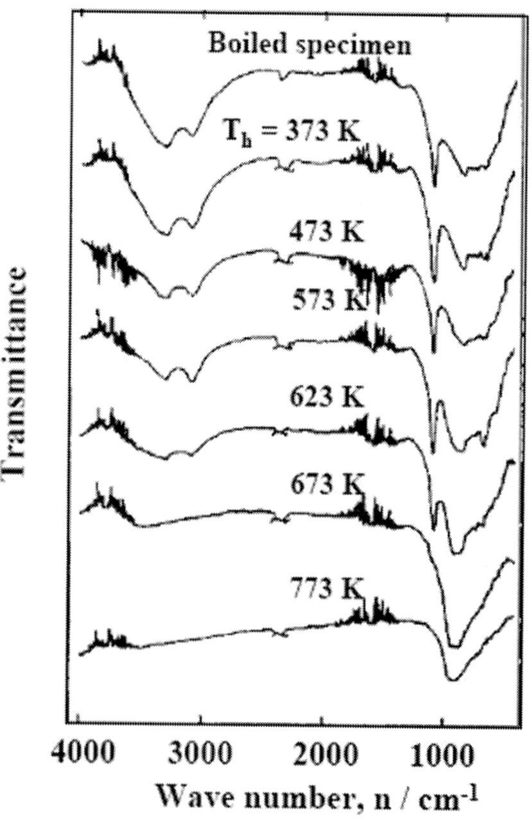


Figure 46. Changes in the FTIR spectra of boiled specimens by heating at different temperatures. Reprinted with permission from *J. Surface Sci. Soc. Jpn.*, **8**, (1987) 279. Copyright (1987), Surface Science Society of Japan.

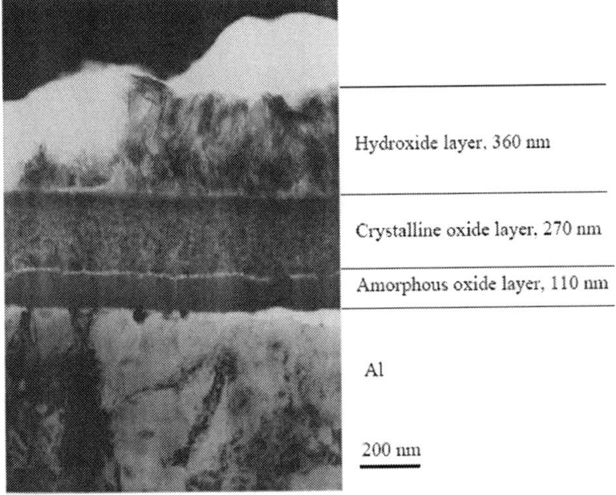


Figure 47. TEM image of the vertical cross section of a barrier type anodic oxide film formed after hydrothermal treatment. Reprinted with permission from *J. Metal Finishing Soc. Jpn*, **38**, (1987) 67. Copyright (1987), Surf. Finishing Society Japan.

## 4. Changes in the Structure of Hydroxide Films by Heating and Anodizing in Acid and Neutral Solutions

Above 673 K, heating of hydroxide film-covered aluminum gives rise to dehydration of the hydroxide, leading to a transformation to oxide. This is ascertained by the analysis of FTIR spectra, which loose absorption peaks at 1100 and 3300 cm$^{-1}$ above 673 K (Fig. 46).[10]

During dehydration, voids are formed in the inner layer of the hydroxide films, and the film becomes thinner.[52] Anodizing of hydroxide film-covered aluminum in neutral solutions causes the formation of anodic oxide films that contain crystalline oxide formed by the dehydration of the hydroxide under high electric fields (Fig. 47).[53–58] Figure 47 shows a TEM image of a vertical cross section of the specimen anodized up to 400 V with a constant current in a neutral borate solution at 313 K after hydrothermal treatment at 373 K for 30 min. The anodic oxide film consists of three layers: an outermost hydroxide layer, an intermediate crystalline oxide layer, and an innermost amorphous oxide layer. The outermost hydroxide layer is formed during hydrothermal treatment and the crystalline and amorphous oxide layers are formed during subsequent anodizing. Small voids are dispersed in the crystalline oxide layer and the voids are highly concentrated at the interface between the crystalline and amorphous oxide layers. The anodic oxide film is considered to grow according to the scheme suggested in Fig. 48.[55,56]

During anodizing, dehydration of hydroxide occurs at the interface between the outermost hydroxide layer and the intermediate layer accompanied by the formation of crystalline oxide and voids,

$$Al_2O_3 \cdot 2.7\,H_2O \rightarrow \gamma\text{-}Al_2O_3 + 2.7\,H_2O + Void \tag{29}$$

The voids become filled with amorphous $Al_2O_3$ newly formed by the reaction of $H_2O$ with $Al^{3+}$ ions that have reached the interface after being transported across the intermediate and innermost layers,

$$2\,Al^{3+} + 3\,H_2O + Void \rightarrow amp\text{-}Al_2O_3 + 6\,H^+ \tag{30}$$

As a result, during anodizing the hydroxide layer becomes thinner with time, and the intermediate layer, which is composed of amorphous and crystalline oxides, thickens. The amorphous oxide in contact with the crystalline oxide is converted to crystalline oxide and voids are formed in the intermediate layer by the volume shrinkage of the oxide, according to

$$amorphous\text{-}Al_2O_3 \rightarrow \gamma\text{-}Al_2O_3 + Void \tag{31}$$

The innermost layer grows by the reaction of Al with $O^{2-}$ ions, which reach the interface between the innermost layer and the aluminum substrate, as

$$2\,Al + 3\,O^{2-} \rightarrow amor\text{-}Al_2O_3 + 6\,e^- \tag{32}$$

At the interface between the intermediate and innermost layers, there is an additional reaction during anodizing, causing the crystallization of amorphous oxide (see Eq. 31).   Overall, the conver-

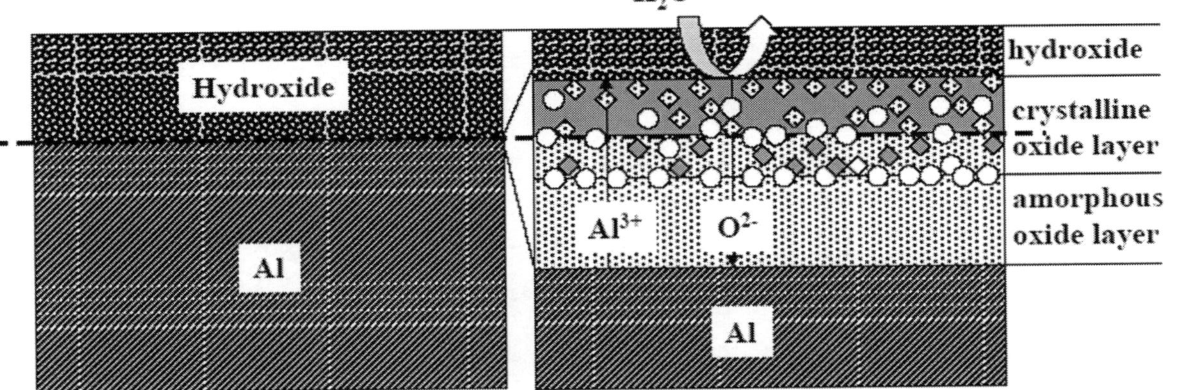


Figure 48. Model of the growth of anodic oxide film during anodizing of aluminum covered with hydroxide film.

sion of amorphous oxide to crystalline oxide leads to an inward movement of the interface during anodizing (Fig. 48b). The inward movement of the interface is accompanied by void transport, leading to an increase in void volume there. The conversion rate of the hydroxide at the interface between the intermediate layer and the innermost layer becomes higher at higher anodizing temperatures, and this enhances the rate of increase in the intermediate layer thickness. This somewhat complicated mechanism for the growth of anodic oxide films after hydrothermal treatment is a significant example of the conversion of hydroxide films to oxide films under high electric fields.

The sum of the thickness of the intermediate and the innermost layers is 380 nm in Fig. 47, suggesting that the anodic oxide film formed after hydrothermal treatment can sustain the high electric field of 0.9 nm/V. This is clearly due to the high field sustainability of the intermediate crystalline oxide layer.

This excellent dielectric property is very attractive in manufacturing aluminum electrolytic capacitors with high capacitances, and the successive processes of hydrothermal treatment and anodizing is widely used in the capacitor manufacturing industry. However, the numerous voids included in the anodic oxide film must be filled with oxide before use of the films as dielectrics in capacitors with high durability.

There are two kinds of voids in anodic oxide films formed after hydrothermal treatment: isolated voids and voids linked by cracks.[59,60] The distribution of crack-linked voids in anodic oxide films can be estimated by re-anodizing (reforming) after immersion in a neutral solution,[57] since solution penetrates into the voids through cracks during immersion and voids are filled with oxide during re-anodizing. Figure 49 shows the changes in anode potential, $E_{ref}$, with time, $t_{ref}$, during reforming with 1 A m$^{-2}$ after im-

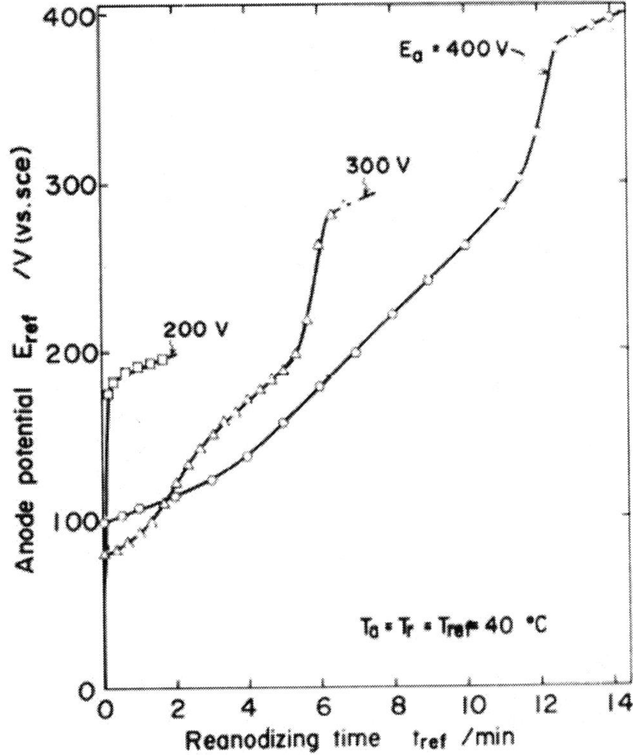


Figure 49. Changes in the anode potential with time during re-anodizing of aluminum subjected to hydrothermal treatment and anodizing. Reprinted with permission from *J. Surf. Finishing Soc. Jpn*, **40**, (1989) 590. Copyright (1989), Surf. Finishing Society Japan.

mersion in a neutral borate solution at 313 K, obtained for specimens anodized to different anode potentials ($E_a$) after hydrothermal treatment. The $E_{ref}$ vs. $t_{ref}$ curves show a 85–100 V jump in $E_{ref}$ at the very initial stage, and then a flat slope at potentials between 85 and 110 V for $E_a$ = 300 and 400 V specimens before a relatively steep slope above these $E_{ref}$ values. The $E_a$ = 200 V specimen does not show the flat segment around $E_{ref}$ = 100 V. When $E_{ref}$ becomes

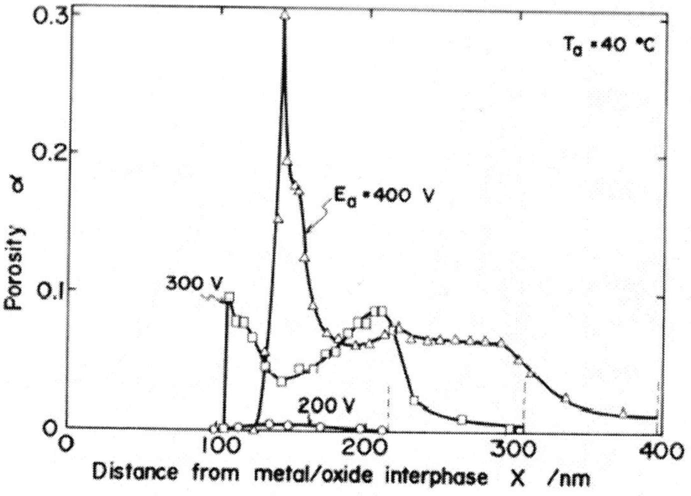


Figure 50. Depth profile of void concentration in anodic oxide films formed on aluminum by hydrothermal treatment and anodizing. Reprinted with permission from *J. Surf. Finishing Soc. Jpn*, **40**, (1989) 590. Copyright (1989), Surf. Finishing Soc. Jpn.

closer to $E_a$, each curve shows a knee point, suggesting the completion of the filling of voids with oxide. Analysis of the $E_{ref}$ vs. $t_{ref}$ curves enables an estimate of the distribution of the linked voids (Fig. 50).

It can be seen from Fig. 50 that there are no voids in the innermost layers of the $E_a$ = 300 and 400 V specimens, and that voids are enriched at the interface between the innermost and intermediate layers and the middle part of the intermediate layer. The high concentration of voids at the interface between the innermost and intermediate layers is ascertained as a white narrow band between the layers in Fig. 47. The peak at the middle parts in the intermediate layer corresponds to the voids at the positions where there was an interface between hydroxide film and the aluminum substrate before anodizing (see the broken line in Fig. 48).

## 5.   Other Combination of Treatments

Coloring of PAOF by filling the pores with dyestuffs and metal particles is a common post-anodizing treatment. Porous type anodic oxide films have large internal surfaces, and easily absorb dyestuff on the pore-wall. Many organic dyestuffs, Alizalin Blue, Alizalin Red-S, Naphtol Green-B, Chromolan Blue-NGG, Anthraquinone, and others, are used to give flashy, vivid colors, gold, yellow, blue, green, etc. Some inorganic dyestuffs are also used, and here the process involves the precipitation of metal salts with low solubility, $Ag_2Cr_2O_7$, PbS, or $Co(OH)_3$ in the pores. The color range of inorganic dyes is more restricted than with organic dyes, but the inorganic dyestuff is more resistant to heat and light.

Electrolytic coloring is generally a method where metal is deposited at the bottom of the pores by applying an alternating voltage.[61] The metal deposition in pores was first reported by Caboni,[62] and then developed by Asada[63] as a process now used around the world for manufacturing architectural components like colored walls, colored window-frame etc. Nickel, Co, Sn, and Cu are mainly deposited to achieve colors in bronze, and from maroon to black. The light fastness of the finishes with electrolytic coloring is far superior to that with organic dyestuffs.

## IV.  ROLE OF ANODIC OXIDE FILMS IN CORROSION PROTECTION OF ALUMINUM

As described in Section I, aluminum has a strong chemical affinity with oxygen to form air-formed oxide films, which are resistant to corrosion. It is not easy to evaluate the protective ability of air-formed oxide films, since their structure and properties strongly depend on the atmosphere where they were formed. Formation of PAOF by anodizing in acid solutions and the subsequent pore sealing are basic processes to improve the corrosion protection ability of aluminum products. As shown in Fig. 39, the rate of dissolution of oxide films in acid solutions decreases greatly by anodizing and pore sealing.

The roles of anodic oxide films in immersion in different solutions and in cathodic polarization will be described in the following Sections.

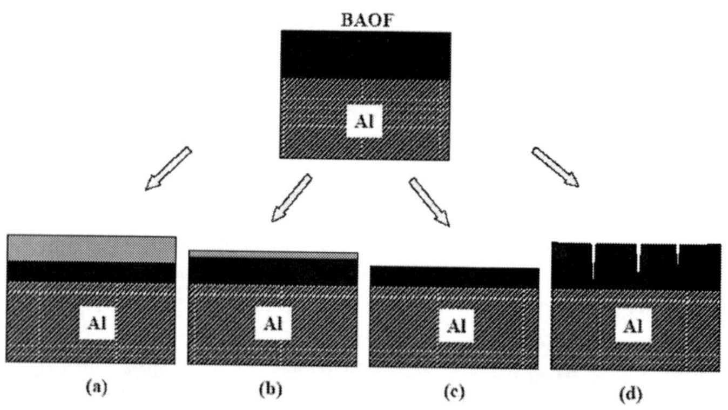


Figure 51. Patterns of deterioration of BAOF during immersion in a variety of solutions: (a) hydration, (b) hydration/dissolution, (c) even dissolution, and (d) uneven dissolution.

## 1.  Immersion of BAOF-Covered Aluminum in a Solution

When BAOF-covered aluminum is immersed in aqueous solutions containing inorganic and organic anions at room temperature, hydration and dissolution of the oxide films occur before the substrate starts to corrode (Fig. 51).[64] Hydration causes the formation of a hydrated layer (Fig. 51a), while the dissolution of the oxide film results in either a uniform film thinning (Fig. 51c) or the formation of cracks and imperfections (Fig. 51d).

In pure water, there is only hydration of oxide occurs due to penetration of water across the BAOF (Fig. 51a). Immersion in phosphate solutions at pH = 7.0 show a variety of behaviors, depending on the concentration. At $10^{-3}$ and $10^{-2}$ M of phosphate concentration, there is no deterioration of BAOF during immersion for up to 120 h, maybe due to adsorption or thin film formation at the surface of the oxide film.[64] At $10^{-1}$ M phosphate, both hydration and dissolution of the oxide occur, leading to decreases in film thickness and the formation of a hydrated oxide layer (Fig. 51b). At 1.0 M phosphate, an even dissolution of oxide occurs at a high rate (Fig. 51c).

In $10^{-1}$ M citrate (pH = 7.0), tartrate  (pH = 6.4)  , and oxalate (pH = 6.4) and in 0.5 M $H_3BO_3$ / 0.05 M $Na_2B_4O_7$ (pH = 7.4) solutions, only uniform dissolution takes place, at rates in the following order

$$oxalate > tartrate > citrate > borate \tag{33}$$

These anions form cationic or anionic complexes with $Al^{3+}$ ions, enhancing the film dissolution,

$$x\, Al^{3+} + y\, anion^{z-} \rightarrow [Al_x(anion)_y]^{3x-yz} \tag{34}$$

In 0.1-M borate, adipate, sulfate, chloride solutions at pH = 7.0, and pure water, there is exclusively hydration of the oxide film proceeds, and the hydration rates decrease in the following:

$$pure\ water > chloride > sulfate > adipate > borate \tag{35}$$

Chloride, sulfate, adipate, and borate ions retard the penetration of water to some extent by adsorption or thin film formation. The behavior of chloride ions with only a weak capability of hydration protection and no dissolution ability for aluminum oxide films is unexpected, as chloride ions are well known as highly aggressive ions in pitting corrosion. Chloride ions are considered to cause the pitting corrosion by passing through the imperfections of the oxide film above the pitting potential.

## 2. Role of Anodic Oxide Films in Pitting Corrosion during Cathodic Polarization

When aluminum covered with air-formed oxide films is polarized cathodically in neutral solutions, the aluminum specimen displays a unique phenomenon, *cathodic corrosion*,[65] due to the amphoteric properties of alumina, which is the main constituent of anodic oxide films.

Cathodic polarization of aluminum in neutral solutions causes hydrogen evolution via water reduction, leading to an increase in pH near the surface,

$$2\, H_2O + 4e^- \rightarrow H_2 + 2\, OH^- \tag{36}$$

Further, the oxide film on aluminum dissolves in an alkaline environment near the surface to expose the aluminum substrate to the solution according to

$$Al_2O_3 + 3\ H_2O + 2\ (OH)^- \rightarrow 2\ Al(OH)_4^- \tag{37}$$

and corrosion of the aluminum substrate occurs too, i.e., the cathodic corrosion, like in

$$Al + 3\ H_2O + (OH)^- \rightarrow Al(OH)_4^- + 3/2\ H_2 \tag{38}$$

Nisancioglu et al. found pitting corrosion of Al covered with air-formed oxide film during cathodic polarization in NaCl solution at potentials more negative than $-1.4$ V (vs. SCE).[66] Despic et al. found that in cathodic polarization of aluminum covered with air-formed oxide films in NaCl solutions below $-1.7$ V (vs. SCE) there is hydrogen evolution, and the amount of hydrogen evolved is larger than would be predicted from the amount of charge passed.[67]

Aluminum covered with BAOF is well known to show *current rectification* in aqueous solutions that enables flowing a large current during cathodic polarization but no current during anodic polarization. The term of *valve metals* for Al, Ta, Nb, and others originates in this phenomenon. Current rectification has been investigated extensively, and has been attributed to p-n semiconductor junction properties of the anodic oxide films[68] and to hydrogen evolution after the migration of protons across the oxide film through the oxide itself or imperfections.[69]

Figure 52 shows anodic and cathodic polarization curves in a neutral borate solution, obtained for aluminum covered with BAOF formed at different anodizing potentials.[70] The anodic current remains only small values during anodic polarization until the potential close to the film formation potential on each specimen is reached, while there is a large cathodic current flowing at a cathodic polarization below minus a few volts. This rectification phenomenon is considered to be due to hydrogen transport, since rectification is not observed when specimens are tested in a vacuum after attaching Ag film to anodic oxide films.

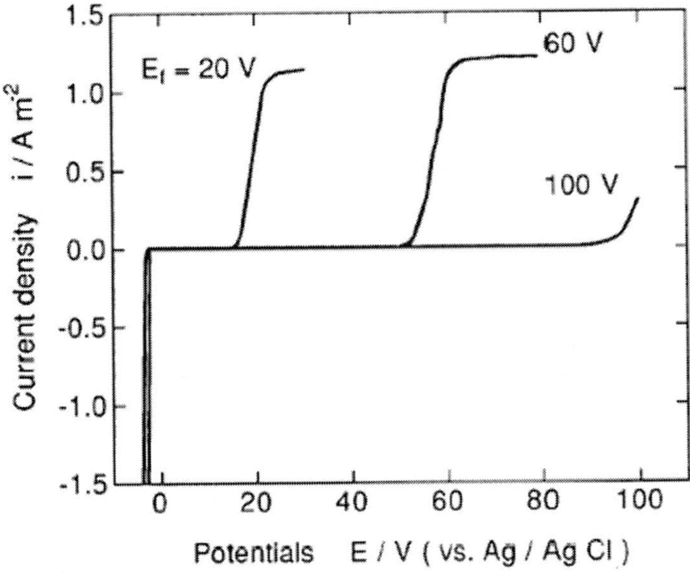



Figure 52. Anodic and cathodic polarization curves in a neutral borate solution, obtained for aluminum covered with BAOF formed at different anodizing potentials. Reprinted from *Corros. Sci.*, **36**, (1994) 677. Copyright (1994), with permission from Elsevier.

Cathodic polarization curves depend slightly on the type of anodic oxide films and the film thickness, showing larger over-potentials on BAOF-covered specimens with higher anodizing potentials (Fig. 53).[71]

Figure 54 shows SEM images of the surface (Fig. 54a-b) and a TEM image of a vertical cross section (Fig. 54c) of PAOF-covered specimens after cathodic polarization up to $-100$ A m$^{-2}$ (Fig. 54a) and $-500$ A m$^{-2}$ (Fig. 54b-c). Pitting corrosion proceeds during cathodic polarization, showing an increase in the size and number of pits at more negative potentials. The pits appear to be covered with anodic oxide film (Fig. 54c).

In a different experiment, a constant cathodic current of $-500$ A m$^{-2}$ was applied to aluminum specimens covered with BAOF to monitor the surface changes in the specimen by AFM.[72] There are

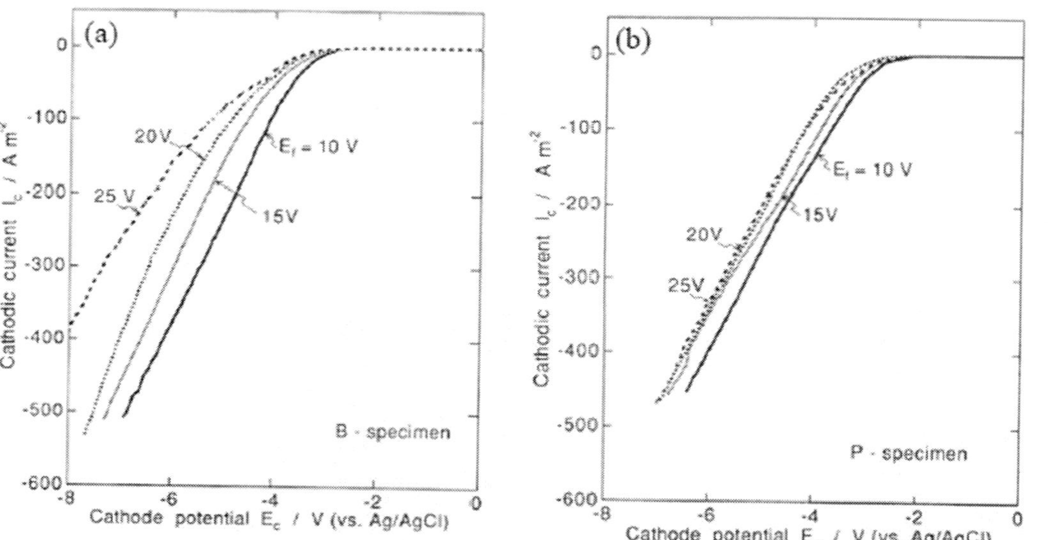


Figure 53. Cathodic polarization curves at 2.9 V/min in a neutral borate solution, obtained for aluminum specimens covered with (a) BAOF and (b) PAOF formed at different anodizing potentials. Reprinted with permission from *Corros. Sci.*, **36**, (1994) 689. Copyright (1994), Elsevier.

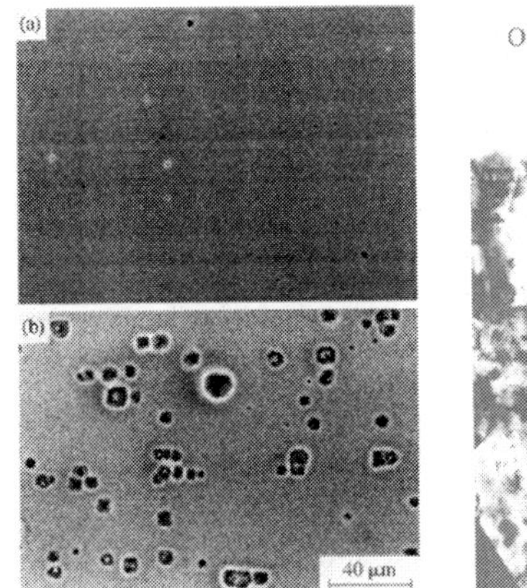

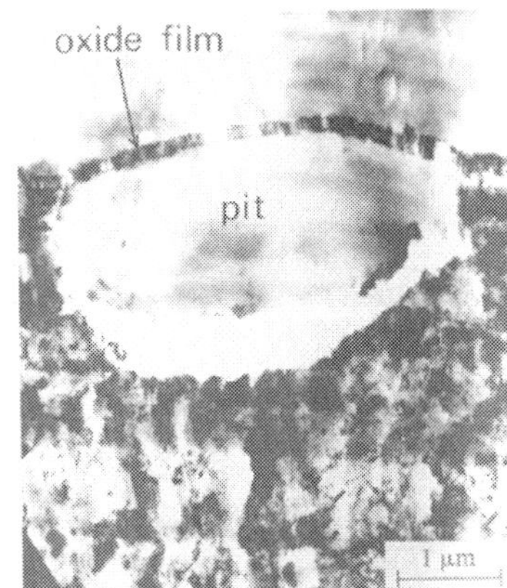


Figure 54. (a), (b) SEM images of the surface and (c) TEM image of the vertical cross section of specimens after cathodic polarization up to (a) -100 A m$^{-2}$ and (b), (c) -500 A m$^{-2}$. Reprinted from *Corros. Sci.*, **36**, (1994) 689. Copyright (1994), with permission from Elsevier.

formation and growth of blisters on the surface during the cathodic polarization, finally resulting in a blowout (Fig. 55).

The mechanism of the pitting corrosion during cathodic polarization can be explained by the transport of H$^+$ across the oxide film at imperfections (Fig.56ab), followed by the reduction to H$_2$ underneath the oxide film (Fig. 56c). Hydrogen gas bubbles become larger with time and the pressure of the H$_2$ gas inside a bubble becomes higher (Fig. 56d) leading to the bubble bursting (Fig. 56e). Finally, there is local dissolution of the aluminum substrate at the area exposed to the solution after a blister has burst, leading to the pitting corrosion.

It must be noted here that pitting corrosion of aluminum occurs also in solutions without chloride ions, and that anodic oxide films play a role in the local dissolution during cathodic polarization.

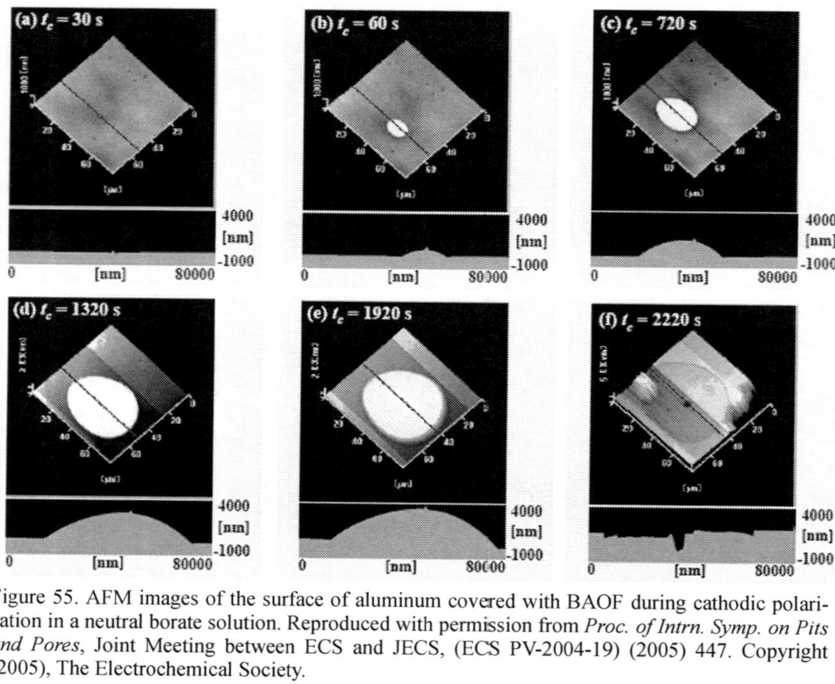

Figure 55. AFM images of the surface of aluminum covered with BAOF during cathodic polarization in a neutral borate solution. Reproduced with permission from *Proc. of Intrn. Symp. on Pits and Pores*, Joint Meeting between ECS and JECS, (ECS PV-2004-19) (2005) 447. Copyright (2005), The Electrochemical Society.

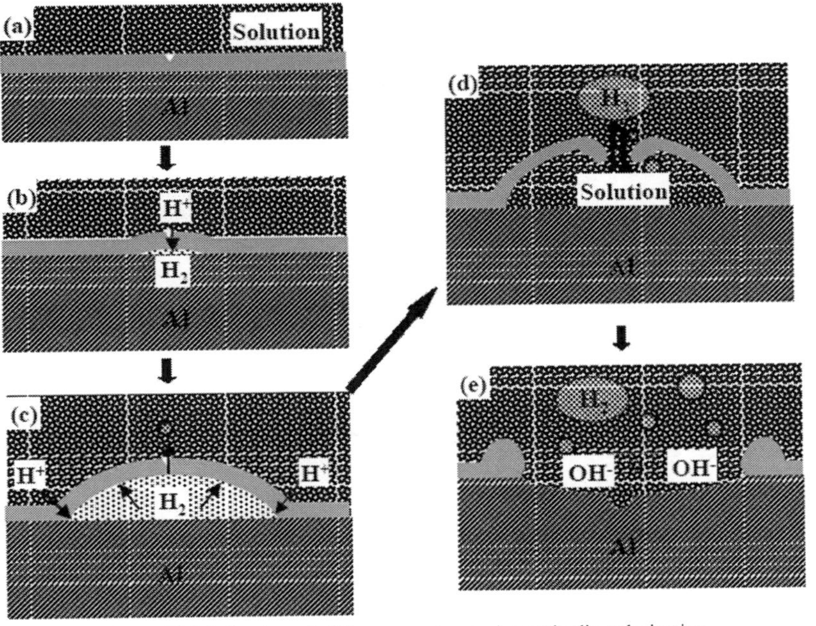


Figure 56. Mechanism of pitting corrosion during cathodic polarization.

## Table 7
### Chemical Composition of Aluminum Alloys.

| No. | Cu | Mg | Si | Fe | Bi | Sn | Al |
|---|---|---|---|---|---|---|---|
| 10 | — | — | 0.1 | 0.3 | — | — | Bal. |
| 13 | — | — | 0.1 | 0.3 | 1.0 | 1.0 | Bal. |
| 15 | — | — | 0.1 | 0.3 | 1.0 | — | Bal. |
| 23 | 5.0 | — | 0.2 | 0.3 | 1.0 | 1.0 | Bal. |
| 63 | 0.3 | 1.0 | 0.6 | 0.3 | 1.0 | 1.0 | Bal. |

## 3. Role of Anodic Oxide Films in the Corrosion of Aluminum Alloys in Alcohol at High Temperatures

When aluminum is immersed in alcohol at high temperature, it dissolves into the alcohol by the alcoxide formation reaction,

$$Al + 3\ ROH \rightarrow Al\text{-}(OR)_3 + 3/2\ H_2 \qquad (39)$$

Newly developed free-machining aluminum alloys, which contain small amounts of tin and bismuth, show a specific corrosion behavior. Extruded aluminum alloys with chemical compositions as indicated in Table 7 were immersed in 2-(2-(2-methoxy ethoxy)ethoxy)ethanol (MEEE) and 2-(2-(2-buthoxy-ethoxy) ethoxy)ethanol (BEEE) for 72 h at 415 K to measure mass changes and rest potential transients.[73] Only the No. 15-specimen shows a mass loss in BEEE, while No. 13 and 15-specimens showed mass losses in MEEE, with a higher corrosion rate on the No. 15-specimen than on No. 13. Figure 57 shows time variations in the rest potential, $E$, of No. 10, 13, and 15-specimens during immersion in MEEE. The No. 10-specimen on which no corrosion occurs shows a slow rise in $E$ after a transient at the initial stage to reach a steady value, suggesting the suppression of corrosion by the growth of a stable thin oxide film. The No. 13-specimen shows a slow continuous drop in $E$ with time to reach a steady low value, and due to the exposure of the substrate by dissolution of the air-formed oxide film. The change in $E$ on the No. 15-specimen, which corrodes heavily, is relatively complex. The $E$ drops till 4 h, and then rises from 4 to 8 h before reaching a high steady value. The drop in $E$ of the No. 15-specimen can be explained in terms of substrate exposure as on No. 13, and the subsequent rise in $E$ is due to enrichment of Sn on the surface of the specimen. Tin particles deposited or remaining on the surface may act as cathodic sites, enhancing pitting corrosion. It is noteworthy that No. 23- and 63-specimens, containing small amounts of Cu or Mg as well as Sn, show no corrosion during 72 h immersion in MEEE and BEEE. Formation of intermetallic compounds between Sn and Mg and segregation of Cu near Sn particles may suppress the activity as cathodic sites.

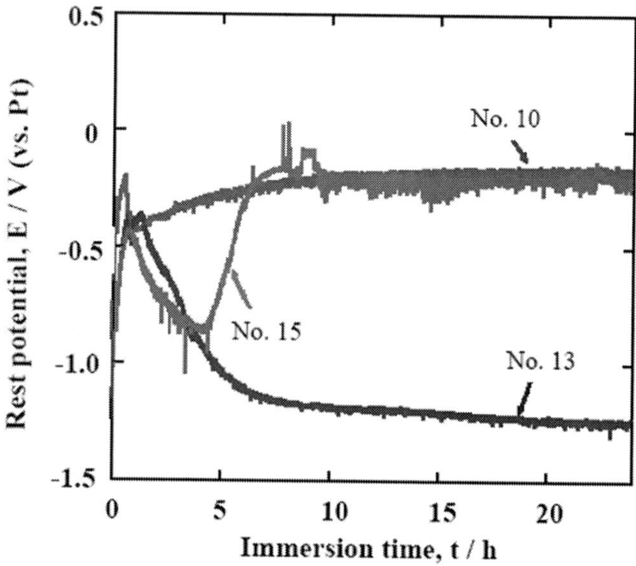

Figure 57. Variations in the rest potential of aluminum alloys during immersion in MEEE at 415 K. Reprinted with permission from *Abstracts of 114th Biannual Meeting of Jpn. Inst. of Light Met.* (2008). Copyright (2008), Jpn. Inst. of Light Met.

Formation of PAOF and the subsequent pore sealing generally suppress the corrosion of aluminum alloys, since pore-sealed PAOF prevents the substrate from exposing to corrosive atmosphere. However, corrosion of aluminum alloys is enhanced by pore-sealed PAOF in some cases. Three specimens of No. 10, 13, and 15, immersed in MEEE at 415 K for 72 h after anodizing in oxalic acid solution and pore sealing in boiling pure water.   Both No.10 and No. 13 specimens showed no corrosion, but substantial mass loss was observed on No. 15. The enhancement of corrosion of No. 15-specimen by anodizing can be explained by crack formation in PAOF, as shown in Fig. 58. Alcohol penetrates through cracks to reach the interface between PAOF and the substrate, and enable to dissolve the substrate along the cracks by the alcoxide formation reaction.

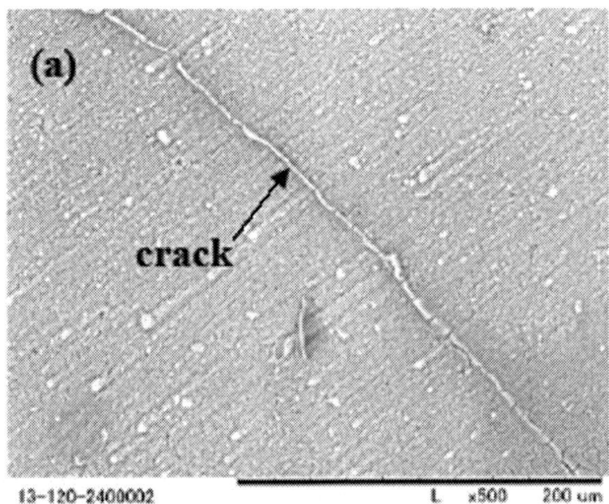


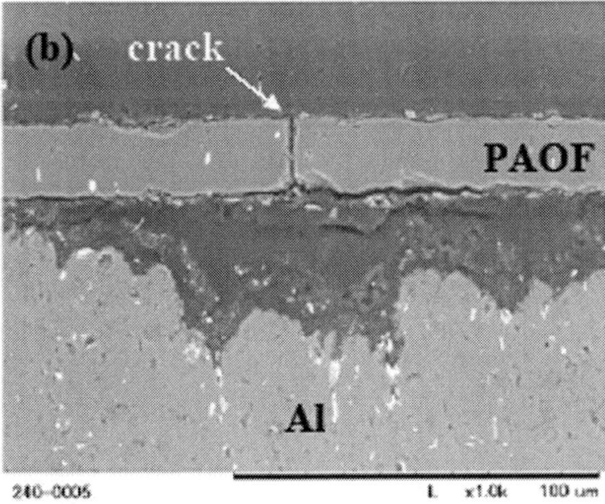


Figure 58. SEM images of (a) the surface and (b) a vertical cross section of a PAOF-covered specimen after immersion for 4 h in MEEE at 415 K. Reprinted with permission from *Abstracts of 114th Biannual Meeting of Jpn. Ins. of Light Met.* (2008). Copyright (2005), Jpn. Inst. of Light Met.

## V.    MICRO- AND NANO-TECHNOLOGIES BASED ON ANODIZED ALUMINUM

The latest decade has seen the development of micro- and nano-technologies have developed in the fields of micro-electromechanical systems (MEMS), nano-electromechanical systems (NEMS), micro total analysis systems (μ-TAS), and bio-technology. There are many investigations on the formation of nano-wires and nano-tubes in pores of porous anodic oxide films (PAOF) on aluminum, and the volume of such investigations is rapidly increasing. One of the reasons why PAOF has attracted investigator interest is that PAOF has the unique morphology of a hexagonal cell structure with nano-scale pores perpendicular to the metal substrate at the center of each cell, and as the cell size can be simply controlled by changing anodizing conditions such as electrolyte solution, temperature, and current density.[1,2] Another reason for the popularity of PAOF is the findings related to the formation of PAOF with highly oriented pores by Masuda et al.[3,74] Masuda et al. found that the degree of orientation of pores in PAOF near the metal substrate becomes higher after longer anodizing, and that highly oriented PAOF can be obtained by re-anodizing after stripping the oxide film formed by the initial anodizing. The Masuda group succeeded to fabricate porous membranes of metal and diamond by using highly oriented PAOF as the templates.[75,76]

Martin et al. pioneered the deposition of materials in the pores of porous membranes like PAOF, and deposited Cd-Te and Cd-Se rods in PAOF pores by AC-electrolysis after stripping the film from the aluminum substrate, where current rectification was established in the resulting system.[77] Martin also deposited poly-pyrrole in PAOF pores by electro- and chemical-polymerization, and suggested the potential for usage as nano-capsules for enzymes.[78] Kyotani et al. succeeded in forming carbon nano-tubes (CNT) in the pores of PAOF by reaction with propylene gas at high temperatures, and proposed that the PAOF containing formed carbon nano-tubes can be used as a membrane for ethanol enrichment from ethanol/water solutions and also as a good electron emitter.[79,80] Chu et al. deposited $TiO_2$ nano-tubes in the pores of PAOF by sol-gel dip coating, and showed that they are excellent candidates as photo-catalytic layers for the decomposition of organic compounds.[81] Mozalev et al. anodized aluminum

/tantalum bi-layers, which had been evaporated on silicon, in phosphoric acid solutions, and showed that tantalum oxide dot arrays on Si are possible, suggesting their application in nano-electronic devices.[82–84]

The authors have been working on micro- and nano-technologies based on Al. In the following, the principles and applications of these techniques will be described.

### 1.    Micro-Technologies with Laser Irradiation

Combining Al anodizing with laser irradiation utilizes two processes: local PAOF removal and PAOF carving (Fig. 59). In the former, PAOF is removed with a relatively large energy laser beam to expose the substrate locally;[44] in PAOF carving, only the outermost layer of the PAOF is removed by carefully controlling the laser power to carve the PAOF.[85]
In the PAOF removal process, an aluminum specimen covered with PAOF (Fig. 59a) is immersed in a dyeing solution (Fig. 59b), and then pores are sealed by immersing the specimen in boiling water (Fig. 59c). After pore-sealing, the specimen is immersed in water or solutions containing metal ions, electrolyte anions, and organic compounds and then irradiated with a pulsed Nd-YAG laser through a quartz window to remove the oxide film locally (Fig. 59d).[85,86] In the film carving process (Fig. 59d), a relatively low energy laser beam is used to remove the outermost film layer, leaving the inner layer of the PAOF.[44,87,88]

In both processes, the laser beam is focused or slightly de-focused on the specimen surface with a convex lens for the laser irradiation (Fig. 60). An XYZ three-dimensional stage is used to control the distance between the lens and the specimen, and for drawing patterns on the surface; in the case of tube-shaped specimens, a $\theta$-stage is also used to rotate the specimen. To draw fine patterns on specimens, a beam expander, an iris diaphragm, and a doublet lens must be set on the beam line.[89] The diameter of the laser beam at the focal plane, $D_f$, can be expressed by the following equation,[90, 91]

$$D_f = 2\lambda / [\pi \sin \{\tan^{-1}(D_0 / 2f)\}] \qquad (40)$$

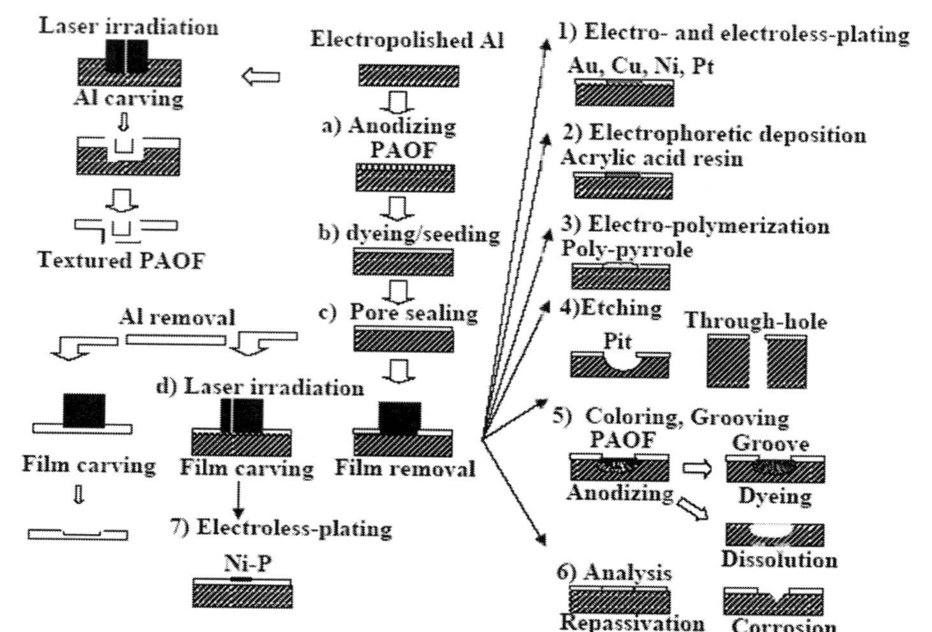



Figure 59. Schematic outline of micro-technologies by laser irradiation.

Figure 60. Physical arrangement of the setup involved in microstructure manufacture with laser irradiation. Reprinted with permission from *J. Surf. Finishing Soc. Jpn*, **56**, (2005) 528. Copyright (2005), Surf. Finishing Soc. Jpn.

where $\lambda$ is the wavelength of a laser beam, $D_0$ the beam diameter before passing the convex lens, and $f$ is the focal distance of the lens. The PAOF removal by laser irradiation can be explained in terms of either laser ablation of the aluminum substrate or thermal shock to the PAOF,[91–94] and dyeing and pore sealing before laser irradiation are significant for enhancing light absorption in the PAOF, leading to film removal without crack formation (Fig. 61). One advantage of laser irradiation in water or in solutions is the potential to avoid specimen contamination from debris produced by the film removal.[95] Debris, tiny flakes of oxides and hydroxides ejected from specimens by the laser irradiation are dispersed into the liquid phase.

After the local removal of PAOF by laser irradiation, metal can be deposited selectively at the film-removed area by electro- or electroless-plating (Fig. 59–1),

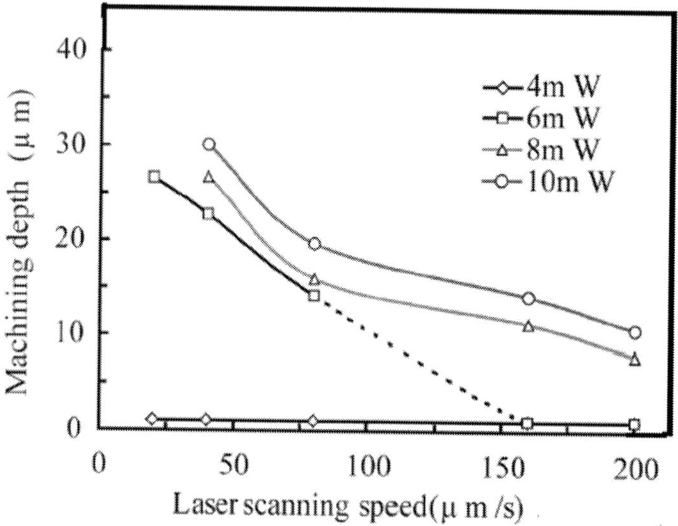


Figure 61. Laser scanning rate vs. depth of channels carved on PAOF for different laser beam powers. Reprinted with permission from *Appl. Phys., A* **88** (2007) 617. Copyright (2007), Springer.

$$Me^{n+} + ne^- \rightarrow Me \text{ (electroplating)} \tag{41}$$

$$Me^{n+} + (n/m)\,Red^{m-} \rightarrow Me \text{ (electroless-plating)} \tag{42}$$

where $Me^{n+}$ represents metal ions and $Red^{m-}$ the reducing reagent. The formation of a relatively thick PAOF (10–20 μm, Fig. 59a) and sufficient pore-sealing (Fig. 59b) are essential in the local electroplating, to avoid metal deposition through imperfections in the PAOF, since the PAOF acts as a template for the local metal deposition.[96–98] Strong adhesion of the metal layer to the substrate can be obtained by laser irradiation in electroplating solutions, since electroplating can be carried out immediately after laser irradiation without specimen transfer via air to a plating solution. This process excluding exposure to the atmosphere may suppress oxide film re-formation.

There is a further advantage of laser irradiation in the electroplating solution for ensuring adhesion of the metal layer, in that metal ions, $Me^{n+}$, included in the solution react with the aluminum substrate and are deposited as minute metal particles during the laser irradiation,[99]

$$3 \, Me^{n+} + n \, Al \rightarrow 3 \, Me + n \, Al^{3+} \quad [ \, Me^{n+} \text{: } Ni^{2+}, Cu^{2+}, Au^{+}] \quad (43)$$

The deposited metal particles may also suppress oxide film re-formation, resulting in good adhesion of metal layers deposited in the subsequent electroplating.

The reaction in Eq. (43) plays a significant role in the local electroless-plating after laser irradiation (Fig. 59-1).[100–103] The deposited metal particles act as catalytic centers during the subsequent electroless-plating, causing the deposition of metal only at the laser-irradiated area. No metal deposition occurs on PAOF that was not irradiated with the laser beam, because there are no catalytic centers here. The authors have performed Ni-P and Cu electroless-plating on pure aluminum and aluminum alloys, using hypophosphite ions as the reducing reagent. In the case of Cu electroless-plating, laser irradiation in a $Ni^{2+}$/hypophosphite solution shows rapid Cu deposition, due to the high activity of the catalytic Ni.[36] More rapid Ni-P deposition is observed on aluminum alloys than on pure aluminum, suggesting that alloying elements exposed by laser irradiation also act as catalytic centers.[103] It must be noted here that in the process the laser irradiation plays the roles of patterning, surface activation, and catalyst deposition prior to the electroless-plating.

In local resin deposition by electrophoretic deposition (Fig. 59-2), the laser-irradiated specimen (Fig. 59c) is immersed in acrylic acid resin nano-particle solutions and anodically polarized to form an acrylic acid resin film by polymerization only at the laser-irradiated area.[100] After electrophoretic deposition, the specimen is heated at 383 K for dehydration and further polymerization of the acrylic acid resin.

In the local electro-conductive polymer deposition by electro-polymerization (Fig. 59-3), the laser irradiated specimen is immersed in a sodium dodecyl-benzene-sulfonate (DBSNa)/pyrrole solution and anodically polarized at room temperature to deposit poly-pyrrole at the laser irradiated area,[105,106]

$$n \; \text{(pyrrole)} + DBS^- \rightarrow \text{(terpyrrole}^+ \cdot DBS^-) + 2\,ne^- \tag{44}$$

Further laser irradiation after PAOF removal causes the formation of micro-pores and through-holes in the aluminum substrate (Fig. 59-4).[93] Laser irradiation in solution rather than air is superior for through-hole formation, since irradiation in solution result in less debris remaining on the specimen surface. Laser irradiation in NaCl solutions under anodic polarization results in hemispherical pores, as a result of electrochemical dissolution.

The process of fabricating grooves on aluminum specimens covered with barrier type anodic oxide films (BAOF) involves irradiation with a laser beam in $H_2SO_4$ or oxalic acid solution while moving the exposed area at a steady rate (Fig. 59–5).[107,108] The specimen is subsequently anodized to form PAOF at the laser-irradiated area before removal of both BAOF and PAOF in a $H_3PO_4/H_2CrO_4$ solution. Local coloring is achieved by immersing the laser-irradiated specimens in dyeing solutions, such as Alizarin Blue, Alizarin Red-S, Naphtol Green-B, and others.

Abrupt destruction of oxide films on metals causes repair of the oxide film in oxide-forming solutions and pitting corrosion of the metal substrate in solutions containing aggressive ions. Analysis of this behavior has been carried out by monitoring the potential- or current-transient after stripping the oxide film mechanically with scratched-,[109] guillotined-,[110] and thin foil breaking electrodes.[111] Oltra et al. first carried out film destruction by laser irradiation to investigate the repassivation of iron,[112] and the authors here have applied the method to anodic oxide films on aluminum.[113,114] They illuminated PAOF-covered aluminum specimens with a laser beam at different potentials in neutral borate solutions containing $Cl^-$ ions (Fig. 59-6), and found that, after PAOF destruction, oxide film formation is followed by pitting corrosion, and the transient strongly depends on the $Cl^-$ concentration and potential.

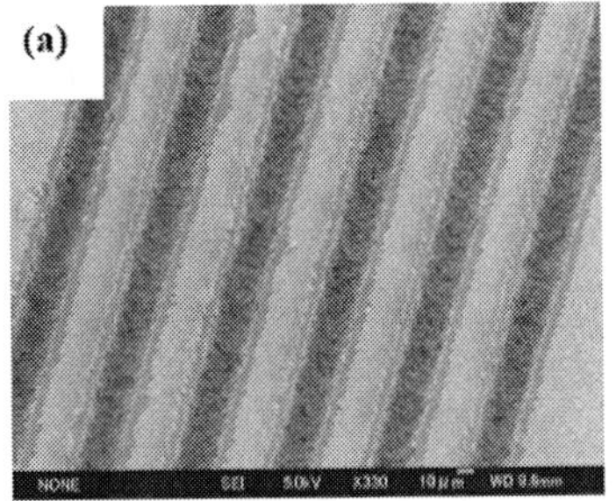
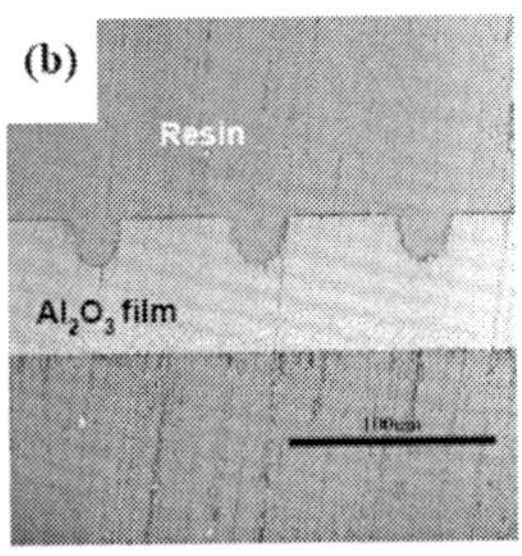



Figure 62. images of micro-channels formed on PAOF by laser carving: (a) surface(SEM), (b) vertical cross section (SCLM). Reprinted with permission from *Appl. Phys., A* **88** (2007) 617. Copyright (2007), Springer.

In the PAOF carving process by laser irradiation, careful control of the laser power and scanning rate is significant. The processes involved here are far more delicate and difficult to achieve than the removal of the whole film described previously. In PAOF carving process, dyeing and pore-sealing are essential steps in carving the PAOF without damage to the metal substrate, as these steps greatly enhance the light absorption enabling a suppression of the intensity of the laser beam reaching the interface between the PAOF and metal substrate. In Fig. 61, the depth of channels formed on PAOF with dyeing and pore-sealing is shown as functions of the laser power and laser scanning rate.[44] The depth decreases with increasing scanning rate at laser powers above 6 mW, while, at 4 mW, 1 µm deep channels are formed at all the scanning rates examined. Figure 62 shows scanning confocal laser microscope (SCLM) contrast images of channels carved on 64 µm thick PAOF by laser irradiation with 6 mW and a scanning rate of 40 µm/s in distilled water.

Local metal deposition on PAOF is possible on surfaces exposed by removing 1 µm of the PAOF (Fig. 63).[87,88] In this process, Pd-seeding of the surface and inner pore wall of the PAOF is carried out during the dyeing (Fig. 64). After Pd-seeding and dyeing, the specimen is subjected to pore-sealing in boiling distilled water (Fig. 64a), this results in a highly crystalline hydroxide layer,

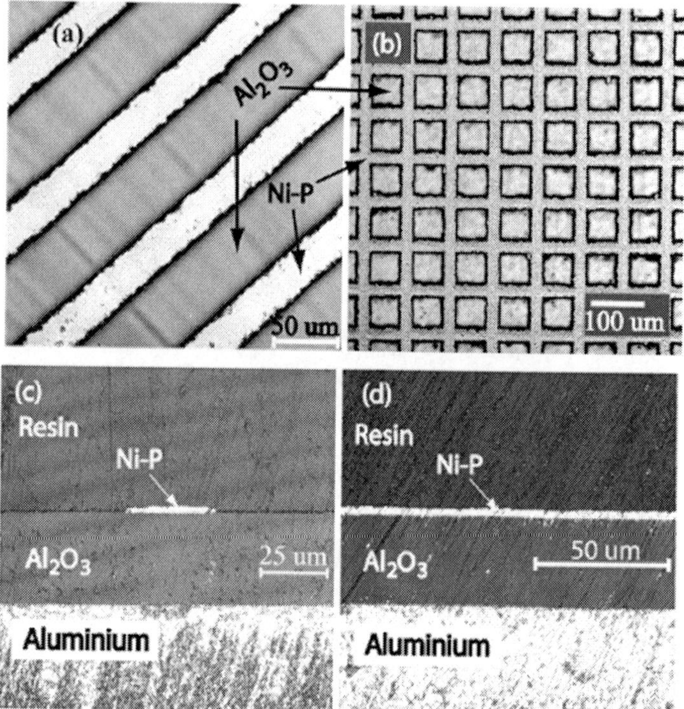


Figure 63. Local Ni-P deposition on PAOF by laser irradiation and electroless-plating; (a) SCLM image of a stripe pattern, (b) SCLM image of a grid pattern, (c) and (d) SEM images of the vertical cross sections of Ni-P deposited specimen (along (c) and across (d) stripes). Reprinted with permission from *Electrochem. Commun.*, **9** (2007) 1596. Copyright (2007), Elsevier.

which does not contain $Pd^{2+}$ ions, at the outermost part of the PAOF. Then, laser carving is carried out at a relatively low energy (2–3 mW) in a Ni-P electroless plating solution (Fig. 64b) to make channels of ca. 1 μm depth. Finally, the specimen is transferred into a hot Ni-P electroless plating solution to allow the local deposition of Ni-P, which occurs only at the carved area (Fig. 64c).

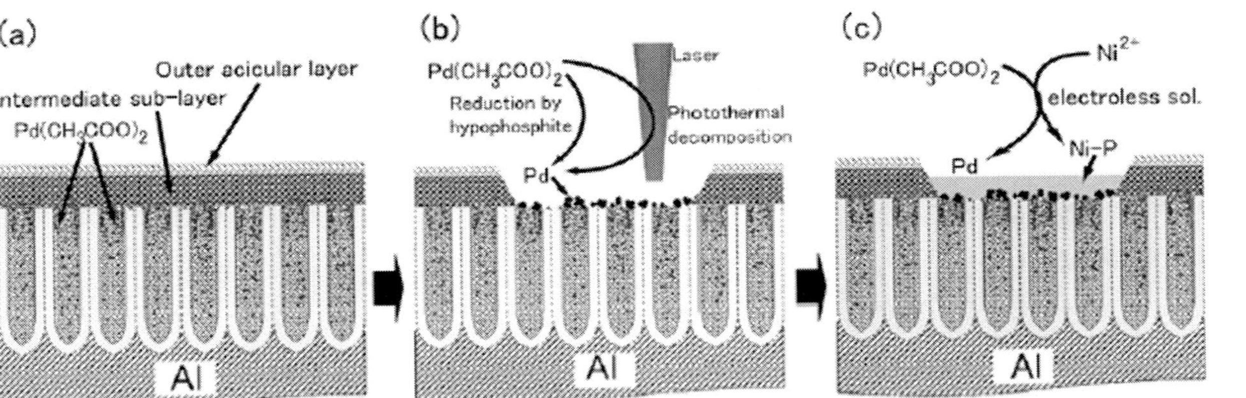



Figure 64. Schematic depiction of the process of local Ni-P deposition on PAOF by Pd-seeding, pore-sealing, laser irradiation, and electro-less-plating. Reprinted with permission from *Electrochem. Commun.*, **9** (2007) 1596. Copyright (2007), Elsevier.

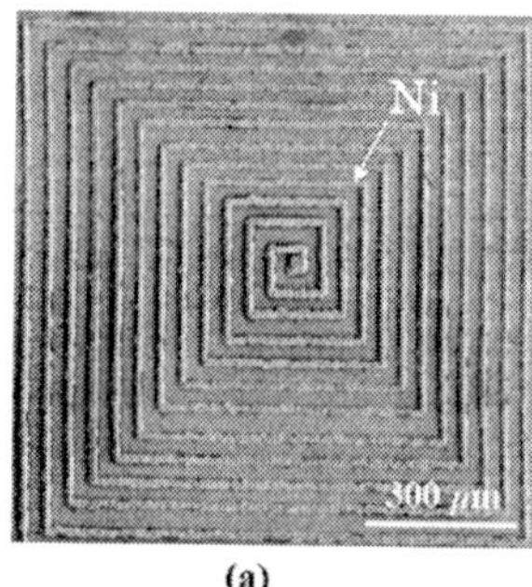


(a)

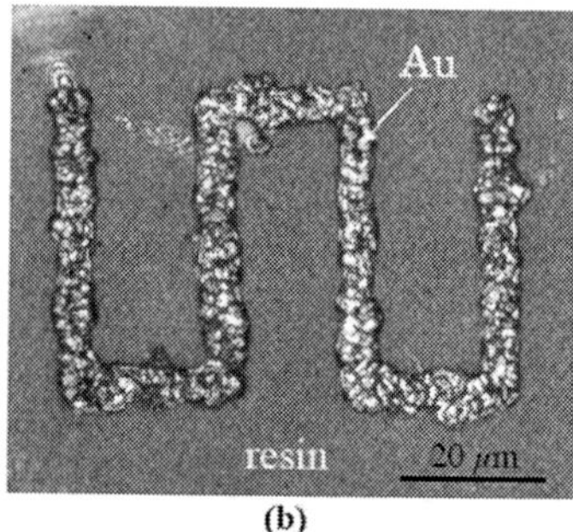


(b)

Figure 65. Prototype of a printed circuit board fabricated by laser irradiation: (a) Ni circuit and (b) Au circuit on an epoxy resin. Reprinted with permission from *J. Surf. Finishing Soc. Jpn,* **56**, (2005) 528. Copyright (2005), Surf. Finishing Soc. Jpn.

## 2.  Applications of Laser Irradiation Techniques

The authors have used the laser technology described above to fabricate printed circuit boards,[93,115-117] plastic injection molds,[114] electrochemical micro reactors,[115] free-standing 3D microstructures,[116] 3D micro manupilators,[101,102] and micro printing rolls,[117] and the following will give brief descriptions of the manufacturing processes and show the resulting structures.

### (*i*)  *Printed Circuit Boards*

In the fabrication of printed circuit boards here, aluminum foil covered with PAOF is subjected to dyeing, pore-sealing, laser irradiation, metal deposition, epoxy resin attachment, and de-attachment of PAOF and metal substrate. The metal deposition is carried out by electroplating, and de-attachment of PAOF and the metal substrate is carried out by dissolving the substrate in alkaline solution. A few to several tens of μm of metal line-width and line-gaps can be obtained on epoxy resin by this process (Fig. 65). The line width and the line-gap are much narrower than those of circuit boards produced by photo-lithography/cu-etching, although the accuracy of the lines is relatively low. Recently, the authors   developed   a   new method for the fabrication of micro

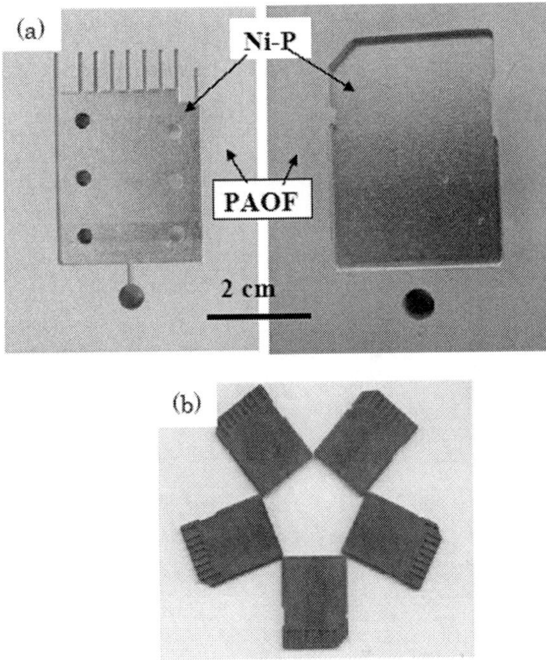


Figure 66. (a) Aluminum molds produced by anodizing, laser irradiation, and Ni-P electroless-plating, and (b) plastic sockets for digital memory cards produced with the aluminum molds.

printed circuit boards by laser removal of Cu layers deposited on epoxy resin plates.[122]

## (ii) Plastic Injection Molds

Fabrication of plastic injection molds here involves electric discharge machining of Al5052 alloy, hard anodizing, laser irradiation, and Ni-P deposition. Plastic injection molds require hard surfaces to prevent wear, and hard anodizing and Ni-P deposition are carried out for this purpose. Thirty μm thick PAOF formed by anodizing at 275 K in 16 mass%-sulfuric acid solution for 140 min gives a 300 Hv micro-Vickers hardness, and a 600 Hv surface is obtained with 30 μm thick Ni-P layers deposited at 353 K over 4 h in electroless plating solutions (Melplate NI875M, pH = 5, Meltex Co.). Figure 66 shows (a) a set of molds produced by the above process and (b) plastic sockets for digital memory cards produced with the molds.[123]

### (iii)    *Electrochemical Micro-Reactor*

An electrochemical micro-reactor fabricated by laser techniques is shown in Fig. 67, the reactor is composed of three components: two aluminum plates and a silicone rubber sheet.[119] One of the aluminum plates has a small depression (2 x 2 x 0.11 mm) with a 5 μm thick Au layer (working electrode: WE) at the bottom, two channels (width: 250 μm, depth: 150 μm), and two through-holes (diameter: 350 μm). The through-holes are connected with fused silica tubes, which function as guides for the inlet and outlet of solution. The other aluminum plate has a small depression with an Au layer (counter electrode: CE) and a through-hole with a nitrocellulose-covered platinum wire (reference electrode: RE). A 100 μm thick silicone rubber sheet is sandwiched between the two aluminum plates held tight with bolts and

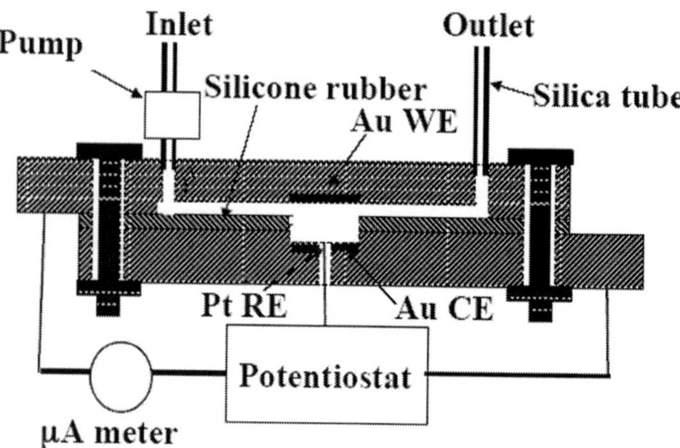


Figure 67. Schematic outline of electrochemical micro-reactor fabricated by laser irradiation. Reprinted with permission from *Electrochim. Acta,* **52**, (2007) 6268. Copyright (2007), Elsevier.

Figure 68 shows the effects of a) potential scanning rate, b) solution flow rate on cyclic voltammograms obtained with the electrochemical micro-reactor in 0.2-M $K_4[Fe(CN)_6]$/0.2-M nuts, to avoid solution leakage from the micro-channels during measurements. The volume of the chamber produced by attaching the two aluminum plates and the silicone rubber sheet is estimated to be about 1.3 µl, which is small enough to be applied to bio-systems where small amounts of solution are examined. $K_3$ $[Fe(CN)_6]$ solution at room temperature. Under stagnant conditions, there is a coupled redox current peak on each cyclic voltammogram (CV), which corresponds to the redox reaction between the $[Fe(CN)_6]^{4-}/[Fe(CN)_6]^{3-}$ ions. The curves have a short potential separation of two peaks, suggesting a high degree of reversibility for the reaction in the micro-reactor. The anodic peak current, $i_{ap}$, in the CV can be expressed by the following equation,[126]

$$i_{ap} = 0.446nFAc\{nF/(RT)\}^{1/2}D^{1/2}v^{1/2} \tag{45}$$

where $n$ is the electron number of the $[Fe(CN)_6]^{3-}/[Fe(CN)_6]^{4-}$ redox reaction, $F$ is the Faraday constant, $A$ the surface area, $c$ the concentration of $[Fe(CN)_6]^{4-}$ ions, $R$ the gas constant, $T$ the temperature, $D$ the diffusion coefficient of $[Fe(CN)_6]^{4-}$ ions, and $v$ is the potential scanning rate. Equation (45) suggests that $i_{ap}$ is proportional to $v^{1/2}$ and $c$. An analysis of Fig. 68a has shown the validity of Eq. (45), suggesting the possibility of quantitative analysis of solutions containing redox species.

In the reactor, cyclic voltammograms under flow conditions show no hysteresis, and there are diffusion limiting currents, $i_L$, depending on the flow rate (Fig. 68b). Assuming that the flow is similar to that in a channel flow electrode, $I_L$ can be expressed as,[127,128]

$$i_L = 1.165\ nFcw(vD^2x^2/b)^{1/3} \tag{46}$$

where $w$ is the electrode width, $x$ the electrode length, and $b$ the half depth of the micro-channel. Analysis of Fig. 68b shows that $I_L$ is proportional to $v^{1/3}$, suggesting the validity of Eq. (46).

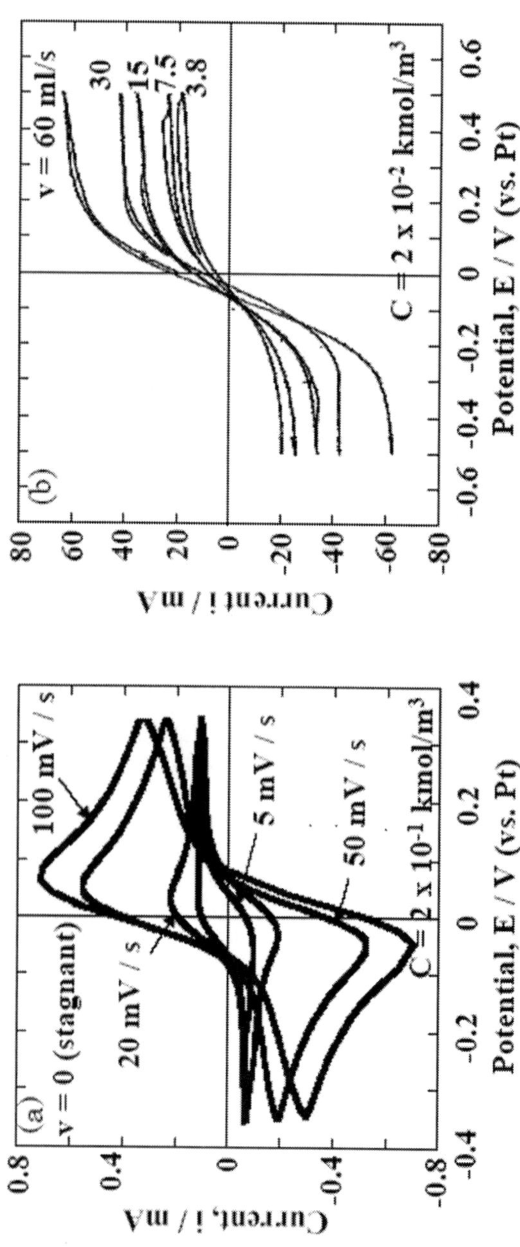


Figure 68. Cyclic voltammograms obtained with the micro-reactor and $K_3[Fe(CN)_6]$ / $K_4[Fe(CN)_6]$ solution under (a) static conditions and b) flow conditions. Reprinted with permission from *Electrochim. Acta*, **52**, (2007) 6268. Copyright (2007), Elsevier.

## (*iv*) *Free-Standing Microstructures*

Fabrication of freestanding microstructures of metals and resins is possible by dissolving PAOF and the aluminum substrate in alkaline solutions after electrochemical deposition at the laser irradiated areas.[120]

Figure 69 shows scanning electron micrographs of three-dimensional microstructures fabricated in this manner. The acrylic acid resin cylindrical network structure (Fig. 69a) was fabricated by electrophoretic deposition on a 2 mm diameter aluminum tube; the flat Nickel mesh (Fig. 69b) was electroplated on aluminum plate; the Ni cylindrical network structure (Fig. 69c), Ni cylinder with gaps (Fig. 69d), and Ni spring (Fig. 69e) all fabricated by electroplating on a 2 mm diameter aluminum tube; and the prismatic network structure (Fig. 69f) used a rectangular 2x2 mm aluminum bar in the manufacture. The authors also succeeded in fabricating 3D microstructures made of platinum.[129] It must be noted that such 3D microstructures are possible only on aluminum substrate where it is possible to form PAOF and perform dissolution in alkaline media. The θ–stage indicated in Fig. 60 made it possible to fabricate the cylindrical and prismatic microstructures, and more complicated structures, like spheres, cones, ovals etc. would be possible by using a 3D free-moving stage.

## (*v*) *Micro-Actuator*

Actuators are devices that transform an input signal into motion. Micro-actuators have been developed as mechano-electric micro systems (MEMS) in this decade, and are mainly made of piezoelectric materials,[130] shape-memory alloys,[131] and electro-conducting polymers.[132,133]

The motion of electro-conducting polymers, for instance polypyrrole (PPy), is based on volume changes due to doping and de-doping with ions as expressed in the following relation,

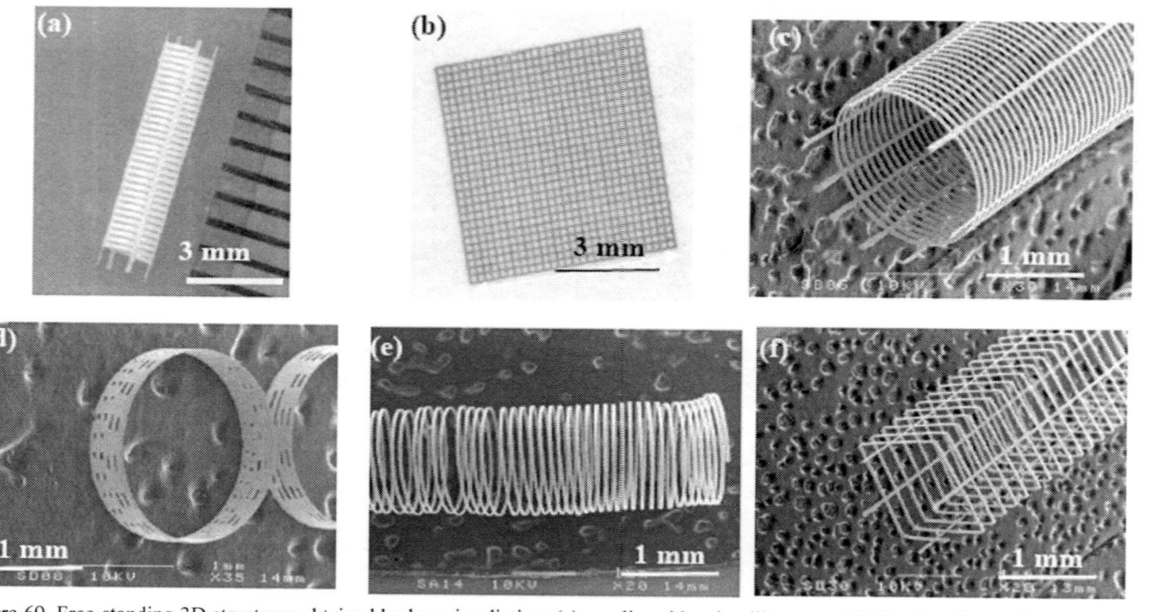

Figure 69. Free-standing 3D structures obtained by laser irradiation: (a) acrylic acid resin pillar cage, (b) Ni micro flat mesh, (c) Ni-pillar cage, (d) Ni ring with gaps, (e) micro-pillar spring, and (f) prismatic cage. Reprinted with permission from *J. Electrochem. Soc.*, **150**, (2003) C567. Copyright (2003), The Electrochemical Society.

$$+ (Na^+)_h \quad \xrightarrow[\text{oxidation}]{\text{reduction}} \quad (47)$$

where $(Na+)_h$ indicates a hydrated sodium ion. The PPy synthesized in dodecyl-benzene-sulfonate (DBS$^-$) solutions includes DBS$^-$ ions as counter ions as shown on the left side of Eq. (47). Cathodic polarization in sodium DBS$^-$ solution reduces the PPy, leading to Na$^+$ ion doping into spaces in the PPy matrix. Sodium ions present as hydrated ions $(Na^+)_h$ in the solution are doped with a water sheath, resulting in expansion of the matrix volume.[134] The reverse, anodic polarization oxidizing PPy, causes a volume shrinkage. A ribbon consisting of two layers, metal and PPy, bends to the metal side by cathodic polarization at the volume expansion of PPy, and to the PPy side when PPy shrinks, the principle of motion of ribbon-type actuators with electro-conducting polymers.

The authors attempted to fabricate three layer 3D actuators after successfully fabricating ribbon type actuators, composed of an acrylic acid resin layer, an Au layer, and a PPy layer. The 3D micro-actuator can be fabricated, using a 2 mm diameter aluminum tube specimen with anodizing (Fig. 70a), laser irradiation (Fig. 70b), Au deposition (Fig. 70c), acrylic acid resin deposition (Fig. 70d), PAOF and metal substrate dissolution, deposition of PPy on the Au layer (Fig. 70e), and attachment to copper wire (Fig. 70f). Figure 71 is a photo of this 3D micromanipulator gripping a 6.5 mg mulite ($3Al_2O_3 \cdot 2SiO_2$) cylinder in a Na-DBS solution. The four fingered manipulator is positioned above the cylinder (Fig. 71a), and then the bottom-ends of the fingers are deflected and the manipulator is moved to enclose the cylinder (Fig. 71b). After the micromanipulator has encloses the cylinder, the bottom-ends of the fingers are bent in to ensure a tight grip (Fig. 71c), and the manipulator gripping the cylinder is lifted up and moved (Fig. 71d).

A two-layer system with Au and PPy would be more flexible actuator than the three-layer system, as the resin layer suppresses the motion. However, the three-layer system gets around the obstacle that PPy loses electro-conductivity by dipping in NaOH solution, a step that is essential for removing PAOF and the aluminum substrate, as shown in Fig. 71.[105] The loss of electro-conductivity has been reported to be due to a reduction or de-polymerization of the PPy.[135–138]

Recently, the authors developed a further process for the fabrication of two-layer systems.[139] In this new process, an aluminum specimen covered with PAOF is subjected to dyeing and pore sealing, and followed by laser irradiation, Au electroplating, and PPy deposition. Finally, the specimen is connected with a Pt plate through a copper cable and immersed in a NaOH solution. The removal of PAOF and the substrate by connecting it to Pt does not cause the loss of electro-conductivity in PPy, as there is no reduc-

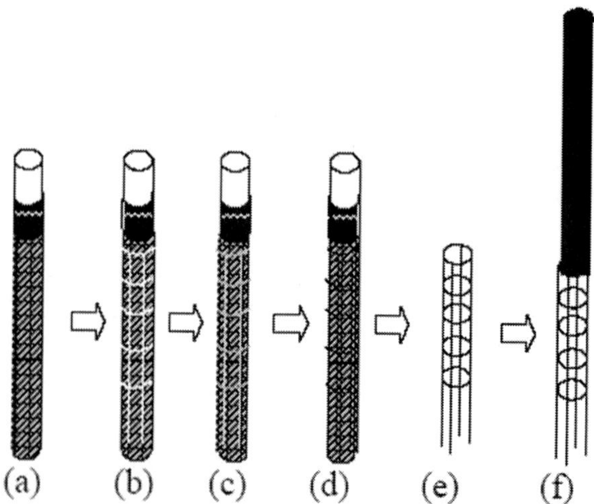



Figure 70. Fabrication process of 3D micro-manipulator: (a) anodizing, coloring, and pore-sealing, (b) laser irradiation, (c) Au deposition, (d) acrylic acid resin deposition (e) PAOF and Al dissolution, and (f) PPy deposition.

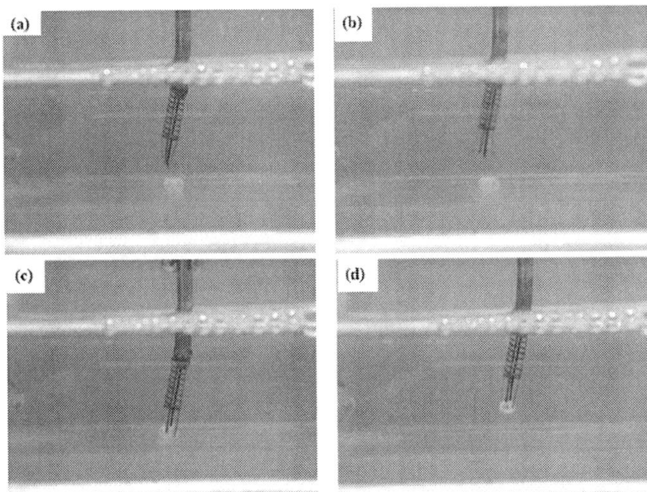

Figure 71. Photos of 3D micro-manipulator (a) suspended over, (b) descending on, (c) holding, and (d) gripping a mulite cylinder in 0.1 M NaDBS solution. Reprinted from *Electrochim. Acta,* **52**, (2007) 4480. Copyright (2007), with permission from Elsevier.

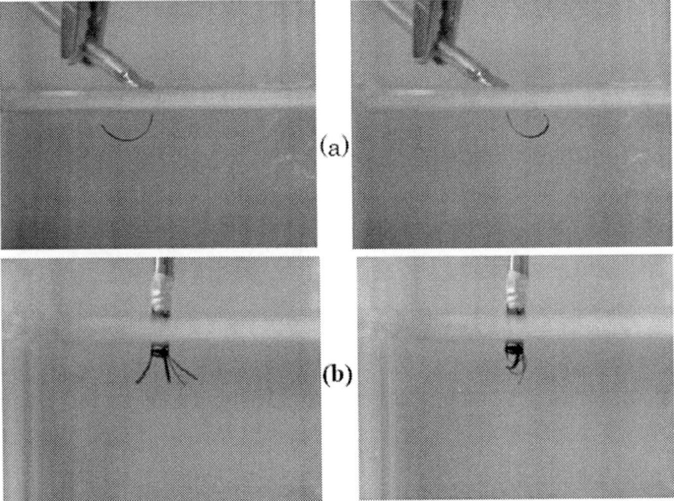

Figure 72. Photos of Au/PPy two-layer (a) ribbon type and (b) 3D actuators. Reprinted with permission from *Abstract of the Biannual Meeting of Surf. Fin. Soc. Jpn,* (2007) Copyright (2007), Surf. Finishing Soc. Jpn.

tion of PPy.[139] The photos in Fig. 72 illustrate the performance of a) ribbon type- and b) 3D-actuators with the Au / PPy two-layer system, showing better flexibility than with the three-layer system.

### (vi)    *Printing Roll*

Offset printing is widely used for mass printing of news papers, books, etc. and is carried out using pre-sensitized (PS) plates. A PS plate is an aluminum plate covered with PAOF and a photosensitive resin. Printing ink easily attaches to the hydrophobic resin surface while water covering the hydrophilic surface of the PAOF protects it from the attachment of printing ink.

A micro-printing roll can be fabricated by a laser irradiation technique as shown in Fig. 73a. A 2mm diameter aluminum tube is anodized to form PAOF and then subjected to dyeing and pore sealing. The tube is irradiated with a pulsed-YAG laser to expose the substrate locally, and then electrophoretic deposition of acrylic acid resin is carried out before heating. A micro-printing roll fabricated thus is shown in Fig. 73b, and the pattern printed on a sheet of paper using the printing roll is shown in Fig. 73c.

Gravure printing is commonly used for long run, high quality printing such as printing of art works, packages, labels, bank-notes etc., and involves the technique that ink remaining in the recessed cells engraved on a copper cylinder is directly transferred to the sheets to be printed (Fig. 74a). Gravure printing rolls, which are made of aluminum, are produced by depositing silver and then copper before engraving patterns on the copper layer, and are re-used by repeating these processes after removing the copper layers (Fig. 74b).

The authors have attempted to develop a method for the fabrication of gravure   printing rolls,   which does not include   an Ag-layer and is re-usable by removing and re-deposition of the Cu-layer.[121] This novel process involves anodizing, laser irradiation, nickel deposition, and copper deposition, and for the re-use, removal of the metal layers from the metal substrate is carried out. Aluminum specimens covered with PAOF are irradiated with a pulsed Nd-YAG laser to form pit arrays at 300 $\mu$m intervals between pits, and then Ni dots are deposited in the pits by electroplating. Finally, a continuous Cu layer is deposited on the PAOF by

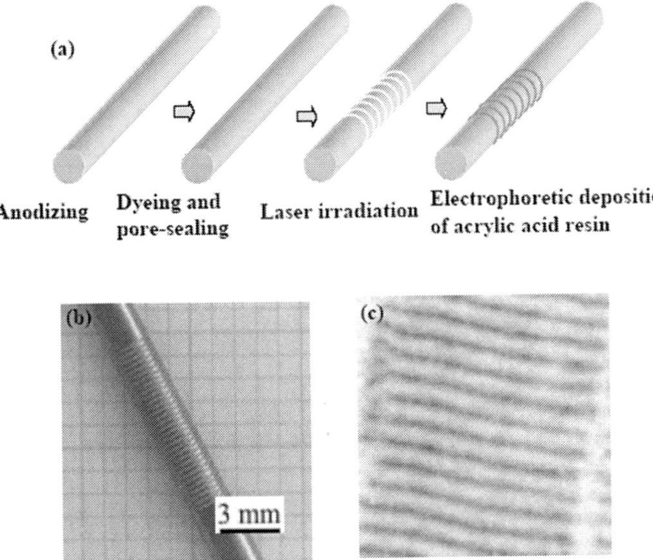



Figure 73. (a) Process of fabrication of micro printing roll, and photos of (b) the micro printing roll fabricated by the above process, and (c) the pattern printed on a sheet of paper using the printing roll in (b).

the growth and coalescence of Cu-domes initiated on the Ni-dots (Fig. 75). Adhesion of the Cu-layer with PAOF is relatively poor, and the Cu layer can easily be detached mechanically. This is an important property for re-use of gravure printing rolls.

## 3.  Micro- and Nano-Patterning with AFM Probe Processing

The scanning tunneling microscope (STM)[140] and Atomic Force Microscope (AFM)[141] were invented by IBM groups, and have led to the development of a number of types of scanning probe micro scopes (SMP): the magnetic force microscope (MFM), the Kelvin force microscope (KFM), the scanning near field microscope (SNOM) and others. These are very useful to characterize the local

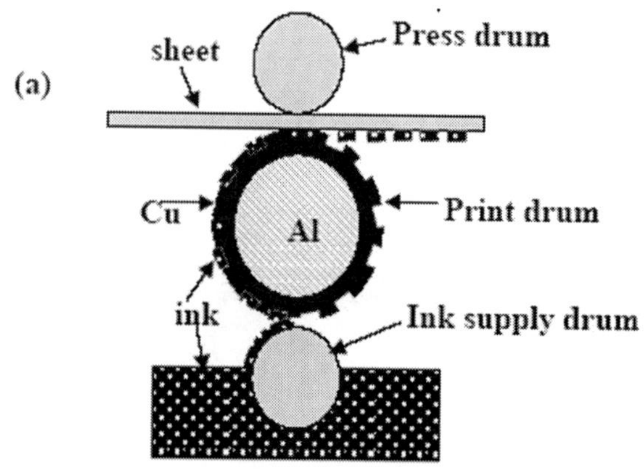


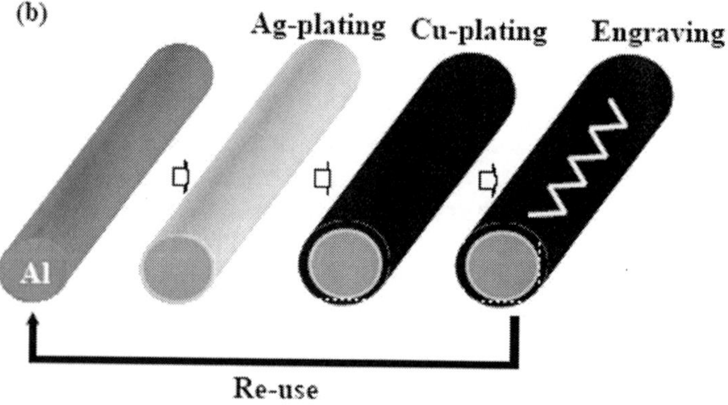


Figure 74. Schematic outline of (a) gravure printing and (b) fabrication of gravure printing drums.

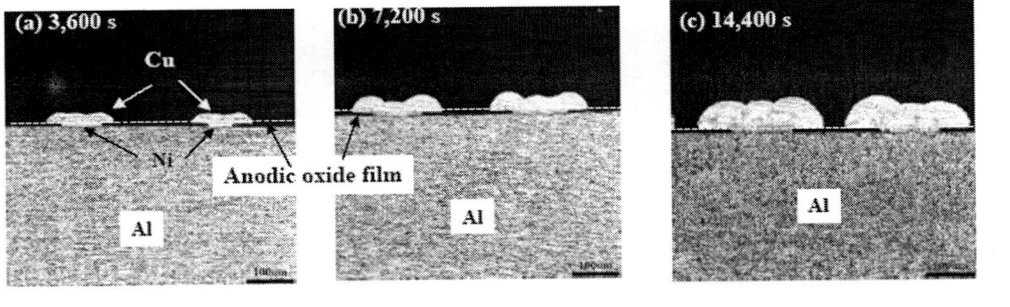


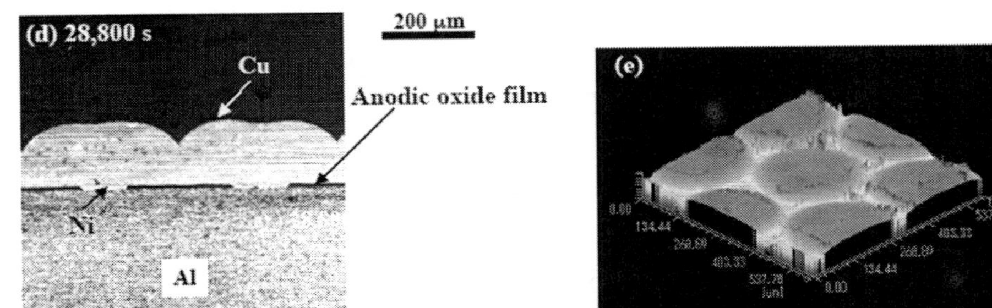


Figure 75. SEM images of vertical cross sections of aluminum specimens at different Cu-electroplating times: (a) $t_c$ = 3,600, (b) 7,200, (c) 14,400, (d) 28,800 s and (e) SCLM height image at $t_c$ = 28,800 s.

morphology and properties of surfaces of solid materials on a nano-scale, and the development of nano-technologies owes much to SPM. In addition to the use in characterization of materials, SPMs can also be used to perform processes with or on materials.[142,143]

The authors developed an AFM probe process on aluminum covered with anodic oxide films in solution, and it will be described next.[144–147] Relatively thin BAOF (10–20 nm) is formed on aluminum by initially anodizing in a neutral boric acid/borate solution, and then, scratching the anodized specimen with an AFM probe before the deposition of metal by electroplating or resin by electrophoresis (Fig. 76). The scratching with the AFM probe is carried out in solutions containing metal ions or nano-sized resin particles. The AFM probe can be used for both in-situ observations and to perform mechanical processes, and the progress of film removal during scratching can be monitored by measuring the rest potential of the specimen. The AFM probe has a pyramidal Si-tip with / without diamond film coating (Fig. 77), as it has been found that silicon-tips coated with diamond film show much higher processing rates and wear much more slowly than tips without coating.

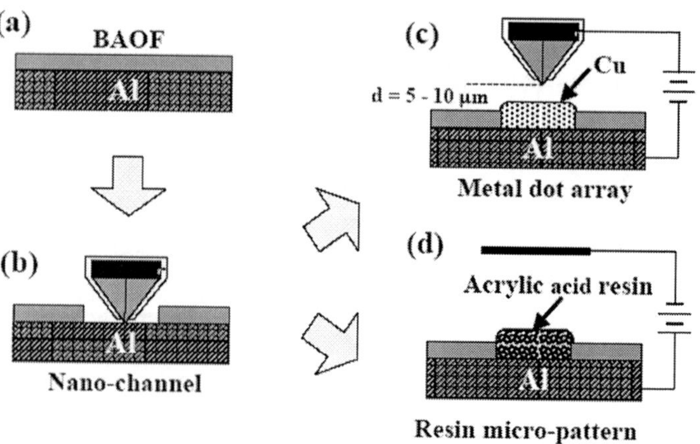



Figure 76. Schematic outline of local deposition of metal and resin by AFM probe processing and electrochemical treatments.

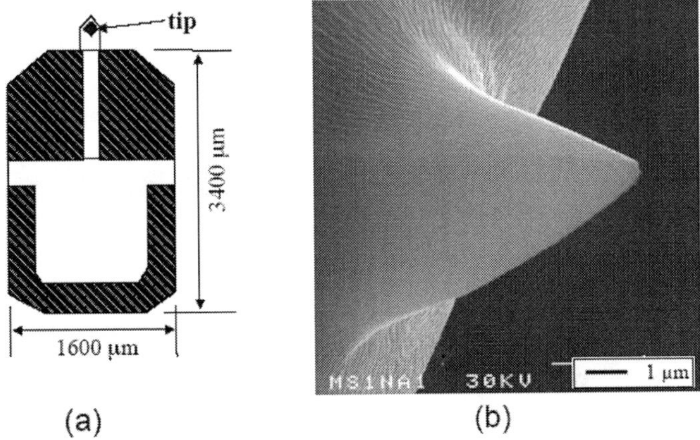



Figure 77. (a) Schematic model of AFM probe and (b) SEM image of the probe tip coated with nitro-cellulose film. Reprinted with permission from *Zairyo to Kankyo*, **52**[1], 12-17 (2003), Copyright (2007), Japan Society Corrosion Engineering.

In the electroplating of metals after film removal by scratching, the surface of all of the AFM probe is covered with a nitro-cellulose film, and this nitrocellulose film is removed only at the top of the Si-tip, which scratches the aluminum plate, prior to the film removal and metal deposition (see Fig. 76b). During metal deposition, the nitrocellulose film-coated AFM probe is used as a counter electrode, and the distance between the counter electrode and aluminum specimen is adjusted to between 2.5–15 μm (Fig. 76c). In the electrophoretic deposition of acrylic acid resins, film-removed aluminum specimens are anodically polarized using a Pt wire as a counter electrode (Fig. 76d).

Figure 78 shows scanning confocal laser microscopic (SCLM) contrast (Fig. 78a) and 3-D height (Fig. 78b) images at the area of a copper dot array fabricated by the repetition of scratching and deposition at different positions. There are five copper dots of ca. 15–20 μm diameter and 3–5 μm height, located at the four corners and center of a 100 μm square. The numbers in Fig. 78a indicate the order of fabrication of the dots. The copper dot at the center of the square is appreciably smaller than the four dots at the corners; this is because   current flows between the probe and   previously deposited copper dots during the deposition of copper at the center of the square.

In the patterning process by resin deposition, aluminum specimens covered with BAOF are immersed in a solution containing acrylic acid resin and melamine resin particles, and then the potential of the specimen is kept at −1.1 V (vs. Ag/AgCl). Scratching with diamond-coated AFM probes was carried out under potentiostatic conditions at −1.1 V with an $F = 100$ μN compression force and at a $v = 50$ μm/s scanning rate. After removing BAOF at different areas, the AFM stage was lowered, and the BAOF was then anodically polarized at different potentials, $E_d$, to deposit acrylic acid-/melamine-resin at the areas where BAOF had been removed. Finally, the specimen was heat-treated at 453 K for 30 min.

Figure 79 shows AFM images of an aluminum specimen covered with BAOF a) after scratching and b) after deposition of the resin by electrophoresis. There is 2 μm long, 80 nm deep, and 800 nm wide groove in Fig. 79a, and a 200–250 nm high and 2.5 μm wide resin wall at the bottom in Fig. 79b.

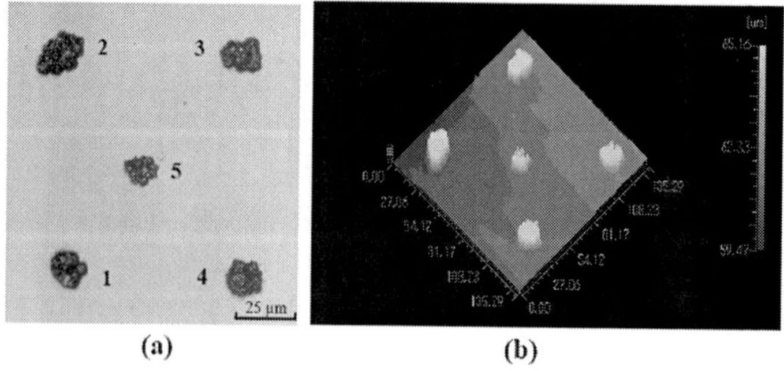

Figure 78. Example of a copper dot array: (a) contrast image and (b) 3-D height image of scanning confocal laser microscopy (SCLM).

## 4. Formation of Composite Oxide Films by the Combination of Anodizing with Other Coating Methods

As described in Section I, BAOF is used as dielectric layers of aluminum electrolytic capacitors, essential devices in many kinds of electronic devices, and their physical and chemical properties determine the performance of the electrolytic capacitor. Recent developments in mobile electronic devices, such as notebook computers, portable telephones, and hybrid and electric vehicles require much smaller electrolytic capacitors with higher electric capacitance. The electric capacitance, $C$, of the aluminum electrolytic capacitor is expressed by the following equation:

$$C = \varepsilon_0 \varepsilon S / \delta \tag{48}$$

where, $\varepsilon_0$ is the vacuum permittivity, $\varepsilon$ the specific dielectric constant of the anodic oxide film, $S$ the surface area, and $\delta$ the film thickness. As the charge accumulated in the capacitor, the product of $C$ and the applied voltage, $V_{appl}$, expresses $Q$, the following equation can be derived

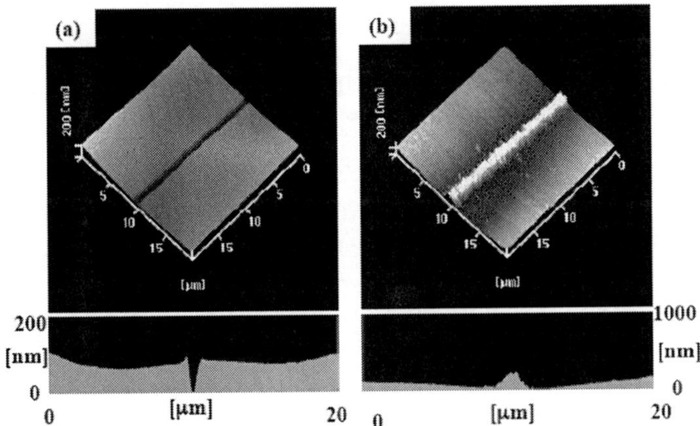

Figure 79. AFM images of aluminum surface after (a) scratching and (b) resin deposition. Reprinted with permission from *Electrochimica Acta*, **53**, (2008) 8118. Copyright (2008), Elsevier.

$$Q = (\varepsilon_0 \, \varepsilon \, S \, / \, \delta) V_{appl} \qquad (49)$$

The $V_{appl}$ should be smaller than the film formation potential, $E_a$, since the application of $V_{appl}$ beyond $E_a$ would lead to film breakdown. Hence, the maximum accumulated charge, $Q_{max}$, in aluminum electrolytic capacitors is expressed in the following equation:

$$Q_{max} = (\varepsilon_0 \, \varepsilon S \, / \, \delta) E_a \qquad (50)$$

The film thickness, $\delta$, is proportional to $E_a$, giving Eq. (51):

$$E_a = \delta / K \qquad (51)$$

where $K$ is the film thickness per unit film formation potential. Substituting $E_a$ with $\delta/K$ into Eq. (50) gives Eq. (52):

$$Q_{max} = \varepsilon_0 \, \varepsilon \, S \, / \, K \qquad (52)$$

From Eq. (52), larger values of $S$ and $\varepsilon$, and smaller values of $K$ give larger $Q_{max}$, and industrially increases in $S$ are achieved by electrochemical etching of the aluminum substrate before anodizing. Another way to increase $Q_{max}$ is increments in $\varepsilon$ or decrements of $K$. These values, however, cannot be changed independently, since they are closely related. Generally, there is a tendency that oxide films with higher dielectric constants have higher $K$-values, as shown in Table 8. The $\varepsilon/K$-values, however, are dif-

### Table 8
### Dielectric Properties of Anodic Oxide Films on Valve Metals.

| Anodic oxide film | $\varepsilon$ | $K$<br>nm V$^{-1}$ | $\varepsilon/K$<br>V nm$^{-1}$ |
|---|---|---|---|
| SiO$_2$ (a) | 3.5 | 0.4 | 8.8 |
| Al$_2$O$_3$ (a) | 9.8 | 1.5 | 6.5 |
| HfO$_2$ (c) | 21 | 1.9 | 11.1 |
| ZrO$_2$ (c) | 23 | 2.0 | 11.5 |
| Ta$_2$O$_5$ (a) | 28 | 1.7 | 16.5 |
| Nb$_2$O$_5$ (a) | 41 | 2.5 | 16.4 |
| WO$_3$ (a) | 42 | 1.8 | 23.3 |
| TiO$_2$ (c) | 90 | 3.0 | 30.0 |

ferent for different anodic oxide films, and anodic oxide films on aluminum have the smallest value in all the oxide films in Table 8. Therefore, the formation of composite oxide films by incorporating valve metal oxides into anodic oxide films on aluminum could offer an avenue to increase $\varepsilon/K$ of anodic oxide films. The authors have developed several methods for the formation of Al-Me (Me: Ta, Nb, Si, Ti) composite oxide films on aluminum:   pore-filling method[148] and a combination of metal organic chemical deposition (MOCVD),[149–150] sol gel dip coating,[151–158] electrophoretic sol gel coating,[159] and liquid phase deposition[160] with subsequent anodizing. In the following, the principles of these processes and their application to tunnel-etched specimens with uneven surfaces will be described.

### (i)  *Pore-Filling Method*[148]

In this process, aluminum specimens are anodized in an acid solution to form PAOF (Fig. 80a), and then immersed in acid solutions on open circuit to widen the pores of PAOF by chemical dissolution (Fig. 80b). The specimens are immersed in solutions containing metal ions to penetrate the solution into the widen pores, and then removed from the solution to dry at room temperature. The dried specimens are heated to decompose the metal complexes and to deposit metal oxide on the pore walls of the anodic oxide films (Fig. 80c). The dipping-heating process is repeated several times before re-anodizing in neutral solutions is carried out in the final step (Fig. 80d). During the re-anodizing, pores fill with new aluminum oxide, and the metal oxide deposited on the pore-walls is incorporated into the newly formed aluminum oxide.

Figure 81 shows the relationship between the reciprocal parallel capacitance, $1/C_p$, and re-anodizing potential $E_f$, for Ti-, Nb-, and Ta-oxide deposited specimens. The slope of the $1/C_p$ vs. $E_f$ curves for $TiO_2$ deposited specimens decreases with increasing $E_f$,

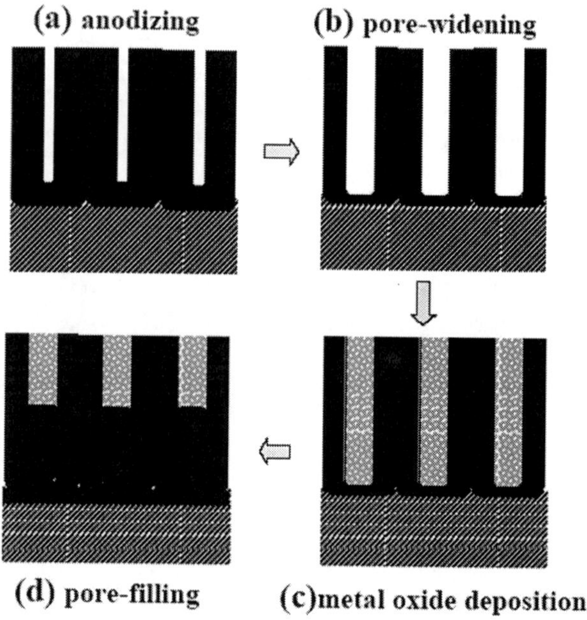


Figure 80. Principle of the pore-filling process: (a) formation of PAOF, (b) pore-widening, (c) deposition of metal oxide, and (d) pore-filling.

while the $1/C_p$ for Nb- and Ta-oxide deposited specimens is proportional to $E_f$. The $C_p$ value for 400 V films incorporating Ti-oxide is 40% higher than that with Nb- and Ta-oxide and without metal deposition.

### (ii) Combination of Metal-Organic Chemical Vapor Deposition with Anodizing[149,150]

In this process, aluminum specimens are covered with Me-oxide by metal organic chemical vapor deposition (MOCVD), where Me-alcoxide vapor is supplied with $N_2$ gas into the specimen chamber and then heated to decompose the alcoxide. The specimen coated with Me-oxide is anodized in a neutral   solution.

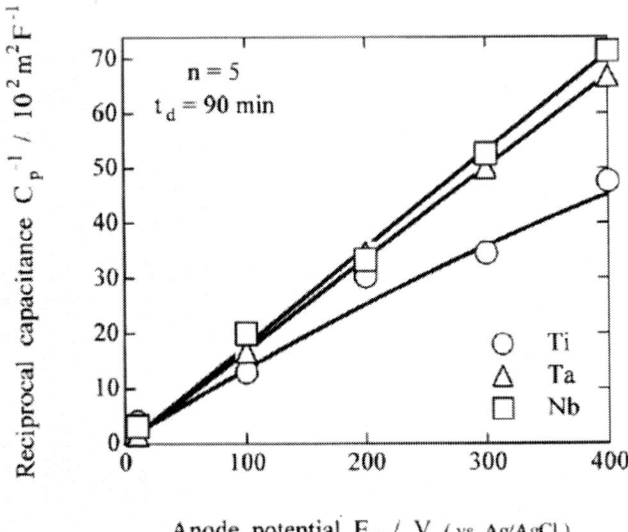


Figure 81. Relationship between the reciprocal parallel capacitance, $1/C_p$, and $E_f$, obtained for Ti-, Nb-, and Ta-oxide deposited specimens with $t_d$ = 90 min and n = 5. Reprinted with permission from *J. Electrochem. Soc.*, **144**, (1997) 2756. Copyright (1997), The Electrochemical Society.

During galvanostatic anodizing, two layer anodic oxide films grow: an outer Al-Me composite oxide layer and an inner alumina layer. Figure 82 shows TEM images of the vertical cross section of DC-etched specimens with MOCVD and anodizing, indicating that the $Ta_2O_5/Al_2O_3$ two-layer films cover the inner surface of tunnel pits of DC-etched specimen uniformly. Impedance measurements showed that $Ta_2O_5$ deposition on DC-etched specimen before anodizing enables increases in capacitance by 50–100 %, compared with only anodized specimens.

### (iii)   *Combination of Sol-Gel Dip Coating and Anodizing*[151–158]

This process involves the deposition of Me-oxide by sol-gel-dip coating of aluminum and anodizing in neutral solutions. The aluminum specimen is dipped in metal-sol, and then removed

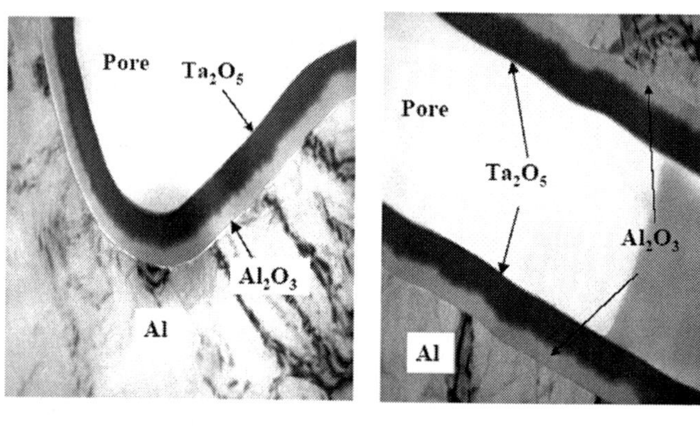



Figure 82. Al-Ta composite oxide films formed in a tunnel pit by the combination of anodizing and metal organic chemical vapor deposition (MOCVD).

at a steady rate with a linear motor system (Fig. 83). Then, the specimen is dried at room temperature and heated at different temperatures in oxygen. The specimen with metal-oxide deposited is anodized in neutral solutions to form composite oxide films. Figure 84 shows TEM images of the vertical cross sections of specimens anodized in 0.5-M $H_3BO_3$ solution up to a) 200, b) 400, c) 600, d) 800, and e) 1,000V after $SiO_2$ coating by sol gel dip coating.

The thicknesses of the $Al_2O_3$ and Al-Si composite oxide layers increase with increasing $E_a$, while the $SiO_2$ layer becomes thinner, disappearing at $E_a$ = 1,000 V. The total film thickness is 660 nm at 1,000 V, indicating 0.66 nm/V of the film thickness per volt of film formation potential. This value is much smaller than that for anodic oxide films formed on aluminum without coating (= 1.5 nm / V). This is due to the high electric field sustainability of Al-Si composite oxide layers. The EDX analysis at the numbered positions in Fig. 84e showed that the concentration of Si becomes lower at more inner positions of the composite oxide layer, while the Al concentration becomes higher. The Si-Al composite oxide film has a 10 % higher film breakdown potential and a 5 % higher capacitance than anodic oxide films formed on specimens without $SiO_2$ deposition.

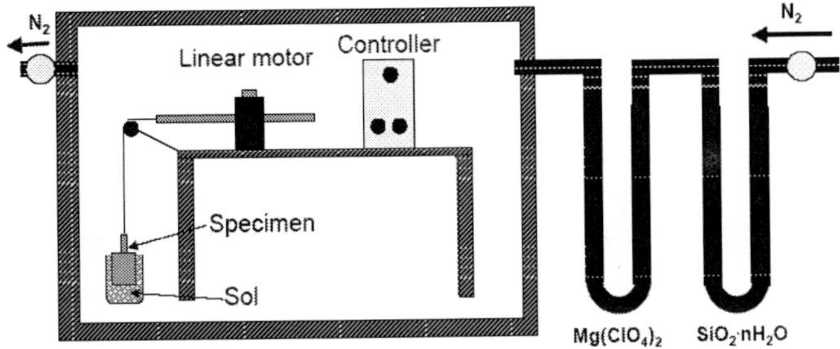



Figure 83. Arrangement of the apparatus for sol-gel dip coating on aluminum.

### (iv) *Combination of Electrophoretic Sol-Gel Coating and Anodizing*[159]

In this process, aluminum specimens are anodically polarized in a solution containing small particles of Me-oxide to deposit the particles on the oxide surface (Fig. 85). The specimen coated with the Me-oxide layer is heated, and anodized in neutral solutions. The electrophoretic sol gel deposition enables a much more uniform deposition of Me-oxide in tunnel pits of DC-etched specimens than it is possible with sol gel dip coating. This can

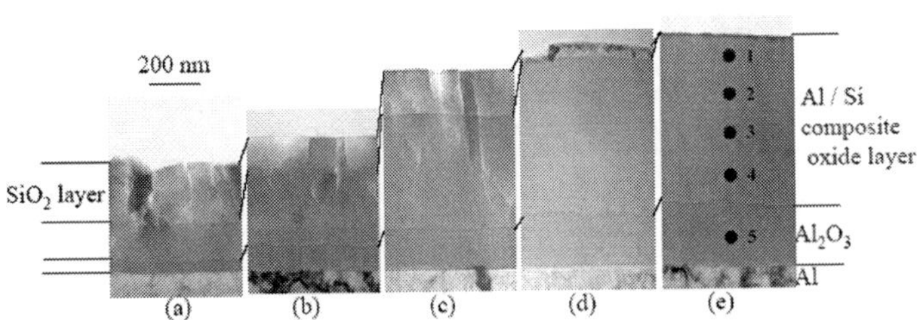



Figure 84. TEM images of the vertical cross sections of specimens coated with $SiO_2$ after anodizing up to (a) 200, (b) 400, (c) 600, (d) 800, and (e) 1000 V. Reprinted with permission from *J. Surf. Fin. Soc. Jpn.*, **54** (2003) 235 Copyright (2003), Surf. Fin. Soc. Jpn.

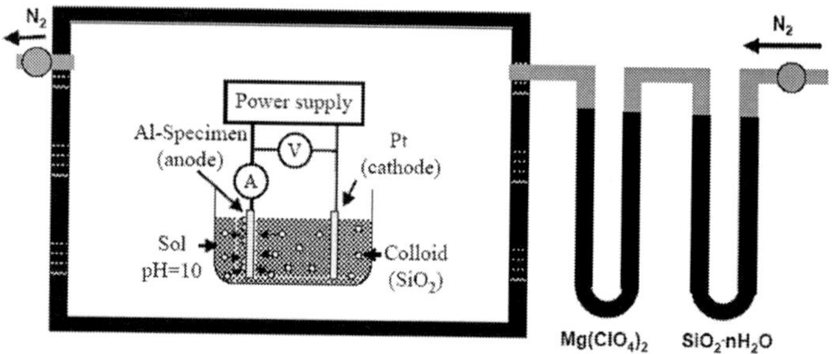



Figure 85. The apparatus for electrophoretic deposition of $SiO_2$.

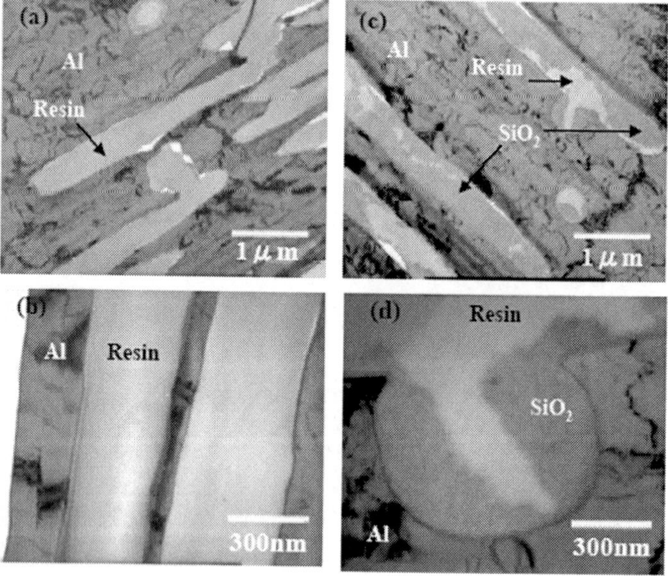



Figure 86. TEM images of cross sections of DC-etched specimens after (a) dip coating (low magnification), (b) dip coating (high magnification), (c) electrophoretic coating (low magnification), and (d) electrophoretic deposition (high magnification). In both processes, heating was at $T_h = 573$ K for $t_h = 1.8$ ks, and in the electrophoretic deposition the coating was carried out up to $V_c = 10$ V. Reprinted with permission from *J. Solid State Electrochem.*, **11**, (2007) 1375. Copyright (2007), Springer.

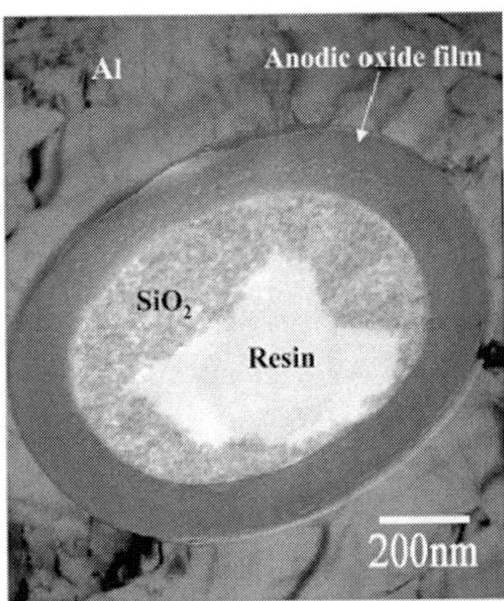


Figure 87. TEM image of cross section of DC-etched specimen after electrophoretic coating and anodizing. Electrophoretic coating was carried out under the conditions of $i_c$ = 0.05 Am$^{-2}$, $V_c$ = 10 V, $T_h$ = 573 K, and $t_h$ = 1.8 ks and anodizing was carried out up to $E_a$ = 100 V. Reprinted with permission from *J. Solid State Electrochem.*, **11**, (2007) 1375. Copyright (2007), Springer

be clearly ascertained in Fig. 86, which shows SiO$_2$ deposition in tunnel pits of DC-etched specimens by dip coating (Fig. 86a-b) electrophoretic coating (Fig. 86c-d). Anodizing after electrophoretic deposition of SiO$_2$ results in the formation of Al-Si composite oxide films, leading to an increase in capacitance (Fig. 87).

### *(v)  Combination of Liquid Phase Deposition with Anodizing*[160]

The liquid phase deposition (LPD) process is based on the hydrolysis of metal fluorides, $MeF_x^{-(x-2n)}$, and the formation of tetra-fluoro-borate:[161]

$$MF_x^{(x-2n)-} + n\,H_2O \rightarrow MO_n + x\,F^- + 2\,nH^+ \qquad (53)$$

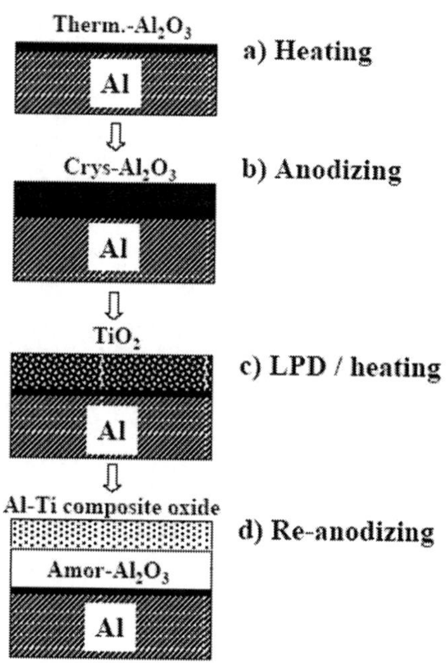



Figure 88. Process of formation of Al-Ti composite oxide film by the combination of liquid phase deposition with anodizing.

$$H_3BO_3 + 4\ F^- + 3\ H^+ \rightarrow BF_4^- + 3\ H_2O \tag{54}$$

The $MO_n$ formed by Eqs. (53) and (54) is deposited on the substrate, and the formation of Al-Ti composite oxide films on aluminum by LPD and anodizing will be described.[160]

Aluminum specimens are heated at 823 K for 10.8 ks in air to form thermal oxide films (Fig. 88a) and then anodized in a neutral borate solution (353 K) with $i_{pa}$ = 50 A m$^{-2}$ up to $E_{pa}$ = 100 V (vs. Ag/AgCl) (Fig. 88b). These steps are essential to prevent the substrate from corroding during LPD, due to the formation of aluminum fluoride. The specimen covered with a thin crystalline $Al_2O_3$ film is immersed in 0.01-kmol m$^{-3}$ (NH$_4$)$_2$TiF$_6$/0.2-kmol m$^{-3}$ H$_3$BO$_3$ solution at 303 K for 7.2 ks at maximum, followed by heat

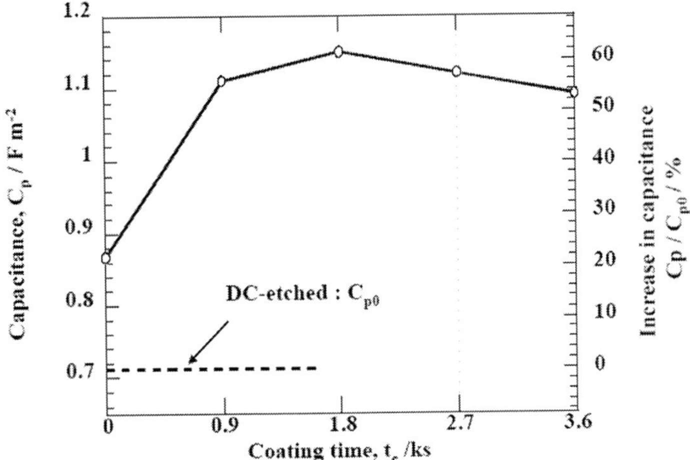



Figure 89. Change in capacitance of Ti-Al composite oxide films formed by LPD and anodizing with $TiO_2$coating time.

treatment at 573 K for 1.8 ks (Fig. 88c). Finally, the specimen is anodized in a neutral borate solution ($T_a = 293$ K) with $i_a = 10$　A $m^{-2}$ up to $E_a = 50$ V (vs. Ag/AgCl) (Fig. 88d). The capacitance of anodic oxide films formed by LPD and anodizing are 60% higher than those formed on non-treated DC-etched specimens (Fig. 89).

## VI.　CONCLUSIONS

This chapter has reviewed the structure and properties of oxide and hydroxide films formed on aluminum, as well as recent developments in micro- and nano-technologies incorporating anodizing of aluminum. Several applications of anodic oxide films in the fabrication of micro-devices are also described. A combination of anodizing with laser irradiation techniques and electrochemical treatments has enabled the development of novel fabrication processes for printed circuit boards, plastic injection molds, electrochemical reactors, and freestanding structures, as well as 3D manipulators and micro-printing rolls. The combination of anodizing with AFM probe processing, and electrochemical treatments enabled micro-patterning with metal and organic resin. The formation of composite oxide films by the combination of MOCVD, sol gel dip coating, electrophoretic sol gel coating, and liquid phase deposition with anodizing has been attempted for the development of new types of aluminum electrolytic capacitors.

At the conclusion it must be emphasized that anodic oxide films on aluminum can be made to play an important role as a template in micro- and nano-technology on aluminum, and that it is still a challenge to form anodic oxide films without imperfections.

## REFERENCES


[1] H. Takahashi and M. Nagayama, *Corr. Sci.*, **18** 911 (1978).

[2] K. Ebihara, H. Takahashi, and M. Nagayama, *J. Metal Fin. Soc. Jpn.*, **33** 156-164 (1982), **34**, 548-553 (1983), **35**, 205 (1984).

[3] H. Masuda and, M. Satoh: Jpn. *J. Appl. Phys.*, **35** L126 (1996).

[4] H. Takahashi and M. Nagayama, *J. Surf. Fin. Soc. Jpn.*, **36** 96-103 (1985).

[5] H. Takahashi, G. Hirose, and M. Nagayama, *J. Surf. Fin. Soc. Jpn.*, **36** 149 (1985).

[6] H. Takahashi, C. Ikegami, M. Seo, and R. Furuichi, *J. Electron Microsc.*, **40** 101 (1991).

[7] A. F. Beck, M. A. Heine, E. J. Caule, and M. J. Pryor, *Corr. Sci.*,**7** 1 (1967).

[8] J. I. Eldridge, R. J. Hussey, D. F. Mitchell, and M. J. Graham, *Oxidation of Metals*, **30** 301 (1988).

[9] G. M. Scamans and E. P. Butler, *Met. Trans. A*, **6A** 2055 (1975).

[10] H. Takahashi, M. Yamagami, R. Furuichi, and M. Nagayama, *J. Surf. Sci. Soc. Jpn.*, **8** 279 (1987).

[11] R. S. Alwitt, *J. Electrochem. Soc.*, **121** 1322 (1974).

[12] H. Takahashi, M. Yamaki, R. Furuichi, and M. Seo, *J. Mineralogical Soc. Jpn.*, (*Koubutsugaku Zashi*), **19** 387-396 (1990).

[13] H. Takahashi, N. Sato, and M. Nagayama, Abs. of the 71st Biannual Meeting of Surf. Fin. Soc., Jpn., (1985).

[14] H. Takahashi, M. Yamaki, and R. Furuichi, *Corr. Sci.*, **31** 243-248(1990).

[15] H. Takahashi and M. Nagayama, *Electrochim. Acta*, **23** 279 (1978).

[16] A. Guntherschultze and H. Betze, *Z. Phys.*, **68** 145 (1931); **71** 106 (1931); **73** 508 (1932); **91** 70 (1934); **92** 367 (1934).

[17] N. Cabrera and N. F. Mott, *Rep. Prog. Phys.*, **7** 163 (1948 – 49).

[18] E. J. W. Verway, *Physica*, **2** 1059 (1935).

[19] M. Nagayama and H. Takahashi, *J. Chem. Soc. Jpn*, (*Nippon Kagaku Kaishi*), **1972** 850 (1972).

[20] H. Konno, S. Kobayashi, H. Takahashi, and M. Nagayama, *Electrochim. Acta*, **25** 1667-1672 (1980).

[21] H. Takahashi, K. Fujimoto, H. Konno, and M. Nagayama, *J. Electrochem. Soc.*, **131** 1856-1861 (1984).

[22] H. Takahashi, K. Fujimoto, and M. Nagayama, *J. Electrochem. Soc.*, **135** 1349-1353 (1988).

[23] K. Shimizu, g. E. Thompson, and G. C. Wood, *Thin Solid films*, **85** 53 (1981).

[24] Y. Xu, G. E. Thompson, G. C. Wood, and B. Bethune, *Corr. Sci*, **27** 83 (1987).

[25] Y. Li, H. Shimada, M. Sakairi, K. Shigyo, H. Takahashi, and M. Seo, *J. Electrochem. Soc.*, **144** 866 - 876 (1997).

[26] H. Shimada, M. Sakairi, and H. Takahashi, *J. Surf. Fin. Soc. Jpn*, **53** 134 (2002).

[27] S. Ikonopisov, *Electrochim. Acta*, **22** 1077 (1977).

[28] F. Di Quarto, S. Piazza, and C. Sunceri, *Corr. Sci.*, **26** 213 (1986).

[29] J. C. Ord, J. C. Clayton, and Brudzewski, *J. Electrochem. Soc.*, **126** 908 (1976).

[30] F. Keller, M. S. Hunter, D. L. Robinson, *J. Electrochem. Soc.*, **100** 411 (1953).

[31] M. Koda, H. Takahashi, and M. Nagayama, *J. Metal Fin. Soc. Jpn.*, **28** 584 (1977).

[32] R. b. Mason and P. E. Fowle, *Electrochem. Soc.*, **101** 53 (1954).

[33] G. Paolini, M. Masoero, f. Sacchi, and M. Paganelli, *J. Electrochem. Soc.*, **112** 32 (1965).

[34] H. Takahashi and M. Nagayama, *J. Chem. Soc. Jpn*, (*Nippon Kagaku Kaishi*), **1974** 453 (1974).

[35] M. Nagayama, K. Tamura, and H. Takahashi, *Corr.. Sci.*, **10** 617 (1970) , *Corr. Sci.*, **12** 133 (1972).

[36] S. Z. Chu, M. Sakairi, H. Takahashi, K. Shimizu, and Y. Abe, *J. Electrochem. Soc.*, **147** 2182-2189 (2000).

[37] H. Takahashi, K. Watanabe, S. Hashimoto, and R. Furuichi, *J. Surf. Fin. Soc. Jpn.*, **41** 423-430 (1990).

[38] K. Watanabe, H. Takahashi, R. Furuichi, and M. Se, *J. Surf. Fin. Soc. Jpn.*, **42** 933-940 (1991).

[39] H. Takahashi, K. Shiga, K. Watanabe, and M. Seo, *J. Surf. Fin. Soc. Jpn*, **44** 542-548, (1993).

[40] J. E. Murphy and C. E. Michelson, Proc. of Sump. on Anodizing Aluminium, Birmingham,   April, p. 3–16 (1967).

[41] H. Takahashi, M. Nagayama, H. Akahori, A. Kitahara, *J. Electron Microsc.*, **22** 149 – 157 (1973).

[42] K. Saito, T. Kikuchi, and H. Takahashi, unpublished work.

[43] M. Koda, H. Takahashi, M. Nagayama, *J. Metal Fin. Soc. Jpn.*, **33** 242-248 (1982), **33**, 614-621 (1982), **34**, 44-50 (1983), **36**, 27-31 (1985).

44  H. Jha, T. Kikuchi, M. Sakairi, and H. Takahashi, *Electrochim. Acta*, **52** 4724 – 4733 (2007).

45  E. P. Short, A. Morita, *Plating and Surf. Fin.*, **72** 102 (1988).

46  A. Dekker and A. Middlehoek, *J. Electrochem. Soc.*, **117** 440 (1970).

47  H. Takahashi and M. Nagayama, *J. Met. Fin. Soc. Jpn.*, **27** 338 (1976), *Corr. Sci.*, **18** 911 (1978).

48  T. A. Renshaw, *J. Electrochem.Soc.*, **108** 185 (1961).

49  K. Kobayashi and K. Shimizu, *J. Electrochem. Soc.*, **135** 908 (1987), J. Jpn. Inst. Light Met., 38, 91 (1988).

50  C. Crevecoeur and H. J. de Wit, *J. Electrochem. Soc.*, **134** 808 (1987).

51  R. S. Alwitt, c. Ortega, N. thorne, and J. Siejka, *J. Electrochem. Soc.*, **135** 2695 (1988).

52  H. Takahashi, M. Yamagami, R. furuichi, and M. Nagayama, Proc. Symp. on Electrochem. Tech. in Electronics, edited by L. t. Romankiw and T. Osaka, *Electrochem. Soc.*, p. 287 (1988).

53  R. S. Alwitt, *J. Electrochem. Soc.*, **114** 843 (1967).

54  T. Kudo and R. S. Alwitt, *Electrochim. Acta*, **23** 341 (1978).

55  H. Takahashi, Y. Umehara, T. Miyamoto, N. Fujimoto, and M. Nagayama, *J. Metal Fin. Soc. Jpn.*, **38** 67 (1987).

56  H. Takahashi, Y. Umehara, and M. Nagayama, *J. Metal Fin. Soc. Jpn.*, **38** 138 (1987).

57  H. Takahashi, Y. Umehara, R. Furuichi, and M. Nagayama, *J. Surf. Fin. Soc. Jpn.*, **40**, 590 (1989).

58  H. Takahashi, K. Takahashi, R. Furuichi, and M. Nagayama, *J. Surf. Fin. Soc. Jpn.*, **40** 1415 (1989).

59  R. S. Alwitt and T. Kudo, *J. Metal Fin. Soc. Jpn.*, **32** 26 (1981).

60  R. S. Alwitt, C. K. Dyer, and B. Noble, *J. Electrochem. Soc.*, **129** 1981 (1982).

61  H. Takahashi, *J. Jpn. Soc. Colour Material*, **62** 607 (1989).

62  C. Caboni, Italian patent, No339232 (1936).

63  T. Asada, Japanese patent, No. 310410 (1963).

64  H. Takahashi, M. Mukai, and M. Nagayama, Proc. of 9th Intern. Congr. on Metal Corros. Vol. 2, p.155, edited by NRC Canada (1984).

65  K. J. Vetter, *Electrochemical Kinetics*, Academic Press, New York (1967).

66  K. Nisancioglu and H. Holtan, *Corr. Sci.*, **19** 537 (1979).

67  A. R. Despic, J. Radosevic, P. Dabic, and M. Kliskic, *Electrochim. Acta*, **35** 1743 (1990).

68  W. Ch. Van Geel, *Physica*, **17** 761 (1951), *Halbleiterprobleme*, **1** 299 (1955).

69  D. A. Vermilyea, *J. Appl. Phys.*, **27** 963 (1956).

70  H. Takahashi, K. Kasahara, K. Fujiwara, and M. Seo, *Corros. Sci.*, **36** 677, (1994).

71  H. Takahashi, K. Fujiwara, and M. Seo, *Corros. Sci.*, **36** 689, (1994).

72 S. Kurokawa, Z. Kato, T. Kikuchi, M. Sakairi, and H. Takahashi, Proc of Intrn. Symp on Pits and Pores, (ECS PV-2004-19) p.447 (2005).

73 Y. Hara, T. Kikuchi, and H. Takahashi, Abstracts of 113rd Biannual Meeting of Jpn. Inst. of Light Met. (2007), 114th (2008).

74 H. Masuda, H. Asoh, M. Watanabe, K. Nishio, M. Nakao, and Tamamura, *Adv. Mater.*, **13** No.3, 189 (2001).

75 P. Hoyer, K. Nishio, and H. Masuda, *Thin Solid Films*, **286** 88 (1996).

76 H. Masuda, M. Watanabe, K. Yasui, D. Tryk, T. Rao, and A. Fujishima, *Adv. Mater.*, **12** No. 6, 444 (2000).

77 J. D. Klein, R. D. Herrick, II, D. Palmer, M. J. Sailor, C. J. Brumlik, and C. R. Martin, *Chem. Mater.* **5**, 902 (1993).

78 C. R. Martin, *Sci.* **266** 1961 (1994).

79 T. Kyotani, L. Tsai, and A. Tomita, *Chem. Mater.*, 7(8) 1427 (1995).

80 T. Kyotani, W. Xu, Y. Yokoyama, J. Inahara, H. Touhara, and A Tomita, *J. Membrane Sci.*, **196** 231 (2002).

81 S. Z. Chu, K. Wada, S. Inoue, and S. Todoroki, *J. Electochem. Soc.*, **149** B321 (2002).

82 A. Mozalev, M. Sakairi, I. Saeki, and H. Takahashi, *Electrochim. Acta*, **48** 3155 (2003).

83 A. Mozalev, G. Gorokh, M. Sakairi, and H. Takahashi, *J. Mater. Sci.*, **40** 6399 (2005).

84 A. Mozalev, I. Mozalev, M. Sakairi, and H. Takahashi, *Electrochim. Acta.* **50** 5065 (2005).

85 H. Takahashi, K. Nukui, and J. Wakabayashi, *J. Jpn. Inst. Metals, Materia Japan* **34** 1276 (1995).

86 H. Takahashi, K. Nukui, M. Seo, Y. Matsumi, and M. Kawasaki: Proc. Symp. on Oxide Films on Metals and Alloys VII.   p.50 (1995), sponsored by Electrochem. Soc.

87 H. Jha, T. Kikuchi, M. Sakairi, and H. Takahashi, *Electrochemistry Communication*, **9** 1596 (2007).

88 H. Jha, T. Kikuchi, M. Sakairi, and H. Takahashi, *Appl. Phys.* **A 88** 617 (2007).

89 T. Kikuchi and H. Takahashi, *J. Surf. Fin. Soc. Jpn*, **56** 528 (2005).

90 *Bisho Kogaku (Micro-Optics) Handbook*, p. 77 (1995), edited by Appl. Phy. Soc. Jpn., Asakura Shoten.

91 C. B. Scruby and L. E. Drain, *Laser Ultrasonics—Techniques and Applications,"* Adam Hilger, (1990).

92 T. Kikuchi, S. Z. Chu, Z. Kato, M. Sakairi, and H. Takahashi, *J. Surf. Fin. Soc. Jpn.*, **50** 697 (1999).

93 M. Sakairi, Z. Kato, S. Z. Chu, H. Takahashi, Y. Abe, and N. Katayama, *Electrochemistry*, **71** 920 (2003).

94 M. Sakairi, S. Yamashita, H. Takahashi, K. Shimamura, and N. Katayama, *J. Surf. Fin. Soc. Jpn.*, **54** 349 (2004).

95 T, Kikuchi, M. Sakairi and H. Takahashi, *J. Surf. Fin. Soc. Jpn.*, **52** 84 (2001).

[96] H. Takahashi, J. Wakabayashi, K. Nukui, M. Sakairi, M. Seo, Y. Matsumi, and M. Kawasaki: Proc. of Intern. Symp. on Aluminium Surface Sci. And Tech. P.344, (1998).

[97] M. Sakairi, J. Wakabayashi, H. Takahashi, Y. Abe, and N. Katayama, *J. Surf. Fin. Soc. Jpn.*, **49** 1220 (1998).

[98] M. Sakairi, J. Wakabayashi, H. Takahashi, Y. Abe, and N. Katayama, *J. Surf. Fin. Soc. Jpn.*, **49** 1227 (1998).

[99] S.Z. Chu, M. Sakairi, I. Saeki, and H. Takahashi, *J. Electrochem. Soc.* **146** 317 (1999).

[100] S.Z. Chu, J. Wakabayashi, Z. Kato, T. Kikuchi, M. Sakairi, and H. Takahashi: Proc. Intern. Symp. of Environmental-Conscious Innovative Material Processing with Advanced Energy Sources (ECOMAP-98) p.224 (1999).

[101] S. Z. Chu, M. Sakairi, H. Takahashi, *J. Electrochem. Soc.* **146** 537 (1999).

[102] S.Z. Chu, M. Sakairi, and H. Takahashi, *J. Electrochem. Soc.* **146** 2876 (1999).

[103] S. Z. Chu, M. Sakairi, H. Takahashi, *J. Electrochem. Soc.* **147** 1423 (2000).

[104] T. Kikuchi, S. Z. Chu, S. Jonishi, M. Sakairi, and H. Takahashi, *Electrochim. Acta*, **47** 225 (2001).

[105] Y. Akiyama, T. Kikuchi, M. Ueda, M. Iida, M. Sakairi and H. Takahashi, *Electrochim. Acta*, **51** 4834 (2006).

[106] T. Kikuchi, Y. Akiyama, M. Ueda, M. Sakairi, and H. Takahashi, *Electrochim. Acta*, **52** 4480 (2007).

[107] H. Takahashi, M. Sakairi, S, Z. Chu, K. Watanabe, T. Kikuchi, K. Miwa, Z. Kato, and S. Jonishi, *J. Jpn. Soc. Light Met.*, **50** 544 (2000).

[108] T. Kikuchi, M. Sakairi, H. Takahashi, Y. Abe, and N. Katayama, *J. Electrochem. Soc.*, **148** C740 (2001).

[109] T. R. Beck, *Electrochim. Acta*, **18** 807 (1973).

[110] G. T. Burstain and R. J. Cinderey, *Corros. Sci.*, **32** 1195 (1991).

[111] G. E. Frankel, B. M. Rush, C. V. Janes, C. E. Farrel, H. J. Davenport, and H. S. Issacs, *J. Electrochem. Soc.*, **138** 643 (1991).

[112] R. Oltra, G. M. Indrianjafy, and R. Roberge, *J. Electrochem. Soc.*, **140** 343 (1993).

[113] M. Sakairi, Y. Ohira, and H. Takahashi, Proc. of Symp. on Passivity and Its Breakdown, co-sponsored by ECS. and ISE. p. 643 (1999).

[114] H. Takahashi, M. Sakairi, and Y. Ohira, Proc. of Intern. Symp. In Honor of Prof. N. Sato, Joint Meeting between ECS. Jpn and ECS (2000).

[115] T. Kikuchi, M. Sakairi, H. Takahashi, Y. Abe, and N. Katayama, *J. Surf. Fin. Soc. Jpn.*, **50** 829 (1999).

[116] H. Takahashi, M. Sakairi, and T. Kikuchi, *J. Surf. Sci Soc. Jpn.*, **22** 370 (2001).

[117] T. Kikuchi, M. Sakairi, H. Takahashi, Y. Abe, and N. Katayama, *J. Surface Finishing Soc. Jpn.*, **54** 137 (2003).

[118] M. Sakairi, S. Moon, H. Takahashi, and K. Shimamura, *J. Surf. Fin. Soc. Jpn.*, **52** 553 (2001).

[119] M. Sakairi, M. Yamada, T. Kikuchi, and H. Takahashi, *Electrochim. Acta*, **52** 6268 (2007).

[120] T. Kikuchi, M. Sakairi, and H. Takahashi, *J. Electrochem. Soc.*, **150** C567 (2003).

[121] T. Saito, T. Kikuchi, M. Sakairi, H. Takahashi, and K. Sato, Abstract of the 212nd Biannual Meeting of ECS (2007).

[122] T. Kikuchi, Y. Wachi, M. Sakairi, H. Takahashi, K. Iino, and N. Katayama, *J. Surf. Fin. Soc. Jpn.*, in press.

[123] S. Kurokawa, Y. Akiyama, M. Yamada, Z. Kato, T. Kikuchi, M. Sakairi, and H. Takahashi, *Benelux Metallurgie*, Vol. 45, No. 1 – 4, 97 – 104 (2007), Proc. of ASST 2006.

[124] K. Sato, A. Hibara, M. Tokeshi, H. Hisamoto, and T. Kitamori, *Anal. Sci.*, **19** 15 (2003).

[125] N. Kitamura, *J. Surf. Fin. Soc., Jpn.*, **54** 402 (2003).

[126] A. L. Bard, L. R. Faulkner, Electrochemical Methods-Fundamentals and Applications, John Wiley & Sons Inc., New York, 1980, p. 218.

[127] V. G. Leivich, *Physicochemical Hydrodynamics*, Prince-Hall, Engelwood Cliffs, NJ. P.112, (1962).

[128] M. Saeki, A. Hishikata, T. Tsuru, and Denki Kagaku, *Electrochemistry*, **64** 891 (1996).

[129] T. Kikuchi, H. Takahashi, and T. Maruko, *Electrochim. Acta*, **52** 2352–2358 (2007).

[130] K. Junwu, Y. Zhigang, P. Taijang, C. Guangming, and W. Boda, *Sensors and Actuators A; Physical*, **121** 156, (2005).

[131] E. Makino, T. Mitsuya, and T. Shibata, *Sensors and Actuators A: Physical*, **88** 256 (2001).

[132] Y. Osada and D. E. De Rossi, *Polymer Sensors and Actuators*, Springer, 2000.

[133] T. F. Otero, I. Boyano, *J. Phys. Chem.*, **B107** 6730 (2003).

[134] A. Bund and S. Neudeck, *J. Phys. Chem.* **B 108** 17845 (2004).

[135] T. Osaka, T. Monma, S. Komaba, H. Kanagawa, *J. Electroanal. Chem.*, **372** 201 (1994).

[136] O. Inganas, R. Erlandsson, C. Nylander, and I. Lundstrom, *J. Phys. Chem. Solids*, **45** 432 (1984).

[137] S. Kuwabata, N. Nakamura, and H. Yoneyama, *J. Electrochem. Soc.*, **137** 2147 (1990).

[138] H. Xie, M. Yan, and Z, Jiang, *Electrochim. Acta*, **42** 2361 (1997).

[139] Ueda, T. Kikuchi, and H. Takahashi, Abstract of the Biannual Meeting of Surf. Fin. Soc. Jpn, (2007).

[140] G. Binnig, H. Rohrer, Ch. Gerber, and E. Weibel, *Phy. Rev. Lett.*, **49** 57 (1982).

[141] G. Binnig, C. F. Quate, and Ch. Gerber, *Phy. Rev. Lett.* **56** 930 (1986).

[142] D. M. Korb, R. Ullmann, and T. Will, *Sci.*, **275** (5303), 1097 – 1099 (1997).

143 H. Sugimura, K. Okiguchi, N. Nakagiri, and M. Miyashita, *J. Vac. Sci. & Tech.*, **B14** (6), 4140 – 4143 (1996).

144 Z. Kato, M. Sakairi, and H. Takahashi, *J. Electrochem. Soc.* **148** C790–798 (2001).

145 Z. Kato, M. Sakairi, and H. Takahashi, *Zairyo-to-Kankyo*, **52** 534–538 (2003).

146 Z. Kato, M. Sakairi, H. Takahashi, *Surface and Coatings Technology*, **169-170** 195 (2003).

147 S. Kurokawa, T. Kikuchi, M. Sakairi, and H. Takahashi, *Electrochim. Acta*, in press.

148 M. Shikanai, M. Sakairi, H. Takahashi, M. Seo, K. Takahiro, S. Nagata, and S. Yamaguchi, *J. Electrochem. Soc.*, **144** 2756 (1997).

149 H. Takahashi, H. Kamada, M. Sakairi, K. Takahiro, S. Nagata, and S. Yamaguchi, Proc. of Intern. Symp. of Dielectric Material Integration for Micro-electronics, sponsored by Electrochem. Soc., 1998, p. 253.

150 E. Sakata, M. Sakairi, H. Takahashi, S. Nagata, and K. Takahiro, Abstract FSISE 2004, (2004).

151 K. Watanabe, M, Sakairi, H. Takahashi, S. Hirai, and S. Yamaguchi, *J. Surf. Fin. Soc. Jpn.*, **50** 359 (1999).

152 K. Watanabe, M, Sakairi, H. Takahashi, S. Hirai, and S. Yamaguchi, *J. Electroanal. Chem.*, **473** 250 (1999).

153 K. Watanabe, M, Sakairi, H. Takahashi, K. Takahiro, S. Nagata, and S.Hirai, *Electrochemistry*, **67** 1243 (1999).

154 K. Watanabe, M, Sakairi, H. Takahashi, K. Takahiro, S. Nagata, and S. Hirai, *J. Electrochem Soc.*, **148** B473 (2001).

155 K. Watanabe, M, Sakairi, H. Takahashi, K. Takahiro, S. Nagata, and S. Hirai, *Electrochemistry*, **69** 407 (2001).

156 K. Watanabe, M, Sakairi, H. Takahashi, and S. Hirai, *J. Surf. Fin. Soc. Jpn.*, **54** 235 (2003).

157 K. Watanabe, M, Sakairi, H. Takahashi, S. Nagata, and S. Hirai, *J. Surf. Fin. Soc. Jpn.*, **55** 471 (2004).

158 S. Koyama, T. Kikuchi, M. Sakairi, H. Takahashi, and S. Nagata, *Electrochemistry*, **75**, 573 (2007).

159 M. Sunada, T. Kikuchi, M. Sakairi, H. Takahashi, and S. Hirai, *J. Solid State Electrochem.*, **11**, 1375 (2007).

160 S. Koyama, Master Thesis, Graduate School of Engineering., Hokkaido University (2008).

161 S. Deki, Y. Aoi, O. Hiroi and A. Kajinami, *Chem. Lett.*, **25** 433 (1996).

*3*

# Electrochemical, Microscopic and Surface Analytical Studies of Amorphous and Nanocrystalline Alloys


Maria Janik-Czachor[*] and Marcin Pisarek[*,**]

*[*]Institute of Physical Chemistry Polish Academy of Sciences, Warsaw, Poland*
*[**]Faculty of Materials Science and Engineering,*
*Warsaw University of Technology*


## I.  INTRODUCTION

To meet the extreme demands of modern technology, functional-ized materials of enhanced specific properties are required. More-over, high resolution methods of characterizing new materials are needed in identifying the factors responsible for such materials unique properties.

This chapter falls under the general heading *Chemistry for Materials Science*, and is aimed at discussing new results, ideas and technologies in the field of the chemical properties of novel materials including (but not limited to) amorphous and nanocrys-talline alloys obtained by rapid quenching, sputter deposition, me-chanical alloying, heavy deformation, electrodeposition and other physical and mechanical methods.

Nonequilibrium processing techniques provide the potential for producing compositionally and structurally graded materials


*Modern Aspects of Electrochemistry*, Number 46, edited by S. –I. Pyun and J. -W. Lee, Springer, New York, 2009.



S.-I. Pyun and J.-W. Lee (eds.), *Modern Aspects of Electrochemistry 46:*     175
*Progress in Corrosion Science and Engineering I,*
DOI 10.1007/978-0-387-92263-8_3, © Springer Science+Business Media, LLC 2009

with optimized properties. And so it is important to discuss means and measures of lowering the reactivity of metastable alloys to reduce (or possibly eliminate) detrimental processes of deterioration such as corrosion/dissolution, or of enhancing their chemical reactivity, durability, and selectivity for catalytic/electrocatalytic processes. It has been recognized that, due to the almost unlimited compositional flexibility of metastable alloys, their properties can be optimized for the requirements of a given application.

Low or even negligible reactivity is required for construction materials, as well as for biomaterials used as implants in orthopedic or dental surgery, for obvious reasons. High reactivity and selectivities are required for efficient electrocatalysts and catalysts. Developing such materials from composition-controlled nonstructured or amorphous alloys by exploiting mechanical, electrochemical, and chemical activation, calls for both fundamental and applied research.

Discussion among specialists from the areas of materials science, physics, chemistry and electrochemistry should contribute towards:

- characterizing the decisive factors, including local changes, introduced during modification and/or activation procedures,
- finding efficient, interdisciplinary ways of tailoring new functional materials for new processes including catalysis and electrocatalysis.

We discuss below the following selected topics:

- the effect of refractory metals on the stability of the passive state of Al-based amorphous alloys;
- the effect of hydrostatic extrusion leading to nanostructurization of austenitic stainless steels on the stability of their passive state;
- the effect of cathodic hydrogen charging on the catalytic activity of Cu-based amorphous alloys.

The above examples illustrate the role of high resolution methods in characterizing local phenomena in electrochemistry and in the chemistry of materials, in general. At the same time, they illustrate the extensive possibilities of modifying the properties of metastable amorphous and nanostructurized materials by chemical or electrochemical pre-treatments.

## II.  STABILITY OF THE PASSIVE STATE OF Al – BASED AMORPHOUS ALLOYS

The passivity of thermodynamically unstable, non-noble metals, such as Fe alloys including stainless steels, Al or Ti and their alloys, is the foundation of their successful practical use. The breakdown of their passivity in chloride-containing media is the main reason for corrosion failures such as pitting or crevice corrosion. For materials science, an understanding of the mechanism of passivity and its breakdown, as well as of the role of alloying elements in improving the stability of the passive state, is essential in the development of new corrosion-resistant materials.

It was recognized long ago that the chemical breakdown of passivity (pitting) results in rapid, highly localized corrosion at local c.d., in a range of from several hundred $mA/cm^2$ up to several $A/cm^2$, which is $10^5$- $10^6$ times higher than that characteristic of passive metal dissolution.[1-6] These results have shown that once pitting corrosion starts, its detrimental effect is strong and leads to a catastrophic dissolution in spite of the small amount of metal dissolved locally. Therefore, prevention of passivity breakdown is crucial.

The interrelation which was found between the composition of the metal substrate, its passivating film and the resistance of the material to pitting suggested that an appropriate alloying of the metal substrate might result in a distinct increase in resistance against pitting.[7-22] Thus, the stability of a passivating film in $Cl^-$-containing media may limit or even exclude pitting in certain systems. Hence, understanding the effect of alloying elements on a passivating film's composition, structure and properties became an essential task for materials scientists interested in the elimination of pitting.

An original suggestion was given by Marcus,[7] who pointed out the effect of Me-Me bond strength in passivity, (see Fig. 1). His discussion aimed at rationalising the role of alloying elements in anodic dissolution and passivation, the essential reactions in the corrosion and protection of metals, respectively.[7] The basis for this

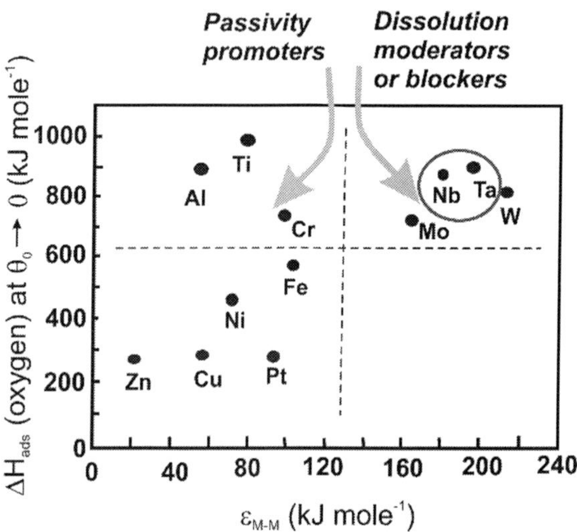



Figure 1. Fundamental factors in the passivation of alloys; metal-metal bond strength $\varepsilon_{m\text{-}m}$ and enthalpy of oxygen adsorption $\Delta H_{ads}$ for different alloying elements.

rationalization is the following: elements that enhance passivation (passivity promoters or passivators) are elements which combine a high metal-oxygen bond strength with a low metal-metal bond strength. The high metal-oxygen bond strength is evidently necessary for the stability of the passive film. The low metal-metal bond strength favours the rapid nucleation and growth of the oxide by facilitating the breaking of the surface metal-metal bonds involved in the early stages of passivation. On this conceptual basis, elements such as Al, Ti and Cr are passivity promoters. In contrast, dissolution moderators or blockers are elements that possess a high metal-metal bond strength: they lower the dissolution rate by increasing the activation energy barrier for the disruption of metal-metal bonds at the surface. They may also have a high metal-oxygen bond strength, but this is not their essential property. Mo, W, Ta and Nb belong to this category of metals. Within the framework of these ideas, it becomes obvious that passivity promoters or passivators will be the main constituents of the passive film, whereas dissolution blockers will tend to remain at

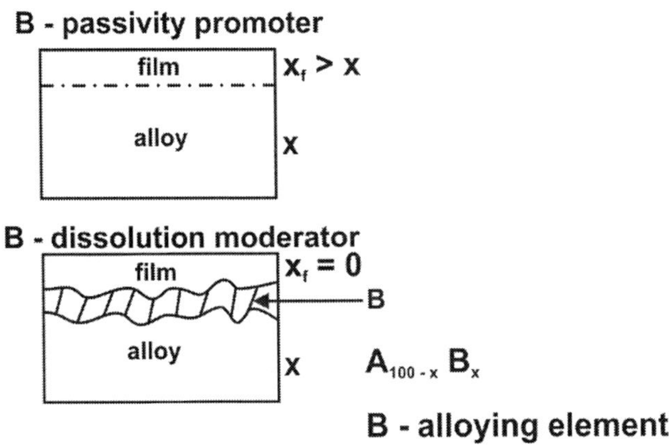



Figure 2. Distribution of alloying elements in the passivating film, and in the film/alloy interphase region after passivation of a binary A-B alloy (schematic). B-, acting as a passivity promoter becomes oxidized and enriched within the passivating film. B-, acting as a dissolution moderator, tends to remain enriched and unoxidized at the film/alloy interface, thus forming a corrosion barrier there.

the film-substrate interface where they will restrain the dissolution process, see Fig. 2.

These ideas made it possible to rationalize the role of alloyed elements in the corrosion resistance of metallic materials. The diagram reported in Fig. 1 shows to which category various metals belong.[7]

This concept can certainly be useful not only for a better understanding of the corrosion behaviour of existing alloys but also in the design of new corrosion resistant materials.

While there are literature data that confirm Marcus' concept for some systems,[6,12-14] the affinities to oxygen of various alloying elements should also be taken into account. Table 1 gives $/\Delta G^0/$ for the elements discussed in connection with Fig. 1. It turns out that though the refractory metals considered above exhibit a high Me-Me bond energy, their affinities to oxygen differ considerably.

Mo, W, Ta and Nb belong to a group of refractory metals with a high Me-Me bond strength: the Me-Me bond strength for all the refractory metals considered is between 170 and 220 kJ/mol. The corresponding value for Al is < 60 kJ/mol.[7] The Gibbs free en-

**Table 1**

**Free Enthalpy of Oxide Formation $\Delta G^0$ for Different Refractory Metals and Aluminum.**

| Oxide | $\Delta G^0$ (kJ/mol) |
|---|---|
| $MoO_3$ | $-\ \ 667.5$ |
| $WO_3$ | $-\ \ 764.2$ |
| $Al_2O_3$ | $-\ 1582.7$ |
| $Nb_2O_5$ | $-\ 1765.9$ |
| $Ta_2O_5$ | $-1911.00$ |

thalpy $\Delta G^0$ of oxide formation are $-667.5$ and $-764.2$ kJ/mol for molybdenum and tungsten oxides, respectively, while that for $Ta_2O_5$ is $-1911$ kJ/mol. It is interesting to note that the corresponding value for $Al_2O_3$ is $-1582.7$ kJ/mol, (see Table 1).

Thus the $|\Delta G^0|$ for molybdenum and tungsten oxide is lower than that for $Al_2O_3$, $Ta_2O_5$, and $Nb_2O_5$. This remains true also when recalculating $|\Delta G^0|$ values for a mole of oxygen in a given oxide. Therefore, the anodic behaviour of Al-Mo and Al-W amorphous alloys may be different than that of the other refractory metals mentioned above. To thoroughly investigate the depth distribution of both components within the protective surface film which would help to differentiate between passivity promoters and the metal dissolution moderators, thick anodic films formed galvanostatically on amorphous alloys were studied.[12-14]

## 1.  Effect of Refractory Metals on the Stability of the Passive State

Amorphous/supersaturated alloys and nanocrystalline alloys with their enormous compositional flexibility and homogeneity, are a particularly interesting and suitable target for investigating the effect of alloying elements on passivity.[22] The effects of composition in a wide concentration range can be studied without any change in the alloy structure.

Table 2 summarizes the effect of refractory metals on the stability of the passive state of Al-based amorphous alloys, listing the breakdown potential $E_{np}$, for different alloys measured under ex-

**Table 2**
**Breakdown Potential $E_{np'}$ for Al-Based Amorphous Alloys with Refractory Metals in a Borate Buffer Solution Containing Cl⁻ Ions.**

| Alloy | Breakdown potential, mV SCE | |
|---|---|---|
| | Borate buffer + 0.1-M NaCl | Borate buffer + 1-M NaCl |
| $Al_{89}Ta_{11}$ | $160 \pm 20$ | $140 \pm 10$ |
| $Al_{74}Ta_{26}$ | $410 \pm 40$ | $390 \pm 40$ |
| $Al_{54}Ta_{46}$ | $800 \pm 70$ | $900 \pm 70$ |
| $Al_{87}Nb_{13}$ | $250 \pm 80$ | $180 \pm 10$ |
| $Al_{74}Nb_{26}$ | $880 \pm 20$ | $570 \pm 20$ |
| $Al_{54}Nb_{46}$ | $1050 \pm 130$ | $1010 \pm 20$ |
| $Al_{85}W_{15}$ | $690 \pm 80$ | |
| $Al_{70}W_{30}$ | $870 \pm 20$ | |
| $Al_{54}W_{46}$ | $1200 \pm 100$ | $1080 \pm 40$ |
| $Al_{85}Mo_{15}$ | $350 \pm 60$ | |
| $Al_{75}Mo_{25}$ | $490 \pm 20$ | $420 \pm 40$ |
| $Al_{60}Mo_{40}$ | $560 \pm 20$ | |
| $Al_{54}Mo_{46}$ | $700 \pm 60$ | |
| Al cryst | from $-800$ to $-650^{a}$ | $-750 \pm 20$ |

$^{a}$Data for crystalline Al also include results from the literature.[23-26]

actly the same experimental conditions.[6,12-14,17,27-28] The data for crystalline Al is also given for comparison.[23-26] One could argue that the crystalline metal[29] is not a good reference for amorphous alloys of a distinctly different structure. However, it is difficult to find a more appropriate reference since pure Al does not exist in an amorphous state. The data show that all the refractory metals cause a shift of $E_{np'}$ towards noble values, thus extending the passivity region by about 1.6 V or more, as compared to the corresponding value for pure crystalline Al.

The effect of Ta compares with that of Mo, whereas the effect of Nb is similar to that of W. The effect of the chemical composition of the alloys is clearly visible in Table 2, in contrast to a suggestion found in the literature[15] implying a predominant effect of defects over chemical composition in general.

## 2.    AES Investigations of Depth Distribution of Refractory Metals within the Passivating Film

Refractory metals are known for their beneficial effect on the passivity of many alloys, both crystalline and amorphous, although the mechanism of their action is still a subject of discussion.[7-14,16-20,22,27-28,30-32] Amorphous alloys thus offer the opportunity of gaining an insight into the role of these elements, even when studying their effects at higher concentrations. In particular, one can check whether the preferential dissolution of Al and/or different transport rates within the passive film of the components determine the anodic behaviour,[21,33] or whether the effect of the refractory metal is confined to that of a metal dissolution moderator rather than a passivity promoter, as suggested by Ph. Marcus.[7] As the previous results[6,13,14] strongly suggest the former effect for Al-W and Al-Mo, the results for Al-Ta and Al-Nb amorphous alloys contradict the general application of the metal dissolution moderators model for the refractory metals.

Al-refractory metal systems including Al-Ta alloys of various concentrations of both components were extensively studied by H. Habazaki et al.[34] and G. Alcala et al.[35] These authors established the ionic transport numbers and relative migration rates of cation species for anodic films formed on Al-Ta alloys whose composition extended from aluminum-rich to tantalum-rich. The authors used various techniques, including ion implanted marker experiments, transmission electron microscopy (TEM), Rutherford backscattering spectroscopy (RBS), and medium energy ion scattering. However, the AES spectra and profiles necessary to establish the depth distribution and chemical state of the metallic components of the anodic layer have been reported only recently.[17,27,28]

For a better understanding of the anodic behaviour of Al-Ta and Al-Nb amorphous alloys and of the role of refractory metals in passivity, surface analytical measurements with the aid of Auger Electron Spectroscopy (AES) combined with $Ar^+$ ion sputtering (depth profiling) were performed on anodized samples to investigate in detail the depth composition of the passive films on Al-Ta and Al-Nb amorphous alloys, and to compare the results with those for Al-Mo/W amorphous alloys. These investigations aimed at gaining an  insight into the beneficial effect of  refractory

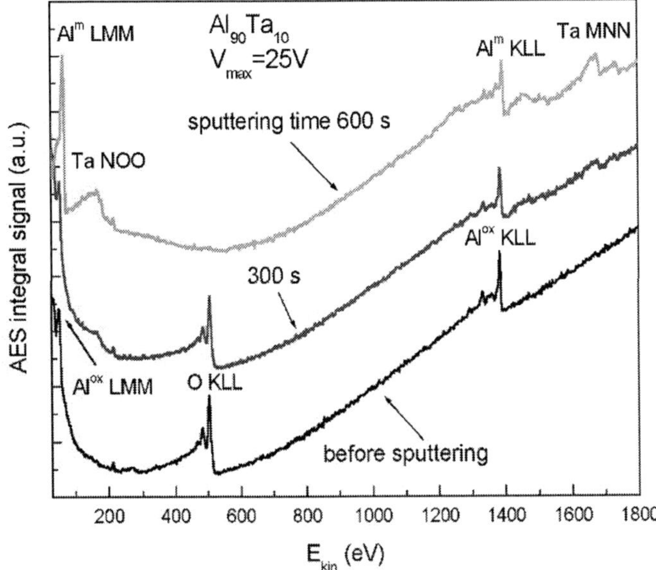



Figure 3. Examples of integral Auger spectra N(E) = f(E$_{kin}$) for Al$_{90}$Ta$_{10}$ amorphous alloy measured during the course of etching of an anodic oxide film formed in a borate buffer at i = 0.5 mA/cm$^2$ up to V$_{max}$= 25 V. Reprinted from *AES and RBS Characterization of Anodic Oxide Films on Al-Ta Amorphous Alloys*, by Z.Werner, A.Jaśkiewicz, M.Pisarek, M.Janik-Czachor and M.Barlak, *Z.Phys.Chem.* **219** (2005) 1461-1479, Copyright (2005) Oldenbourg Wissenschaftsverlag GmbH.

metals.[6,12-14,17,20,27,28,36] The chemical state of surface species was identified by using a high resolution scanning Auger microanalyzer – a Microlab 350. The energy resolution of the analyzer is continuously variable between 0.6% and 0.06%. The conversion of the Auger signals into the atomic concentration of the components was undertaken by using sensitivity factors S$_f$ from the Avantage database software[37] and by adopting a NLLSF (Non-Linear Least Square Fitting) procedure to deconvolute the Al$^m$ and Al$^{ox}$ Auger signals emerging from the anodic oxide films and from the metal substrate beneath.

Figure 3 shows examples of survey spectra for an anodized Al-Ta amorphous alloy. Disappearance of the O (KLL) peak in the

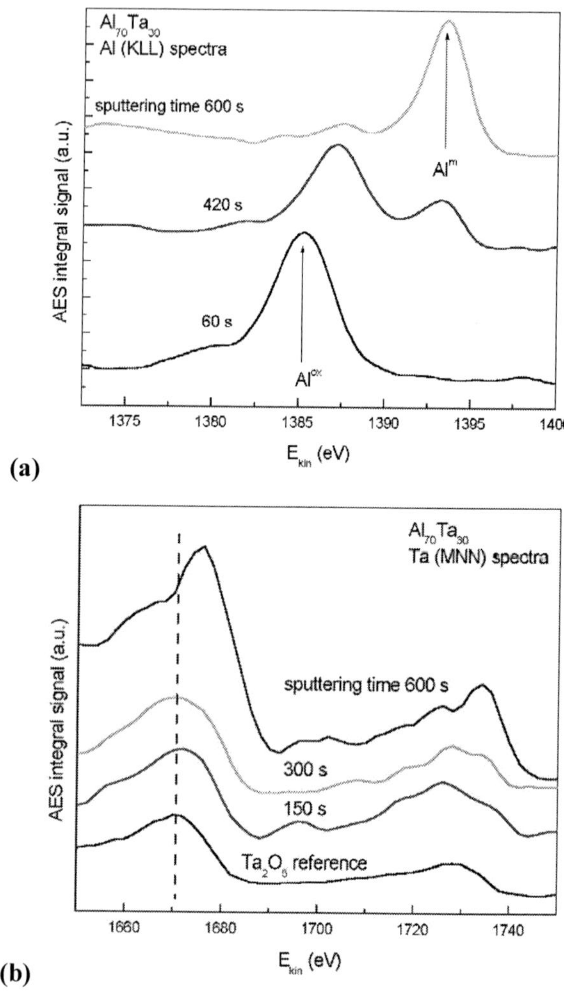



Figure 4. High resolution AES spectra showing changes in the position and shape of (a) Al (KLL) and (b) Ta (MNN) Auger signals during the course of etching the anodic layer from an oxidized state (passive layer) to a metallic state (substrate). Reprinted from *AES and RBS Characterization of Anodic Oxide Films on Al-Ta Amorphous Alloys*, By Z.Werner, A.Jaśkiewicz, M.Pisarek, M.Janik-Czachor and M.Barlak, *Z.Phys.Chem.* **219** (2005) 1461-1479, Copyright (2005), Oldenbourg Wissenschaftsverlag GmbH.

course of sputtering is well visible, reflecting analytical results at various depths of the sample: within the anodic oxide film and, eventually, within the substrate.

Figures 4a and 4b present the evolution of high-resolution Ta and Al spectra, respectively, during the course of $Ar^+$ etching of the anodic layer. In Fig. 4b the Ta spectra are compared with the reference $Ta_2O_5$ spectrum to show that as long as the spectrum is measured within the anodically oxidized layer, Ta remains in an oxidized state. The same is true for Al, Fig. 4a, where separation of the $Al^m$ and $Al^{ox}$ signals is well visible. An *intermediate state* of $Al^{3+}$ ions at the interface region between the anodic film and the substrate was revealed with AES: at 420s of sputtering the film itself was already very thin, so that the $Al^m$ signal appeared. Moreover, the position of the $Al^{ox}$ signal was shifted to a higher energy than that typical of $Al^{ox}$ ($\sim$ 1392 eV instead of 1386 eV, see Fig. 4a). These results suggest that there is an intermediate electronic state of alumina cations within the interphase region at the metal/oxide film interface, and that these cations are neither $Al^{3+}$ nor $Al^m$.

Figure 5a presents a typical composition profile for $Al_{90}Ta_{10}$ after anodization up to 25V. Three distinct domains can be distinguished within the profile: I - the uppermost of Al oxide; II - an anodic oxide film containing a *mixture* of Ta and Al oxides; III - the substrate containing $Al^m$ and $Ta^m$ only. AES results thus indicate that with this alloy composition, Ta is completely absent in the surface layer I. This is not the case at Ta $\geq$ 30% at, as seen from Fig. 5b, where tantalum is already present in the uppermost monolayers of the anodic oxide film.

Apparently, during anodic oxide film formation both components migrate towards the surface as $Al^{+3}$ and $Ta^{+5}$, whereas $O^{-2}$ migrates towards the substrate. The resultant depth distribution of both metallic components is probably the result of a tendency of $Al^{+3}$ to migrate faster forward than $Ta^{+5}$ [35,36] within the high electric field existing within the anodic layer. The final effect for low Ta concentration in the substrate is the formation of a thin $Al_2O_3$ film without any tantalum (compare Fig. 5a), whereas at a sufficiently high Ta concentration only a gradient of Ta exists in the uppermost part of the film (see, Fig. 5b). As a consequence of such a redistribution of elements within the profile of $Al_{90}Ta_{10}$, a *relative enrichment* of Ta may occur within the inner part of the

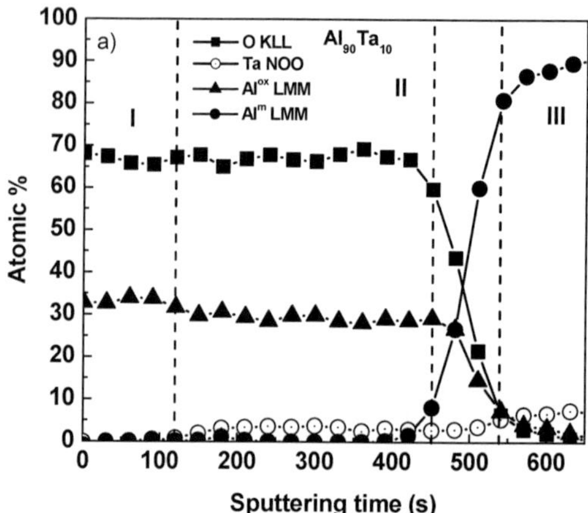

**(a)**

**(b)**

Figure 5. Composition profiles of Al-Ta AA with an oxide layer formed during anodic polarization in borate buffer. (a) $Al_{90}Ta_{10}$ ($V_{max}$ = 25 V); regions I, II, and III correspond to the uppermost surface layer of pure alumina, principal mixed oxide, and substrate, respectively. (b) $Al_{70}Ta_{30}$ ($V_{max}$ = 70 V); only two regions are distinguishable: principal mixed oxide I and substrate II. An intermediate region between the oxide and the substrate is not numbered.

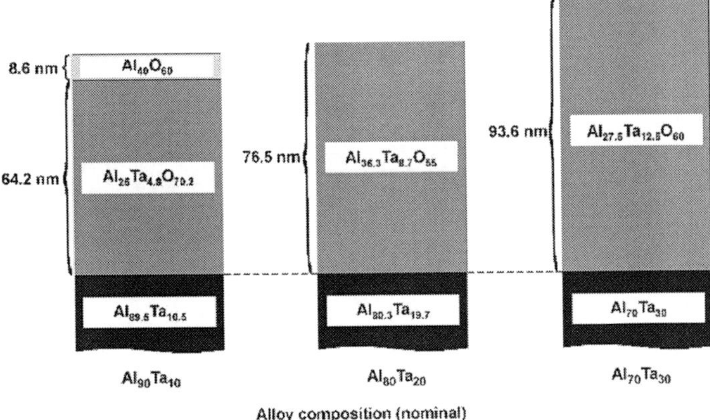


Figure 6. Composition and thickness of anodically oxidised surface layers ($V_{max}$ = 70 V) on amorphous Al-Ta alloys as deduced from SIMNRA analysis of RBS spectra.

film which, however, does not exist within the films formed on higher Ta alloys. The details are discussed elsewhere.[27] It thus seems that the role of tantalum in enhancing the stability of the passive state of Al-Ta amorphous alloys is mainly as a passivity promoter which undergoes oxidation and takes part in anodic film formation.

Rutherford Backscattering Spectroscopy (RBS) results, providing an average composition of the sublayers existing on the substrate, confirm the above statement, (see Figs. 6 and 7, and Table 3).

To verify the results of composition profile analysis obtained by ion sputtering and AES spectroscopy, samples of amorphous alloys, oxidized in air and anodically oxidized up to different voltages, were examined using RBS [38] with a beam of 1.7 MeV alpha particles. Particles backscattered from the sample at an angle of 170° were detected with a Si(Li) detector of 14 keV energy resolution. The spectra were recorded at a total charge incident on the sample amounting to 10 μC. Usually, the beam was incident at an angle of 0° to the sample normal, but, for oxidized layers formed with up to 25 and 50V anodic voltage, an angle of incidence of 60° was used to increase depth resolution.

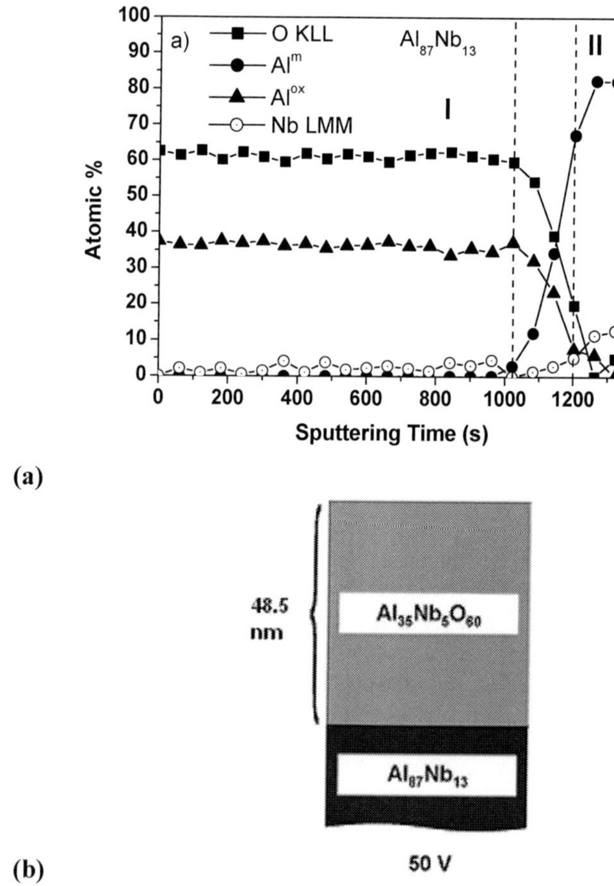



Figure 7. AES composition profiles of Al-Nb AA covered with an oxide layer formed during anodic polarization ($V_{max}$ = 50 V) in borate buffer solution (a). Composition and thickness of an anodically oxidised surface layer (50 V) on an amorphous Al-Nb alloy as deduced from SIMNRA analysis of RBS spectrum (b).

The spectra obtained were simulated using SIMNRA code. The compositions and thicknesses of individual sub-layers forming the sample seem to be reliable to within a few percent, one should bear in mind that RBS measurements do not provide direct information on the geometrical thickness of the analysed layers, but on their *surface atomic density,* i.e., the number of atoms in the layer behind the unit surface area within a thickness of a few micrometers. The thicknesses generated by the simulation code are originally expressed in units of $10^{15}/cm^2$ of *average atoms* forming the layer and their conversion into geometrical thicknesses requires information about specific densities, which is not always known. In our case, the thicknesses of the anodically oxidized layers (shown in Figs. 6 and 7) were calculated under the following assumptions:[27]

- The oxidized layers are dense, i.e., they exhibit neither voids nor pores,
- The layers are composed of a mixture of aluminum trioxide and tantalum pentoxide in the proportion identical to that of the substrate composition,
- The specific densities of the layer components are: $Al_2O_3$ 3.97 $g/cm^3$, $Ta_2O_5$ 8.2 $g/cm^3$,
- The specific densities of mixtures vary linearly with composition between the limiting values for the pure oxides.

Under these assumptions, the conversion coefficients between the thickness of the anodically oxidized layers expressed in units of $10^{15}$ of average atoms and the geometrical thickness are: $Al_2O_3$ 0.856 Å/$10^{15}$; $Al_{90}Ta_{10}$ 0.99 Å/$10^{15}$; $Al_{80}Ta_{20}$ 1.09 Å/$10^{15}$; and $Al_{70}Ta_{30}$ 1.14 Å/$10^{15}$.

Table 3 compares the Ta/(Ta+Al) atomic ratio within the anodic film to the substrate. The average Ta (Ta+Al) atomic ratio within the film is higher than in the substrate for $Al_{90}Ta_{10}$ only, where the superficial, uppermost $Al_2O_3$ sublayer exists. For the other two alloys the above parameter is nearly the same for both film and substrate.

**Table 3**

**Ta/(Ta+Al) Atomic Ratio in the Substrate and in the Anodic Film as Determined from RBS Measurements.**

|  | $Al_{90}Ta_{10}$ | $Al_{80}Ta_{20}$ | $Al_{70}Ta_{30}$ |
|---|---|---|---|
| Substrate | 11 | 20 | 30 |
| Anodic film | 16 | 20 | 32 |

Hence, Ta may become enriched in the inner part of the film, thus forming a corrosion barrier as a dissolution moderator at sufficiently low Ta concentration in the substrate. However, the latter effect is rather small, as the superficial $Al_2O_3$ upper layer I is thin ($< 20\%$ of the total film thickness only and, shrinking with Ta content in the alloy[28]), thus leaving behind only a small portion of excessive Ta in layer II. This subject is discussed in more detail elsewhere.[27]

The absence of a pure alumina surface layer in anodic films on high Ta alloys disagrees with the results of Habazaki et al.[34-35] who have found such a layer even at 39% at. Ta content in the substrate. While this difference between the data obtained in different laboratories may be a result of differences in the experimental procedure applied, Habazaki's result points to the role of Ta as a dissolution moderator under certain conditions. More research is necessary to clarify this.

Following the discussion on the role of refractory metals in the passivity of Al-based amorphous alloys, one has to note that the recent results for Al-Nb alloys seem to deny the idea of Nb as a metal dissolution blocker, (see the example in Fig. 7). The composition profiles of the anodic film grown on an $Al_{74}Nb_{26}$ amorphous alloy at $i = 0.5$ mA/cm$^2$ up to various voltages have shown that Nb is present from the very beginning of the depth profiling procedure, even at low Nb concentration (13 at.%), as can be seen in Fig. 7a. Moreover, RBS analysis leading to an average composition of the surface sublayers confirmed that no superficial $Al_2O_3$ layer was present within the upper part of the anodic oxide film; see the example in Fig. 7b obtained for anodically treated $Al_{87}Nb_{13}$ samples under similar experimental conditions as those in Fig. 7a.

It is worth recalling here the early results for Al-Mo and Al-W alloys, obtained with a AES 500 Ribber spectrometer.[6,13,14] Figure 8 shows partial Mo and W profiles for Al-Mo and Al-W samples, containing 15% or 45–50% at. of the refractory metal, anodized up to 50V. The prominent feature of these profiles is the absence of any Mo or W signal up to $\sim 60$ min. of sputtering, suggesting that only aluminum and oxygen are present in the upper part of the film. This is true for both low refractory metal concentration (15 at.%) and $\sim 3$ times higher concentration (45–50 at.%). Apparently, both Mo and W tend to remain unoxidized during the

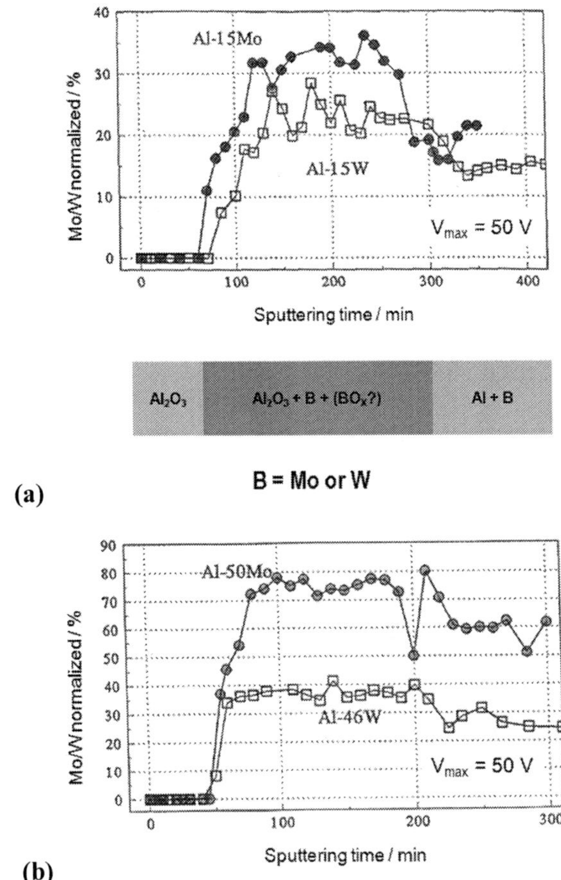


Figure 8. Partial composition profiles of Al-Mo and Al-W amorphous alloys with an oxide layer formed during anodic polarization ($V_{max} = 50$ V) in borate buffer solution. Mo/W was normalized (%) with respect to the total amount of metallic components (B divided by $Al^{ox} + Al^m + B$). (a) Al-15B (b) Al-46/50B. The scheme below shows the following regions: the uppermost surface layer of pure alumina, principal oxide mixed with B, and substrate.

anodization process. The absence of Mo or W in the upper part of the film and the enrichment of Mo or W in the inner part are clearly visible. These results confirm that we do have a relative enrichment in refractory metal in the inner part of the film. The size of the Mo- or W- depleted zone is practically independent of the final voltage $V_{max}$.[6,14] The enrichment level seems to depend on the Mo or W content in the alloy.[6] The findings concerning the depth distribution of Mo and W within the anodic film and near the interphase oxide/substrate region confirm that Mo and W act as dissolution blockers in the Al-Mo and Al-W alloys.[13,14] These results are in line with those of Habazaki who found that Mo remains unoxidized during anodization at low and moderate voltages.[30-32]

- Ta and Nb in Al-based alloys become oxidized during anodization, and participate in the formation of the surface oxide film. They act as *passivity promoters*.

- Mo and W in Al-based alloys tend not to undergo oxidation during the anodization process. They accumulate within the inner part of the film, acting as *dissolution moderators*.

- Various refractory metals contribute to an anodically-formed oxide film in various ways, thus protecting the alloy underneath according to different mechanisms. The mechanism operative in a specific system depends on the relative values of the free enthalpy of oxide formation of all the alloy components and on their metal-metal bond strength.

## III.  EFFECT OF HYDROSTATIC EXTRUSION LEADING TO NANO-STRUCTURIZATION ON THE STABILITY OF THE PASSIVE STATE OF AUSTENITIC STAINLESS STEELS

In recent years, a number of methods for refining the structure of metals by severe plastic deformation (SPD) have been developed. Some of those methods permit grain refinement to a nanometric level. Numerous investigations show that metals having such a structure are characterized by a number of specific properties. However, their corrosion properties are not well known.

Nanostructured materials initially attracted the attention of scientists and engineers because of their specific physical and mechanical properties.[39-42] Although they are metastable systems, the degree of homogeneity of nanomaterials is higher, than for common 'as received' materials. One of the methods which has been developed for successful nanostructurization is that of hydrostatic extrusion (HE), which enables bulk nanocrystalline materials to be produced.[43] Many studies have been carried out on the physical and mechanical properties of nanostructured Al, Al-based alloys, Ti, Cu, Ni and stainless steel.[44-48] However, their corrosion properties, including the stability of the passive film on such technically important materials as HE stainless steel, have not been studied fully. Grain refinement combined with the severe plastic deformation introduced during the HE process may cause drastic changes in corrosion resistance. In particular, the overall surface area of grain boundaries is multiplied and their composition and structure may change. There is much in the literature concerning the passivity and pitting[49-59] of metals and alloys, including austenitic stainless steels. However, in order to gain insight into the effect of hydrostatic extrusion on passivity, investigations carried out under the same experimental conditions should be compared.

The principles of the HE process are quite simple and are shown in Fig. 9.

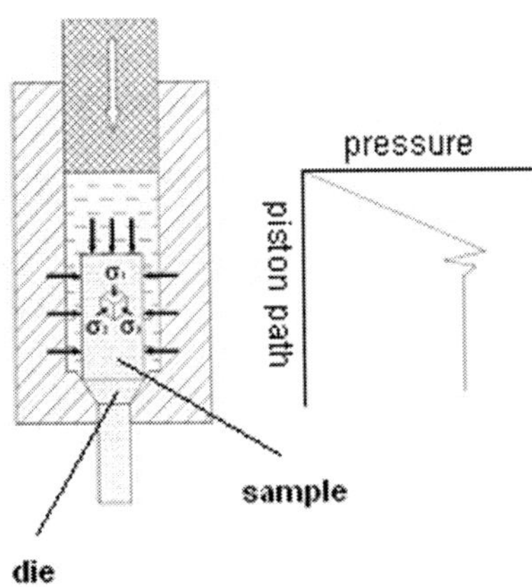


Figure 9. Scheme of HE apparatus.

A billet of the test material is extruded through a die located in the operational chamber. During the hydrostatic extrusion process the piston does not contact the material being extruded and in the deformation zone of the die, the cone, the material flows on a film of lubricant (hydrodynamic lubrication). The very low friction enables severe plastic deformations in the range of 2.0 to be obtained.

Hydrostatic extrusion is one of the methods used to produce nanostructured materials and to thereby obtain a homogenous microstructure throughout the entire volume of the processed material. Crucial to an understanding of the results presented below is that the HE samples were cut perpendicular to the deformation direction and that only the central part of the hydroextruded stainless steel rod was used in the investigations. Details are given elsewhere.[60]

The next chapter compares the results of potentiodynamic experiments on two steels, 316 and 303, in $Cl^-$-containing electrolyte. 316 stainless steel contains only 0.05 at.% S, whereas the S content of 303 stainless steel is of an order of magnitude more. Both steels contain similar amounts of Cr or Ni. The detailed composition of both steels is given in Table 4.

## 1.    Anodic Behavior of As-Received and HE Processed Materials in $Cl^-$ Containing Solutions

Figure 10 summarizes the potentiodynamic polarization measurements for mechanically polished and air-exposed samples of 316 and 303 in borate buffer solution +0.1 M NaCl, after HE and in the as-received state. The current density within the passive region for the as-received material is low, from 5 up to 10 $\mu A/cm^2$, which is comparable with the findings of Yang and Mcdonald.[61] The sudden increase in the current density by more than a factor of 50 at potentials above $E_{np'} = 0.33 \pm 0.02\ V_{NCE}$ (i.e., $0.37 \pm 0.02\ V_{SCE}$) for 316 and $E_{np'} = 0.28 \pm 0.02\ V_{NCE}$ ( i.e., $0.32 \pm 0.02\ V_{SCE}$ ) for 303 is due to pit nucleation.

316 HE stainless steel in the same solution exhibits excellent reproducibility of the potential independent current density within the stable passivity region, having a value of $42 \pm 1\ \mu A/cm^2$, some 4 to 8 times higher than that for the as-received material. The pitting potential $E_{np'}$ is not well reproducible for the HE material. For

**Table 4. Chemical Composition of Austenitic Stainless Steels 303 and 316.**

| Steel | | Fe | Cr | Mn | Ni | Mo | Ti | Cu | Si | P | S | C |
|---|---|---|---|---|---|---|---|---|---|---|---|---|
| 303 | Wt.% | balance | 17.5 | 2.0 | 8.5 | 0.4 | – | 0.6 | 1.0 max | 0.2 max | 0.35 max | 0.15 max |
| | At.% | balance | 18.3 | 2.0 | 7.9 | 0.2 | – | 0.5 | 1.9 max | 0.4 max | 0.6 max | 0.7 max |
| 316 | Wt.% | balance | 17.2 | 1.8 | 11.4 | 2.1 | 0.3 | – | 1.0 max | 0.045 max | 0.03 max | 0.08 max |
| | At.% | balance | 18.2 | 1.8 | 10.7 | 1.2 | 0.3 | – | 1.9 max | 0.08 max | 0.05 max | 0.4 max |

The chemical composition of both materials is quite similar, although 303 stainless steel contains approximately ten times more S, five times more P, two times more C, and six times less Mo.

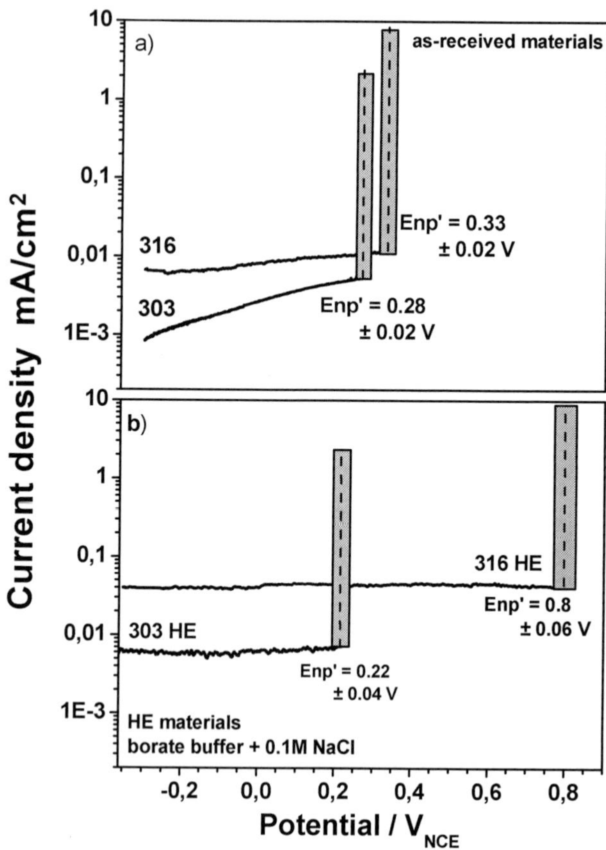



Figure 10. Potentiodynamic polarization curves for type 303 and type 316 stainless steels in a borate buffer + 0.1M NaCl solution: before (a) and after HE (b).

316 HE breakdown of passivity occurs at and above $E_{np'}$ (HE) = $0.80 \pm 0.06$ $V_{NCE}$ (i.e., $0.84 \pm 0.06$ $V_{SCE}$), (see Fig. 9). Apparently, HE leading to homogenization of the substrate material and the passivating film, resulted in an increase in its resistance to pit nucleation, in spite of its metastable state and the large residual

stresses, etc. 303 HE stainless steel exhibits a different behavior, as shown in Fig. 10.

HE 303 steel in the same solution (in contrast to the behavior of the as-received material) exhibits a stable, potential independent, current density of about 6 $\mu A/cm^2$, which is thus higher than for the as-received material. It undergoes a breakdown of passivity at $E_{np'} = 0.22 \pm 0.04\ V_{NCE}$. These results suggest that the changes introduced by HE treatment may affect the structure of each steel in a different way, leading to an extension of the stable passivity region in 316 stainless steel, and to a slight reduction of it in 303 HE. As will be discussed below, structural and surface analytical investigations helped towards an understanding of these differences.[60,62-64]

- The values of the apparent breakdown potential $E_{np'}$ suggest that HE Type 303 tends to be less resistant to pitting in buffered $Cl^-$ solution than the as-received material. The reason for such behavior will be discussed below.
- Electrochemical accelerated laboratory tests — anodic potentiodynamic polarization curves — have shown that, compared to the as-received material, HE Type 316 exhibits a current density 4 to 8 times higher than in the passive region. The values of the apparent pitting potential $E_{np'}$ suggest that HE Type 316 is more resistant to pitting in buffered $Cl^-$ solution than the as-received material. This is in direct contrast with the findings for Type 303 stainless steel.

## 2.  Microstructure Examination and Image Analysis

In order to understand the electrochemical tests results, and in particular the various effects of the HE process on the stability of the passive state, including the breakdown of 316 and 303 stainless steels in $Cl^-$ containing electrolyte, microstructural and image analysis results should be carefully considered, as follows: Light microscopy reveals a typical microstructure of austenite in type 303 in the as-received state; deformation twins and many small non-metallic inclusions are visible. The inclusions were identified as MnS, see Section III.3. Typical grain size is in the range of 20-40 $\mu m$. More details are given elsewhere.[60,62-64] Light microscopy reveals that the microstructure of as-received type 316 stainless

 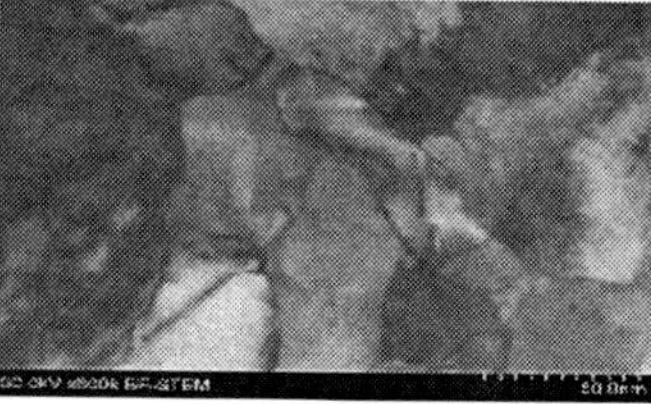

Figure 11. Microstructure of 303 austenitic stainless steel after hydrostatic extrusion; SEM-BF (bright-field) images.

steel is similar to that of type 303, with grain sizes in the same range. Sporadic non-metallic inclusions are present in the material.

Figure 11 presents a SEM-BF image of 303 stainless steel after hydroextrusion. Evidently, HE resulted in a strong refinement of the microstructure. Elongated nano-grains (*domain structure*) of ~ 10–20 nm width are clearly visible. The image reveals deformation twins in the primary austenite, which intersect each other at various angles, forming characteristic blocks – a specific structure at the nanometric scale.

For comparison Fig. 12 shows transmission electron micrographs of 316 steel after nanostructurization. The results suggest that the structure of the HE material consists of very small crystallites, equal in size to those in 303 HE. There are also

Figure 12. Transmission electron micrograph of 316 stainless steel after hydroextrusion. In the upper insert the selected diffraction pattern (SAD) is shown.

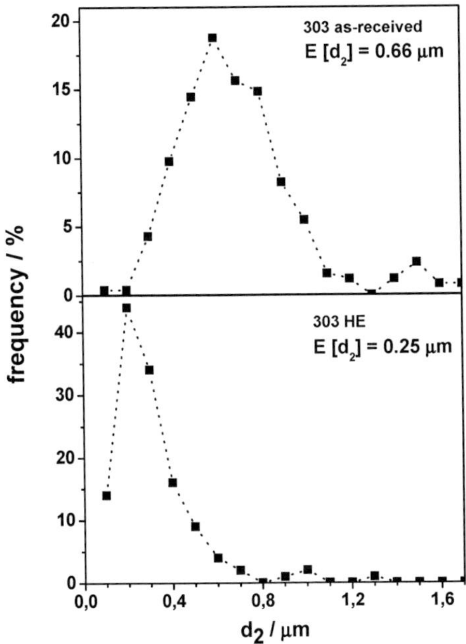



Figure 13. Results of image analysis performed for inclusions in 303 stainless steel after the HE process. See
text for details.

domains, suggesting the presence of nano-grains. Thus, the HE
procedure resulted in an average reduction of grain size by a factor
of ~ $10^2$ or more. The upper-right corner of the image in Fig. 12
shows the selected area diffraction pattern (SAD) of 316 HE steel.
One can see characteristic strong diffraction rings consisting of
*spot-like* reflections, with the reflections being distinctly stronger
in certain directions, suggesting a textured HE material. Detailed
discussions are given elsewhere.[62]

The amount of impurities in 316 stainless steel is low (see Table 4). Therefore, non-metallic inclusions in this steel are rare, and
do not play an important role in the breakdown of passivity. Ex-
aminations of 303 stainless steel's microstructure revealed that
there are a large number of nonmetallic inclusions both within the
grains and at the grain boundaries. Figure 13 shows the results of

image analysis for the inclusions in as-received material. On the surface area selected for the stereological analysis, about 260 inclusions were considered. The mean value of the equivalent diameters $E(d_2)$ of the inclusions in as-received steel is 0.66 μm.

The number of inclusions in 303 HE was much higher than that in the as-received material. About 130 inclusions with an equivalent diameter from 0.1 to 1.6 μm were considered. A rough estimate is that the number of inclusions after HE increased by a factor of 16 per unit area. The mean values of the equivalent diameters $E(d_2)$ of the inclusions are: 0.25 mm in HE pretreated material (compared with ~ 0.66 μm in the as-received material), (see Fig. 13). Careful inspection of the 303 HE samples suggest that inclusions with a diameter below 0.1 μm are also present, but, are below the resolution of the image analysis program used in this investigation. Despite the drawbacks of the method applied, it is evident that the HE pretreatment reduces the size of inclusions by a factor of 3 (compare Fig. 13), and apparently increases their total number per unit volume.

- Microscopic and TEM experiments have revealed that HE resulted in significant change in the structure of type 303 and 316 stainless steels. Grain size was reduced by a factor of ~ $10^3$, and a strongly textured material was obtained.
- TEM and stereological image analysis have revealed that HE resulted in a reduction in the size of nonmetallic sulfide inclusions in Type 303 stainless steel by a factor of 3 or more, thus increasing their number within the matrix.

## 3.  Local AES and SAM Analyses of the Inclusions in 303 Stainless Steel

A high resolution Scanning Auger Microprobe—Microlab 350 (Thermo VG Scientific) equipped with an FEG-tip (Field Emission Electron Gun) was used for the local AES analysis. The Microlab 350 was used to monitor the surface morphology (SEM) and local chemical composition, utilizing the Scanning Auger Microscopy (SAM) functions of the instrument with a lateral resolution of about 20 nm. The detailed lateral distribution of elements was examined utilizing the Auger line scan and mapping function.

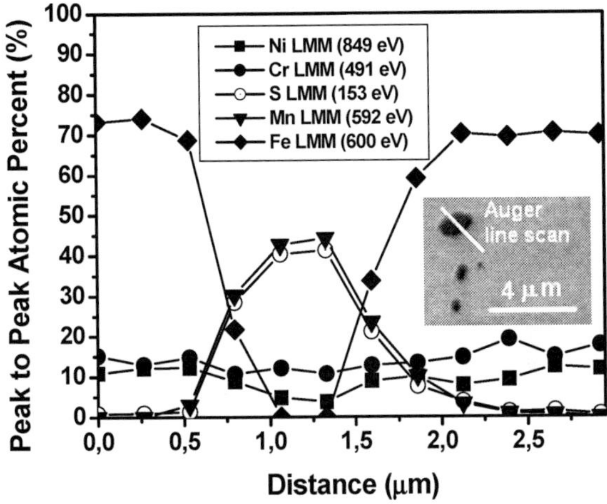


Figure 14. Auger line analysis recorded across an MnS inclusion in 303 austenitic stainless steel. Reprinted from M.Pisarek, P. Kędzierzawski, T. Płociński, M. Janik-Czachor, K. J. Kurzydłowski, "Characterization of the effects of hydrostatic extrusion on grain size, surface composition and the corrosion resistance of austenitic stainless steels", *Mater.Charcter.* **59 (9)**, 1292, Copyright (2008) with permission from Elsevier.

To obtain a semi-quantitative distribution of elements within a typical inclusion, a line scan analysis was performed across the inclusion/steel-interface (see Fig. 14). The line scan indicates that the inclusion consists mainly of sulphur and manganese in equal proportions, but also contains some iron and nickel. Thus, the Auger line analysis taken across an MnS inclusion may suggest that some Fe and Ni are minor components of it, though one cannot exclude the possibility that the inclusion is sufficiently thin to allow attenuated signals from the matrix to reach the detector.

Our SAM measurements point to the role of discontinuities in the passivating film on the MnS inclusions in 303 HE stainless steel (see the example in Fig. 15). The distribution image of the Auger O KLL signal from the surface shows unambiguously that there is a lack, or at least a depletion, of oxygen within these

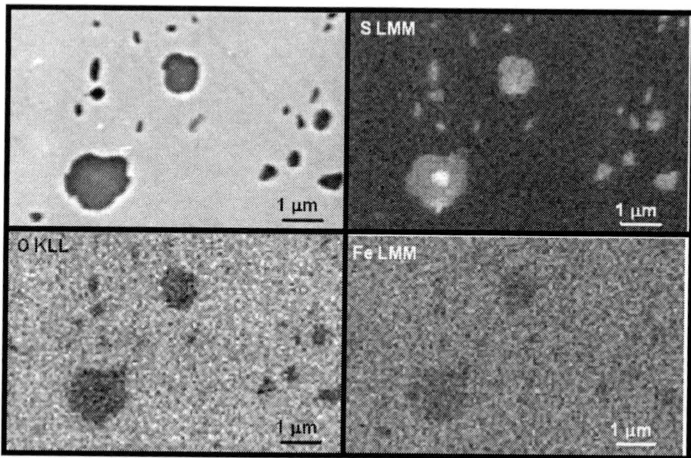


Figure 15. Distribution images of elements: Fe (LMM), S (LMM), O (KLL) on the surface of 303 HE stainless steel after HE process, suggesting a distinct depletion of oxygen on the surface of sulfide inclusions. Reprinted from M. Pisarek, P. Kędzierzawski, T. Płociński, M. Janik-Czachor, K. J. Kurzydłowski, "Characterization of the effects of hydrostatic extrusion on grain size, surface composition and the corrosion resistance of austenitic stainless steels", *Mater.Charcter.* **59 (9)**, 1292, Copyright (2008) with permission from Elsevier.

surface regions, where the S LMM Auger signal infers the presence of manganese sulfide inclusions. It is therefore clear that these regions and their boundaries with the matrix tend to be more prone to Cl⁻ attack than is the matrix itself.

There is a lot of discussion in the literature on the role of nonmetallic inclusions[18, 27-31] in the deterioration of corrosion resistance in stainless steels. There is a general agreement that $MnS_x$ inclusions are the most damaging.[65-73] The main reasons for their detrimental role, pointed out by different authors, are:

- their chemical instability in acidic solutions,[65]
- the presence of voids or crevices at the inclusion/matrix boundary,[66-68]
- the absence or lower stability of the passivating film formed on the inclusions locally[69] as compared to the passivating film on the matrix.

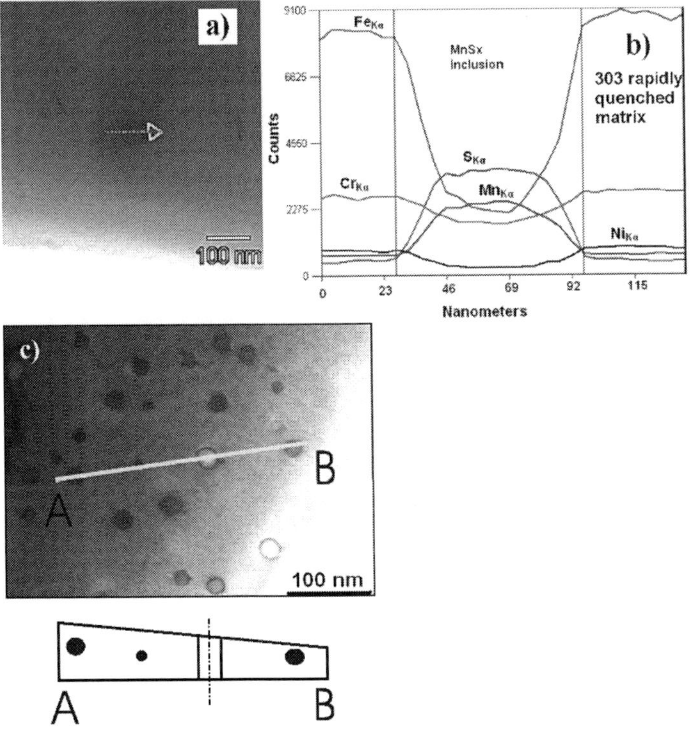



Figure 16. (a) TEM image of MnSx inclusion in a rapidly quenched 303 stainless steel ribbon after additional thinning. The edge of the resultant 303 foil is visible at the lower part of the image. (b) EDS line scan across the inclusion, marked by an arrow in (a). (c) A number of MnSx inclusions within the 303 foil near its edge with the diameters below 30 nm (black circles). Two inclusions were removed completely during anodic thinning (open "empty" circles). A scheme of a cross-section of the foil along the line A-B is shown below (c).

Our above results suggest that the last reason, in general, is the most significant:

- SAM microanalytical investigations of Type 303 stainless steel revealed local discontinuities within the passivating oxide film over the sulfide inclusions at the surface before and after HE. Such discontinuities provide for direct contact between the inclusions and the aggressive environment and apparently facilitate the breakdown of passivity.

- The larger density of local discontinuities within the surface oxide film due to the presence of more sulfide inclusions in the HE Type 303 steel compared to the as-received material is most probably responsible for the low stability of 303 HE in Cl⁻ containing solution.

An interesting example of a joint application of electrochemistry and high resolution techniques for characterization of modified material is given below in connection with a local chemical analysis of very small nonmetallic inclusions occurring within the 303 stainless steel matrix, (see Fig. 16).

Rapid quenching, which is a commonly used procedure for amorphization or nanostructurization of materials, was used in an attempt to obtain a grain refinement in 303 stainless steel. It has been found that rapid quenching from $1530^0$C at a rate of cooling from 20 up to 50 rpm/min. did not result in any distinct grain refinement. However, it did reduce the size of the MnS inclusions by a factor of $10^2$ or more (thus ~ 2 orders of magnitude more than the HE procedure did, compare Fig. 13). It was possible to visualize the inclusions within the rapidly quenched matrix by TEM, after a procedure of thinning the samples electrochemically in an acidic solution (95% concentrated acetic acid + 5% perchloric acid) or, alternatively, by the ion sputtering technique.

An example is given in Fig. 16c. A large number of small, round inclusions with a diameter of 20–30 nm are clearly visible. It is worth nothing that SAM (Scanning Auger Microscopy) analysis failed to obtain composition maps of the inclusions within the matrix. Signals from the austenite matrix only were detected. Apparently the inclusions on the surface were removed completely during the thinning procedure. EDS analysis was then performed to analyze the inclusions located *within* the thin austenite foil (compare the Scheme below, Fig. 16c). The information depth of EDS is ~ 1 μm or $10^3$ times larger than that of AES, and, therefore, it was possible to detect the inclusions within the foil. On the other hand, by using a thin foil one was able to eliminate possible strong signals from the matrix (Fe, Ni, Cr) which could adversely interfere with the Mn and S signals.

An example of such an analysis is presented in Fig. 16b. A line scan across the inclusion (Fig. 16a) confirmed, in agreement with the AES data for 303 HE, that the inclusion is made up of Mn and S. Some signals from Ni(K$\alpha$) and Cr (K$\alpha$) apparently originating from the matrix were also detected within the inclusion area, (compare scheme underneath Fig. 16c).

The above discussion shows that the advances in EDS microanalytical technique enabled chemical analysis of highly dispersed different phases within a metal matrix to be obtained. However, EDS does not provide true surface information, in contrast to that obtained by AES.

The surface oxide film on 316 HE seems to be fully continuous which is compatible with the fact that MnS inclusions are absent there. To determine the internal composition of the oxide film covering the metal matrix, the depth composition profiles were measured on 316 and 303 stainless steels using AES and $Ar^+$ ion sputtering technique.

## 4.    AES Investigations of the Surface Oxide Film on HE Materials

The internal composition of the oxide film covering the metal matrix was deduced from composition profiles.[62-63] Figure 17(a)-(b) shows partial compositional profiles (the relative Cr/Fe atomic concentration) of air-formed films on Type 303 and 316 stainless steels before and after the HE process, as measured with AES combined with $Ar^+$ ion sputtering. O (KLL) composition profiles are also given. The boundaries between the film and the substrate are marked by the dotted lines. As seen from Fig. 17, the film on the as-received materials is removed after ~ 36s of sputtering, which corresponds to a thickness of about 3.6 nm, based on a sputtering rate of 0.1 nm/s. Similar composition profiles are shown in Fig. 17 for the HE processed materials. The film thickness seems to be the same for as-received Type 316, as-received Type 303 and Type 316 HE, but, seems larger (about 5.6 nm) for Type 303 HE. As highlighted in Fig. 17, the main difference here is a larger enrichment of Cr for the HE processed materials than for the as-received ones. The maximum Cr/Fe concentration ratio within the film on HE steels reaches ~ 0.5, whereas it approaches only ~ 0.4 for both steels in the as-received state. Additional details are

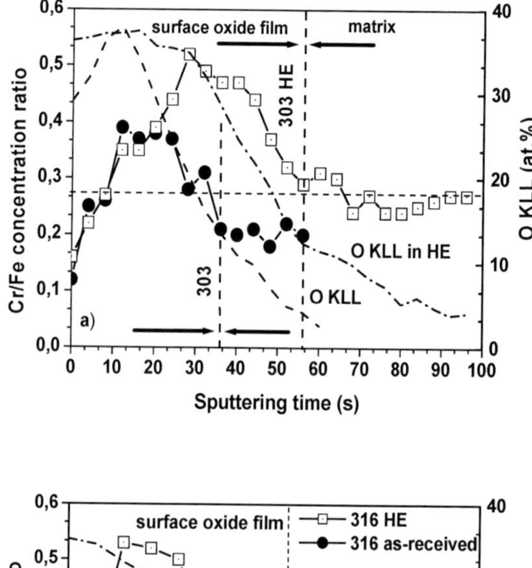



**(a)**

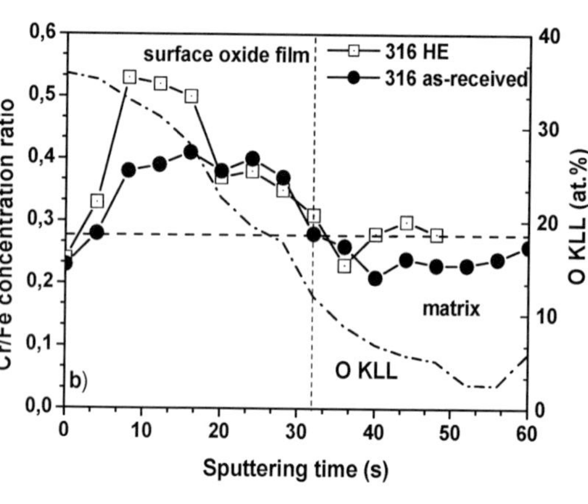



**(b)**

Figure 17. Comparison of Cr/Fe concentration ratio vs. sputtering time for Type 303 (a) and 316 (b) stainless steels before and after the HE process. The plots show the depth distribution of Cr/Fe within the surface oxide film and in the underlying matrix. The oxygen profiles are also marked by a dotted line. Dashed vertical lines mark the oxide/matrix boundaries. Dashed horizontal lines mark the Cr/Fe concentration ratio within the bulk matrix. Reprinted from M. Pisarek, P. Kędzierzawski, T. Płociński, M. Janik-Czachor, K. J. Kurzydłowski, "Characterization of the effects of hydrostatic extrusion on grain size, surface composition and the corrosion resistance of austenitic stainless steels.", *Mater.Charcter.* **59 (9)**, 1292, Copyright (2008) with permission from Elsevier.

given elsewhere.[62-63] Apparently, hydrostatic extrusion may enhance the migration of components to the surface of the metastable matrix, which contains a large density of grain boundaries and accumulated internal stresses, thus resulting in a higher enrichment of chromium within the surface oxide film on the HE substrate.[62] Some differences in the degree of chromium enrichment in the surface film due to different pretreatment of samples have already been reported.[74]

The above AES results suggest that Cr acts as *passivity promoter* in austenitic stainless steels, being enriched and oxidized within the film.

- There are distinct differences in the chemical composition of the post-HE air-formed films. Type 303 HE and Type 316 HE surfaces are covered with an air-formed film enriched with Cr to a greater extent than films present on the as-received material.

- Apparently the greater enrichment of Cr in the oxide film on HE pretreated 303 stainless steel is not sufficient to eliminate local film discontinuities caused by the presence of a large number of MnS inclusions and so it does not improve the steel's resistance to pit nucleation.

- The general conclusion from this investigation is that nanostructurization (via hydrostatic extrusion) and the following *homogenization* of the material appears inefficient in improving resistance to pit nucleation in steels containing MnS inclusions—in this case, type 303 stainless steel. Even after a reduction in size, the MnS inclusions do not undergo passivation; thus there exists a large density of locations susceptible to pitting attack in a chloride-containing environment.

## IV.  CATALYTIC ACTIVITY OF Cu-BASED AMORPHOUS ALLOYS MODIFIED BY CATHODIC HYDROGEN CHARGING

Materials scientists are interested in the preparation and characterization of new functional materials, including new catalysts for technically important reactions.

In this chapter we show how a detrimental process of material degradation can be successfully applied to obtain new functional materials exhibiting appreciable catalytic activity and high selectivity. Thus, the above processes provide useful methods of metal alloys modification for practical applications in catalysis.

The dehydrogenation of aliphatic alcohols to produce corresponding carbonyl compounds is one reaction requiring an efficient catalyst. Cu powder is known to catalyse such processes, but easily becomes deactivated, mostly due to sintering. In order to improve performance, supported Cu catalysts, usually Cu on an oxide support, are used. The best examples of these are the binary and ternary industrial Cu-metal oxide systems used in methanol synthesis.[75-77]

Amorphous alloys may be considered as model systems[4] and are known as precursors for a number of efficient catalysts for technically and ecologically important reactions.[78-80] Specifically, they have been found to display outstanding properties in ammonia synthesis[81] and have been suggested for use in global carbon dioxide recycling.[82]

Amorphous Cu-Zr and Cu-Hf alloys are known as precursors for efficient and selective supported catalysts for the dehydrogenation of alcohols.[83-84] The transformation of these precursors into a $Cu/ZrO_2$ or $Cu/HfO_2$ catalyst requires a modification procedure. Chemical and electrochemical methods have proved to be efficient here.[84-85] In particular, cathodic hydrogen charging with subsequent air exposure has appeared to be highly efficient.[84,87-89] This method, moreover, has appeared promising for Cu-Ti amorphous alloy, which had seemed to be *totally inactive*, and unsuitable for the activation processes discussed above.

To illustrate the problems involved with the activation of Cu-Ti alloy for catalytic application the results of Molnar et al.[90] should be considered, (see Fig. 18). Those results concern an attempt to activate Cu-Ti amorphous alloys prepared by mechanical alloying. Such alloys offer a high specific surface area, which is beneficial to heterogeneous catalysis. Unfortunately, as one can see from Fig. 18, it was impossible to activate the Cu-Ti powders, in contrast to those of Cu-Hf and Cu-Zr.

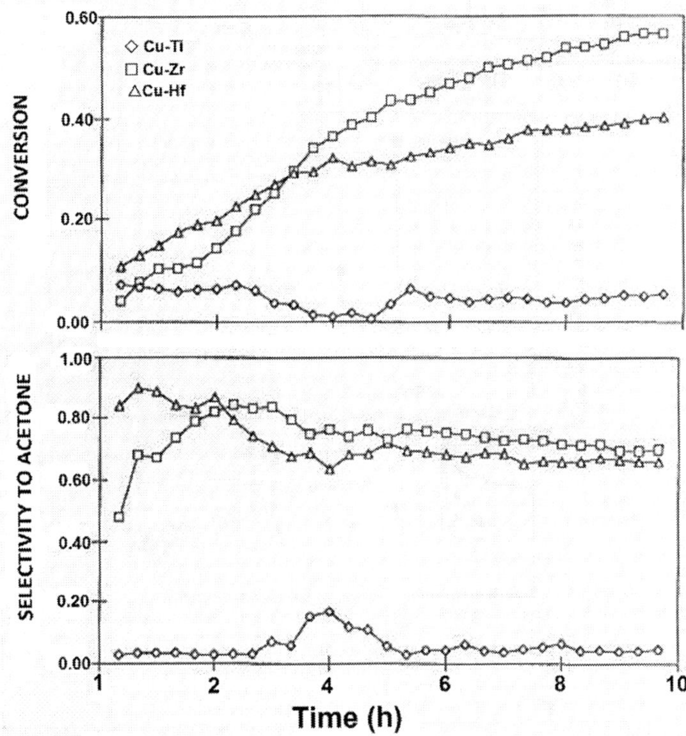


Figure 18. Activity and selectivity of Cu-M powders in the dehydrogenation of 2-propanol (523K, 20 ml/min hydrogen). M = Ti, Zr or Hf.

In order to successfully enhance Cu-Ti amorphous alloys for catalytic purposes, the mechanism of the catalytic reaction should be understood. Moreover, the structure and composition of the *active sites* on the surface of a catalyst, and their functions, assessed on the basis of surface analytical investigations, should be considered.

Most earlier experimental observations, including the results of surface science studies[17] and data acquired by catalytic[92-93] and kinetic investigations,[94-95] are in harmony with the so-called carbonyl mechanism depicted in Scheme 1 for 2-propanol. With respect to the surface active sites involved in the process

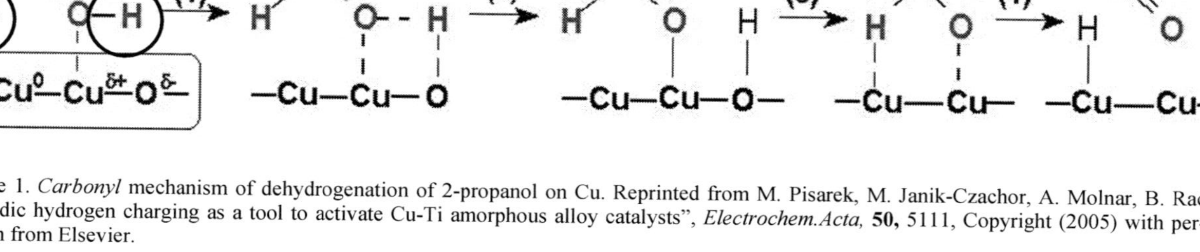

Scheme 1. *Carbonyl* mechanism of dehydrogenation of 2-propanol on Cu. Reprinted from M. Pisarek, M. Janik-Czachor, A. Molnar, B. Rac "Cathodic hydrogen charging as a tool to activate Cu-Ti amorphous alloy catalysts", *Electrochem.Acta*, **50**, 5111, Copyright (2005) with permission from Elsevier.

the results are somewhat contradictory. The rate of dehydrogenation was found to be proportional to the concentration of exposed $Cu^0$ sites.[96-98] In other studies, in contrast, the coexistence of metallic and oxidized copper was shown to be necessary for catalytic activity.[99] In fact, surface oxygen was found to greatly enhance the ability of Cu to dissociatively chemisorb methanol and ethanol in single crystal studies.[100-103]

Both the latest results[95] and earlier observations[104] show that the rate-determining step in the dehydrogenation of secondary alcohols is the breaking of the O–H bond.

Detection of surface alkoxide species on copper[91,100-101,105-106] is considered to be definitive proof that O–H bond breaking occurs prior to the breaking of the $\alpha$ C–H bond. This is despite the fact that the $\alpha$ C–H bond in secondary alcohols is somewhat weaker than the O–H bond. It is supposed that this may be due to the favourable orientation of the hydroxyl group on the surface.[95]

The catalytic action of the copper catalyst is the result of partial charge transfers between the various entities on the surface and in the adsorbed molecule according to the electronegativity of individual atoms. The carbonyl mechanism, in fact, was originally suggested to interpret the dehydrogenation of alcohols on oxide catalysts such as $Cr_2O_3$,[107-108] and may be viewed as the weakening of the corresponding 3 bonds in the aliphatic alcohol through interactions with 3 surface entities $Cu^0$, $Cu^{\delta+}$ and $O^{\delta-}$, as shown in Scheme 1.

Hence, the goal for materials scientists is to transform a stable Cu-Ti amorphous alloy, which is highly resistant to activation, into a modified material where the above 3 entities exist, in a suitable configuration on the surface ($Cu^0 + Cu^{\sigma+} + O^{\sigma-}$). In an attempt to achieve this, cathodic hydrogen charging followed by air exposure was used as a modification method.

Our own recent results have shown that the procedure developed by Gebert et al.[109-110] was efficient in transforming Cu-Hf amorphous alloys into stable, efficient (~ 90% conversion) and selective (~ 95%) catalysts.[88] After hydrogen charging the Cu-Hf samples were exposed to air for a prolonged time to desorb hydrogen. This allowed the second element to undergo oxidation, thereby enhancing segregation phenomena. Our preliminary results[89] suggest that the same procedure is promising for Cu-Ti, as well.

**Table 5**

**Effect of Hydrogen Charging on Catalytic Behavior of Cu-Based Amorphous Alloys.**

| Precursors | Treatment | Conversion (%) | Selectivity (%) |
|---|---|---|---|
| Cu-Zr | High pressure hydrogen charging | ~ 50 | ~ 85 |
| Cu-Hf | Cathodic hydrogen charging | ~ 90 | ~ 98 |
| Cu-Ti[a] | Cathodic hydrogen charging | ~ 60 | ~ 99 |

[a]Only cathodic hydrogen charging is efficient in the activation of Cu-Ti precursor

## 1.  Catalytic Activity, Porosity and Specific Surface Area of Cu-Based Catalysts

Table 5 compares the conversion and selectivities in dehydrogenation of 2-propanol to acetone obtained on modified Cu-based catalysts. Only the most efficient methods were compared for each alloy under consideration. Although the conversion level is lower for Cu-Ti than for Cu-Hf catalysts, yet, the selectivities are superior for the Cu-Ti catalyst.

Table 6 shows the effect of hydrogen charging on the catalytic activity of Cu-Ti amorphous alloys, as well as, on their BET specific surface area. Catalytic activity increases with charging time

**Table 6**

**Average H/M Atomic Ratio, BET—Specific Surface Area, Porosity and Conversion of 2-Propanol on Cu60-Ti40 Amorphous Alloy After Cathodic Charging at –1 mA/cm$^2$ Over Varying Charging Time from 27 h Up to 96 h.**

| Cu60-Ti40 $i \approx 1$ mA/cm$^2$ (0.1 M H$_2$SO$_4$) | Atomic ratio H/M | BET surface (m$^2$/g) average[a] | Porosity (cm$^3$/g) average[a] | Conversion (%) after 3 h | Conversion (%) after 5 h |
|---|---|---|---|---|---|
| t = 27 h | 0.39 | 2.3 (~230) | 0.005 (0.5) | ~11 | ~10 |
| t = 50 h | 1.0 | 2.7 (~270) | 0.006 (0.6) | ~22 | ~18 |
| t = 70 h | 0.86 | 1.1 (~110) | 0.002 (0.2) | ~26 | ~27 |
| t = 96 h | 1.0 | 1.7 (~170) | 0.003 (0.3) | ~60 | ~60 |

[a]Average value calculated with respect to the total weight of the ribbon. The values of BET surface and porosity estimated for the modified layer of the material only are given in parentheses. True values of BET specific surface area and porosity (in parentheses) are estimated for the modified region only. Those values were estimated taking into account that the cathodic hydrogen charging modified only ~ 1% of the Cu60-Ti40 ribbon thickness[11].

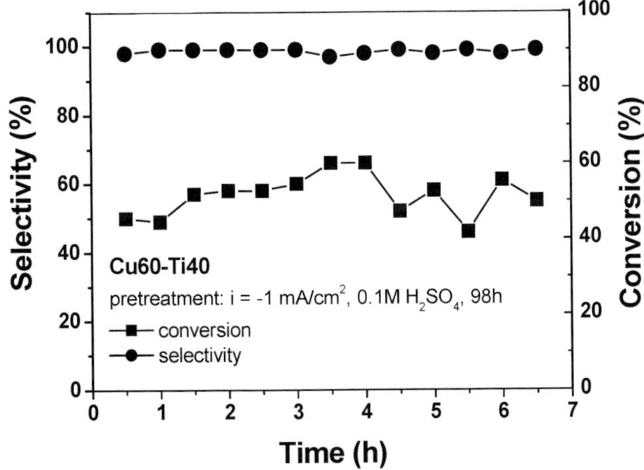


Figure 19. Catalytic activity and selectivity to acetone for dehydrogenation of 2-propanol on a Cu60-Ti40 amorphous ribbon after hydrogen charging in 0.1-M H$_2$SO$_4$ at low current density. Reprinted from M. Pisarek, M. Janik-Czachor, A. Molnar, B. Rac "Cathodic hydrogen charging as a tool to activate Cu-Ti amorphous alloy catalysts.", *Electrochem.Acta*, **50**, 5111, Copyright (2005) with permission from Elsevier.

up to conversion of ~ 60%, while the H/Metal atomic ratio (H/M) also increases from 0.39 to ~ 1.0, which is not in harmony with BET specific surface area and porosity data, as can be seen from Table 6. This subject will be discussed in more detail below.

The selectivity to acetone of the Cu-Ti catalyst, which reached about 100%, is superior to that obtained for modified Cu-Hf[111] or Cu-Zr. Figure 19 shows an example of the stability of the catalytic activity of hydrogen-charged Cu-Ti and its selectivity, reaching 99% over a prolonged testing period, at a conversion reaching ~ 60%. Figures 20 and 21 show BET specific surface area and porosity vs. cathodic hydrogen charging time for Cu-Ti and, for comparison, for Cu-Hf amorphous alloys.

Although BET areas may not necessarily be relevant to catalysis, the rather small values (a few m$^2$/g or less) are indicative of a small concentration of active sites. The average BET areas are small and decrease with hydrogen charging for Cu-Ti; neverthe-

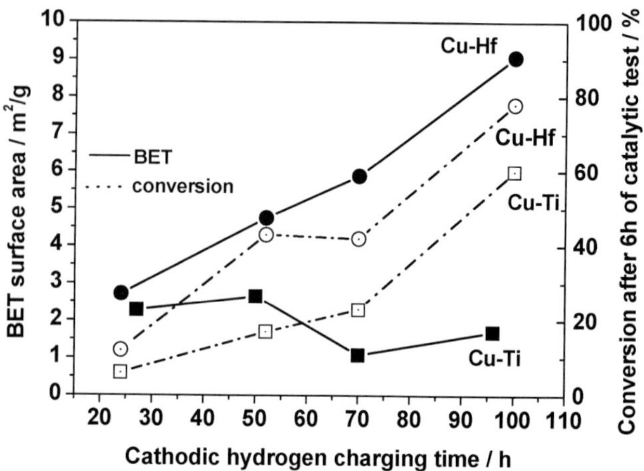


Figure 20. BET specific surface area (estimated with respect to the total weight of the modified ribbon) vs. time of hydrogen charging of Cu-Hf and Cu-Ti alloys at low current density in 0.1 M $H_2SO_4$. Corresponding conversion data are also given.

less, catalytic activity increases. Inspection of the average pore diameter, moreover, reveals that it oscillates between 8.6 nm and 7.2 nm for a hydrogen charging time varying from 27 h up to 98 h.[85,111] Thus, the average porosity characteristic is not in harmony with the catalytic data either.

It is important to emphasize that for this modified precursor surface $Cu^0$ could not be detected, even after hydrogen treatment. It may be argued that the sensitivity and accuracy of $N_2O$ titration are not high enough to measure the surface concentration of $Cu^0$ accurately. However, for Cu-Zr and Cu-Hf alloys and powders in earlier studies[90,112] activities could be correlated with the $Cu^0$ concentration measured by the same method. Furthermore, even if small amounts of undetected $Cu^0$ are present on Cu-Ti, this definitely cannot account for the activities, which are comparable to those determined for Cu-Zr and Cu-Hf.

The BET surface area and porosity (estimated with respect to the total weight of the modified ribbon) given in Table 6 and in Figs. 20–21 require an additional comment. Our attempt to ex-

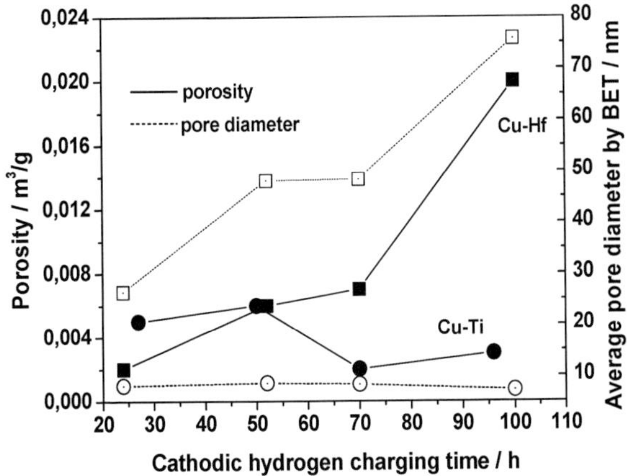



Figure 21. Average porosity (estimated with respect to the total weight of the modified ribbon) of the surface of the precursors vs. time of hydrogen charging of Cu-Hf and Cu-Ti alloys at low current density in 0.1 M $H_2SO_4$. Corresponding average pore diameter data are also given.

amine the changes in a cross-section of Cu-Ti amorphous alloy introduced by cathodic hydrogen charging (and peeling off the Cu, see the comment to Fig. 21, below) was not successful. The hydrogen modified layer was thin ($< 1\mu m$) as compared to the thickness of the ribbon, and not homogenous in thickness, so it was not possible to estimate its thickness unambiguously. Apparently, hydrogen concentration is maximized at the very thin surface layer as titanium hydride, which then blocks further entry of hydrogen into the bulk. This type of blocking effect by hydrogen is known for other metals.[113] Hydrogen charging, therefore, affects primarily the surface layer, (in the case of Cu60-Ti40 ~ 1% of the total thickness of the ribbon) which is relevant to catalysis. Certainly, the specific surface area and porosity also change only locally; thus the corresponding *average* figures for the modified layer given in Table 6 should be, in fact, multiplied by a factor of about $10^2$, as indicated in parentheses in Table 6.

As Ti is known for its tendency to readily form hydrides,[114] it is thus very probable that, upon cathodic charging, hydrogen penetration into the bulk results in the formation of $TiH_2$. Because the hydride has a higher volume, swelling takes place, inducing phases separation and crystallization and the migration of Cu to the surface. Since the catalytic results for the above material are not in harmony with those of their BET specific surface area and porosity measurements, an attempt was made to carefully examine the morphology and surface composition to identify the decisive factors responsible for the increased catalytic activity of the Cu-Ti precursor after activation with hydrogen charging. Considering the large difference between the behaviour of Cu-Zr[83-84] and Cu-Ti, we conclude that Cu-Ti alloy is more stable, and that only a very thin surface layer is involved in the modification.

## 2.  Surface Characterization of Cu-Based Catalysts

High-resolution local characterization is necessary to gain insight into the mechanism of catalytic action of modified Cu-based alloys. The local chemical and morphological changes of the catalyst at the nanometer scale during modification of the catalyst precursor need to be monitored. Scanning Auger microscopy (SAM) is a suitable technique for this purpose because it is surface sensitive. It offers information limited to a depth of 1 nm or less.[86,115-116] Simultaneously, a lateral resolution of ~20 nanometers can be obtained. An Auger microprobe analyzer, a Microlab 350 (Thermo Electron), was employed to monitor subtle changes in surface morphology (scanning electron microscopy, SEM) and local chemical composition, utilizing Auger electron spectroscopy (AES).

The Microlab 350 has a Schottky type field emission electron source, which consists of a single crystal tungsten wire coated with zirconium oxide semiconductor material. This provides a fine electron beam with a diameter of several nanometers, which is suitable for local excitation of the process of Auger electrons emission.[117] Detailed lateral distribution of elements was examined utilizing the Auger line scan function. The chemical state of surface species was identified using the high-energy resolution of the Auger spectrometer (the energy resolution of the spectrometer is continuously variable between 0.6% and 0.06%) and the appropriate standards. Details are given elsewhere.[85,89,111]

It was clear from the previous results[85,88,111] that the catalytic activity of Cu-Hf increases with hydrogen charging time  because

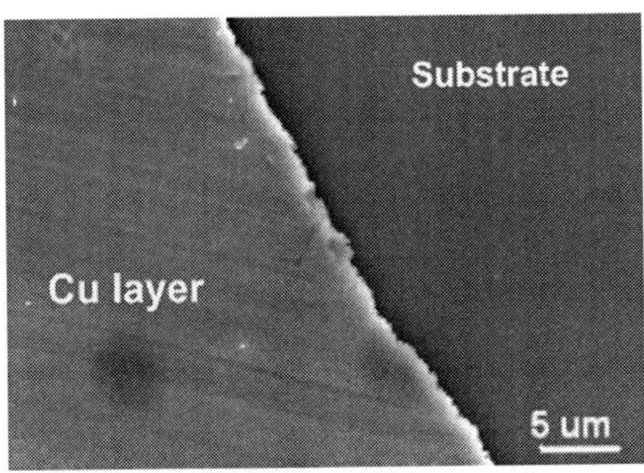


Figure 22. Morphology of a 60Cu-40Ti amorphous glassy ribbon after cathodic hydrogen charging in 0.1-M $H_2SO_4$ at low current density for 98 h. Cu layer peels off and does not contribute to the catalytic activity.

of the development of specific surface area and porosity. In contrast, it was still unclear what was the decisive factor increasing the catalytic activity of Cu-Ti, although a similar unique behaviour of the Cu-Ti system had already been observed.[87] Additional data acquired by surface analytical techniques are presented below to further characterize the working CuTi catalysts.

Figure 22 shows a typical morphology of Cu60-Ti40 amorphous alloy after hydrogen charging at $i = -1$ mA/cm$^2$ in 0.1-M $H_2SO_4$ for 96 h. Here the picture is completely different than that observed for the Cu-Hf amorphous alloy,[87,89] where a typical Cu catalyst on a $HfO_2$ support was formed during hydrogen charging. A rather flat Cu layer (Cu segregated at the surface during cathodic hydrogen charging) covers part of the surface. Upon cathodic charging hydrogen penetrates into bulk resulting in the formation of $TiH_2$. Since the hydride has a higher volume swelling takes place inducing phases separation and crystallization and the migration of Cu to the surface.

This peculiar type of Cu segregation usually resulted in a partial peeling of the Cu surface layer, thus exposing the remaining oxide support underneath. As a major part of the Cu layer peeled off the surface, it did not participate in the catalytic process.

A closer look at both areas after tilting of the sample revealed a surprising picture. The Cu surface layer (Fig. 23) is all cracked

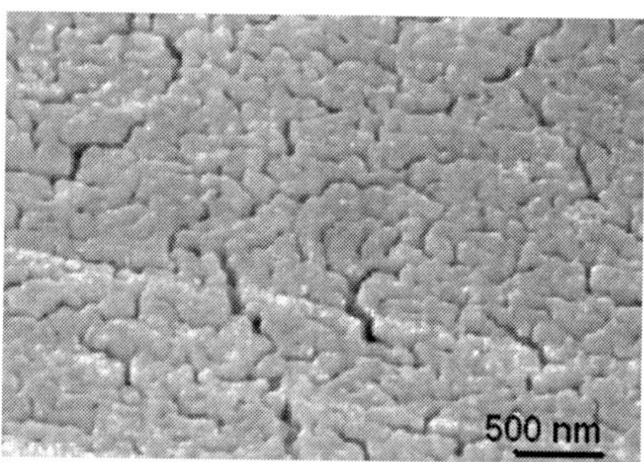


Figure 23. High resolution SEM image of Cu layer after cathodic hydrogen charging at $i = -1$ mA/cm$^2$ in 0.1-M H$_2$SO$_4$ during 98 h. The image demonstrates details of a typical morphology with small cracks developed in the process of hydrogen charging.

and contains micro-voids, which may be easy paths for hydrogen penetration during continuation of the cathodic pretreatment or during the catalytic test reaction. The substrate image (Fig. 24) reveals small, highly porous clusters on the surface, with diameters of 20–100 nm. It is, however, obvious that still smaller ones, below the resolution limit of the Microlab 350, are also present.

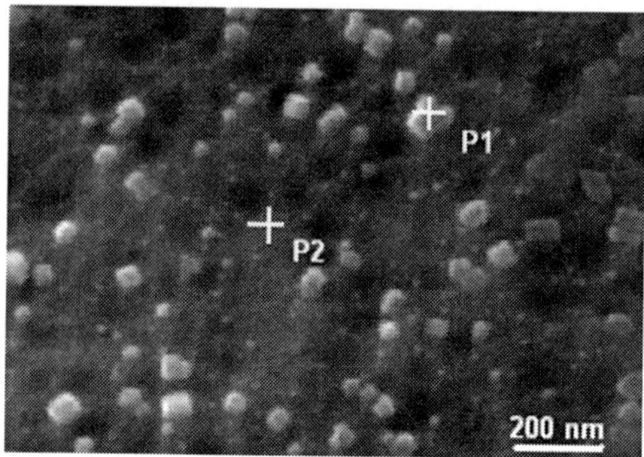


Figure 24. High resolution SEM image of typical substrate beneath a Cu layer after cathodic hydrogen charging at $i = -1$ mA/cm$^2$ in 0.1 M H$_2$SO$_4$ during 96 h. The picture shows very small metal clusters on the oxide support. The points of Auger local chemical analysis: P1 – cluster, P2 - substrate are marked on the SEM image.

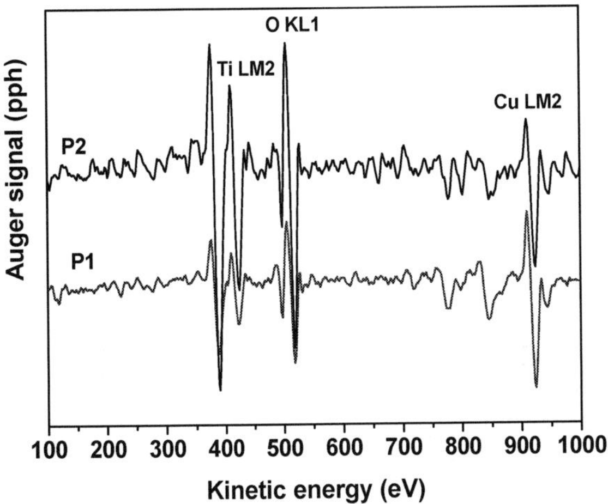


Figure 25. Local differential Auger spectra recorded on a cluster (P1) and on the porous substrate (P2) after sputtering for 10s, see Fig. 24 .

Local Auger spectra presented below (Fig. 25), taken at a cluster (P1) and at the substrate (P2), of the SEM image (Fig. 24) reveal the following typical features:

- the cluster consists of Cu, O and some amount of Ti; the corresponding peak to peak heights ratios in the differential spectrum are Cu(LM2)/Ti(LM2) = 2.5, O(KL1)/Cu(LM2) + Ti(LM2) = 0.6 for P1; for the other clusters the corresponding values varied from 2.3 to 6.7 for pph ratio Cu(LM2)/ Ti(LM2);

- the background consists of Ti and O and a small amount of Cu; the corresponding peak to peak height ratios in the differentiated spectrum are Cu(LM2)/Ti(LM2) = 0.7, O(KL1)/ Cu(LM2) + Ti(LM2) = 0.9 for P2; for the other locations the corresponding values varied from 0.5 to 0.8 for pph ratio Cu(LM2)/Ti(LM2).

These results illustrate the chemical inhomogeneity of the catalysts surface, which apparently facilitates its selectivity for acetone in the catalytic reaction.

Thus, it is clear that all three components (Cu, Ti, O) are present in different proportions both on the surface and in the clusters. The latter are enriched in Cu (to different degrees depending upon the cluster),[89] which certainly is important for their catalytic efficiency. Moreover, the substrate is more oxidized than the clusters.

## 3. Modified Mechanism of Dehydrogenation of Aliphatic Alcohol on Cu-Ti Catalysts

The results of our investigations have shown that the mechanism of dehydrogenation of aliphatic alcohols proposed for Cu powder, and apparently valid for modified Cu-Hf amorphous alloy precursor, does not apply to Cu-Ti catalysts. Thus we propose its modification, as follows:

While $Cu^{\delta+}$ and $O^{\delta-}$ centres play their role according to the previously proposed mechanism (Scheme 1, steps 1 and 2), the identification of the actual surface species in step 3 to split off a second hydrogen atom needs further consideration. In fact, the actual surface site responsible for the removal of this hydrogen was not discussed in earlier studies. The only exception is a work comparing the characteristics of Cu and ZnO in the dehydrogenation of ethanol.[118] Here, a Cu site is suggested to participate in the final step as depicted in Scheme 1.

The unique behaviour of Cu-Ti suggests that ionic Ti centres may play a role in the final step of dehydrogenation. The complexity of the actual catalyst surface is a strong indication that a mechanism similar to that occurring on oxides requiring the close proximity of all three components may be operative here. The main feature of this mechanism is the removal of the β hydrogen by a metal ion centre (Scheme 2), which is similar to an enolic-type mechanism.[107,118] Then, the final product is formed as a result of a simultaneous hydride transfer as suggested in Ref. 85, 89, 111 and 118.

Alternatively, α-hydrogen removal by a surface Ti site may also be possible (Scheme 3). Both suggestions are based on the known tendency of Ti to form stable hydrides readily. The third mechanism, however, seems less probable than the second for the reason discussed elsewhere[107,111,119] for oxide-type catalysts:

- Detrimental processes of material degradation by hydrogen embrittlement occurring during cathodic hydrogen charging appeared to be a useful method of functionalization of Cu-based amorphous alloys precursors for catalytic applications.

- The only successful method to transform Cu-Ti into an active catalyst is cathodic hydrogen charging. This is the only procedure found so far which can induce the changes in structure, composition and morphology necessary for effective performance as a catalyst.

- The electrochemical process of cathodic hydrogen charging proved to be an efficient method of modifying so far catalytically inactive Cu-Ti into efficient (average conversion attained ~ 60%) and selective (99%) catalysts for the dehydrogenation of aliphatic alcohols.

- Combined high resolution microscopic and local surface analytical investigations of Cu-Ti catalysts modified by cathodic hydrogen charging provided a new insight into the catalytic mechanism of the dehydrogenation of 2-propanol. Accordingly, a modification of the previously suggested mechanism was proposed.

- High resolution SAM investigation show that modified Cu-Ti catalysts do not consist of two well-separated phases. Highly porous, Cu-Ti-O particles (enriched in Cu) on a Ti-O-Cu support exist there. A modification of the previously suggested catalytic reaction mechanism was thus required. The proposed modification involves $Cu^{\delta+}$, $O^{\delta-}$ and $Ti^{\delta+}$ entities as possible active centres. The above results provide an insight into a possible engagement of the catalyst's support itself in the catalytic reaction.

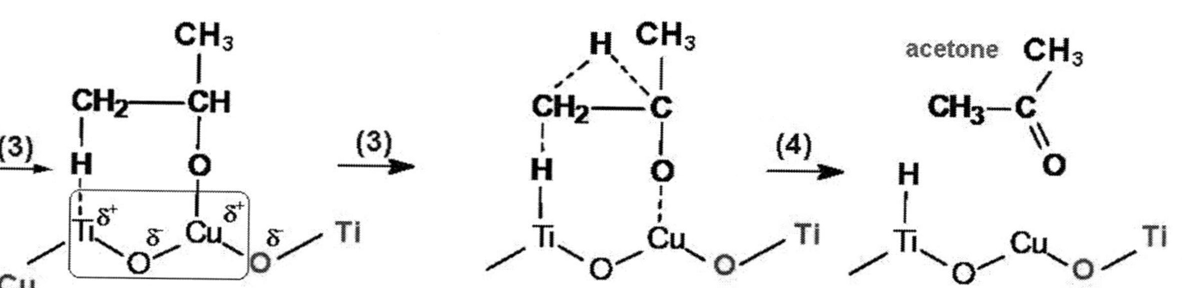

Scheme 2. *Enolic-type* of mechanism suggested for dehydrogenation of 2-propanol on Cu-Ti catalyst. Reprinted from M. Pisarek, M. Janik-Czachor, A. Molnar, B. Rac "Cathodic hydrogen charging as a tool to activate Cu-Ti amorphous alloy catalysts", *Electrochem.Acta*, **50**, 5111, Copyright (2005) with permission from Elsevier.

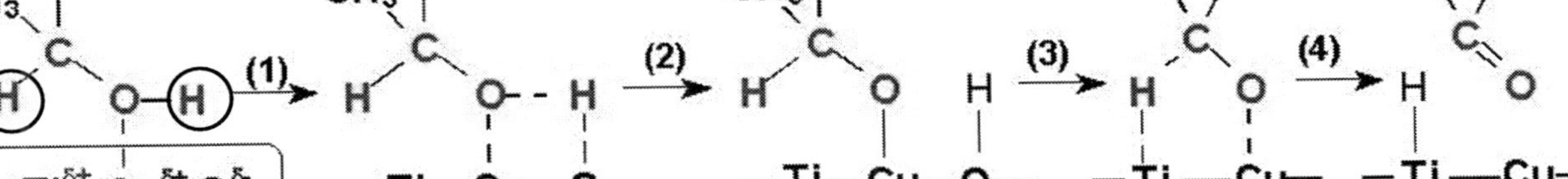

Scheme 3. *Carbonyl* type of mechanism conceivable for dehydrogenation of 2-propanol on Cu-Ti catalyst. Reprinted from M. Pisarek, M. Janik-Czachor, A. Molnar, B. Rac "Cathodic hydrogen charging as a tool to activate Cu-Ti amorphous alloy catalysts", *Electrochem. Acta*, **50**, 5111, Copyright © 2005 with permission from Elsevier.

## V.  SUMMARY AND CONCLUSIONS

In this chapter we have shown on three examples that a combination of electrochemistry, microscopy, and surface analysis helps to identify the factors responsible for a specific function of the modified amorphous or nanocrystalline alloys. Figure 26 illustrates the information depth and lateral resolution provided by different experimental techniques (and by different instruments) used in the investigations of metastable materials discussed above. While light microscopy and SEM are useful mainly in examining morphological details at the surface, AES and SAM provide important chemical information confined to a few top atomic monolayers. X-ray microanalysis, RBS and XRD pattern in fact provide bulk information, but for very thin samples they are a source of surface information. Electrochemistry, on the other hand, provides average information on the rate of surface reactions. Depending upon the goal, a combination of these methods is useful for materials functionalization and/or for characterizing the topography and chemistry of their surfaces.

Analysis of the existing data suggests that the changes occurring in the metastable materials in contact with an electrolyte or with air – leading to their functionalization – may extend deeper, and laterally wider, than those one would expect for crystalline materials. Therefore, metastable materials, including amorphous and nanocrystalline ones, are suitable as model systems and provide good targets for functionalization. Their potential in these areas is still not sufficiently recognized and exploited. One has to consider, however, that the local and in-depth effects are confined to only nanometers or less. Therefore, for future progress in materials science, electrochemical and high resolution surface analytical and microscopic investigations are both crucial and promising.


## ACKNOWLEDGMENTS

The authors are grateful to Prof. Koji Hashimoto, Prof. Andrzej Szummer and Prof. Arpad Molnar for their inspiration to undertake research in the area of *chemistry for materials science*. Thanks are also due to all the co-authors who contributed their

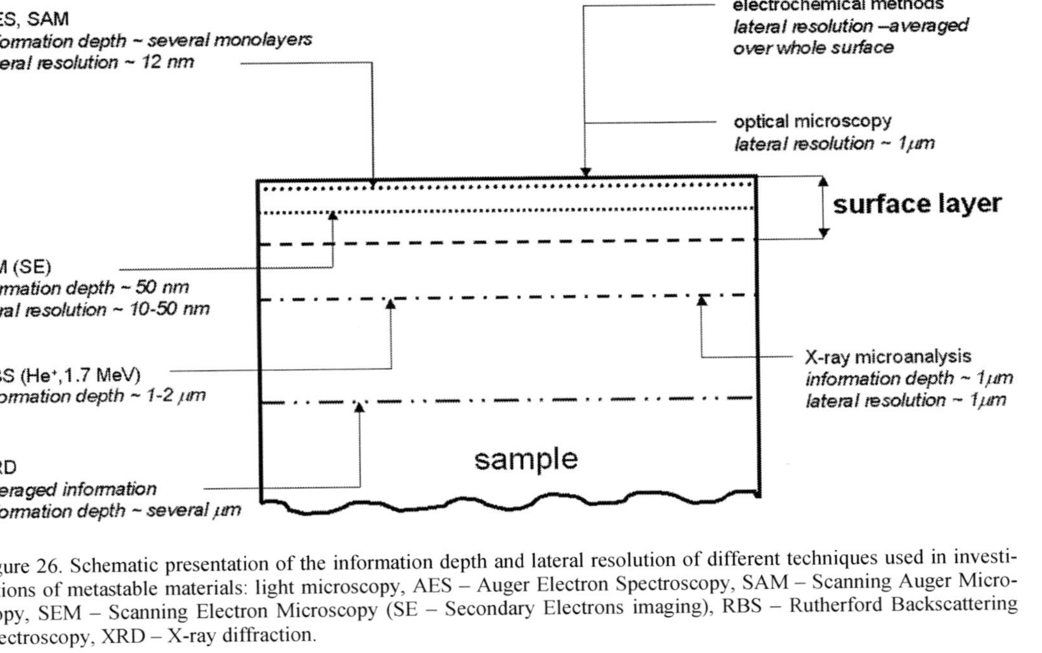


Figure 26. Schematic presentation of the information depth and lateral resolution of different techniques used in investigations of metastable materials: light microscopy, AES – Auger Electron Spectroscopy, SAM – Scanning Auger Microscopy, SEM – Scanning Electron Microscopy (SE – Secondary Electrons imaging), RBS – Rutherford Backscattering Spectroscopy, XRD – X-ray diffraction.

knowledge and experience to the joint papers quoted in this chapter.

Surface characterizations were performed using a Microlab 350 located at the Physical Chemistry of Materials Center of the Institute of Physical Chemistry, PAS and of the Faculty of Materials Science and Engineering, WUT.

# REFERENCES


[1] H.J. Engell and N.D. Stolica, *Z. Phys. Chem. N.F.*, **20** (1959) 113.

[2] H.J. Engell and G. Herbsleb, *Z. Phys. Chem. N.F.*, **215** (1960) 167.

[3] H.Kaesche, *Z. Phys. Chem. N.F.*, **34** (1962) 87.

[4] M. Janik-Czachor, Z. Szklarska-Smialowska, *Brit. Corr. J.*, **4** (1969) 138.

[5] K. J. Vetter and H. H. Strehblow, *Localized Corrosion* NACE 1974, p.240.

[6] A. Wolowik, thesis, *Polish Academy of Sciences*, Warsaw 2001.

[7] Ph. Marcus, *Corr. Sci.*, **36** (1988) 2155.

[8] H. Habazaki, M. A.P eaz, K. Shimizu, P. Skeldon, G. E. Thompson, G. C. Wood, and X. Zhou, *Corr. Sci.*, **38** (1999) 2053.

[9] X. Zhou, G.E. Thompson, P. Skeldon, G.C. Wood, K. Shimizu, H. Habazaki; *Corr. Sci.*, **41** (1999) 1599.

[10] H. Yosioka, H. Habazaki, A. Kawashima, K. Asami, and K. Hashimoto,; *Corr. Sci.*, **32** (1991) 313 and 327.

[11] X. Zhou, H.Habazaki, G. E.Thompson, and G. C. Wood, *Corr. Sci.*, **38** (1996) 1563 and X. Zhou, G.E. Thompson, H. Habazaki, K. Shimizu, P. Skeldon, G.C. Wood, *Thin Solid Films*, **293** (1997) 327.

[12] M. Janik-Czachor, A. Wolowik, and Z. Werner, *Corr. Sci.*, **36** (1994) 1921.

[13] A. Wolowik, M. Janik-Czachor, and A. Szummer; Proc. Symp., Passivity and its Breakdown (Eds. P.M. Natishan, H.S. Isaacs, M. Janik-Czachor, V.A. Macagano, P. Marcus, M. Seo), The Electrochemical Society, NJ, Vol. **97-26** (1998) 533.

[14] A. Wolowik and M. Janik-Czachor; *Mat. Sci. Eng. A*, **267** (1999) 301.

[15] E. Sikora, B.A. Shaw, and T. Miller, Proc. Symp. *Passivity and its Breakdown* (Eds. P.M. Natishan, H.S. Isaacs, M. Janik-Czachor, V.A. Macagano, P. Marcus, M. Seo), The Electrochemical Society, NJ, Vol. **97-26** (1998) 654.

[16] B. Shaw, W.C. Moshier, R. Went, P. Miller, and E. Principe; 5th International Symposium on Electrochemical Methods in Corrosion Research, EMCR'94, Lisbon, Portugal, 1994.

[17] M. Janik-Czachor, A. Jaskiewicz, P. Kedzierzawski, and Z. Werner, *Mat. Sci. Eng. A;* **358** (2003) 171.

[18] G.S. Frankel, R.C. Newman, C.V. Johnes, and A. Russak, *J.Electrochem. Soc.*, **140** (1993) 2192.

[19] M. Janik-Czachor, *J. Electrochem. Soc.*, **128** (1981) 513C.

[20] H. Habazaki, P. Skeldon, G. E. Thompson, and G. C. Wood; *Phil. Mag. B*, **71** (1995) 81 and H. Habazaki, K. Shimizu, P. Skeldon, G. E. Thompson, G. C. Wood, and X. Zhou, *Corr. Sci.*, **37** (1997) 731.

[21] R. Kirchheim, B. Heine, S. Hofmann, and H. Knote; *Corr. Sci.*, **31** (1990) 191; *Corr. Sci.* **29** (1989) 889.

[22] K. Hashimoto, P.-Y. Park, J.-H. Kim, H. Yoshioka, H. Mitsui, E. Akiyama, H. Habazaki, A. Kawashima, K. Asami, Z. Grzesik, and S. Mrowec, *Mat. Sci. Eneg. A* **198** (1995) 1.

[23] H. Kaesche, *Korrosion der Metalle*, Springer Verlag, Berlin, 1976, p. 264.

[24] J. B. Bessone, D. R. Salians, C. E. Mayer, M. Ebert, and J. W. Lorenz, *Electrochim. Acta*, **37** (1992) 2283.

[25] H. Bohni and H.H. Uhlig, *Corros.Sci.*, **9** (1969) 329.

[26] F. Mansfeld, Y. Wang, and H. Shih, *Electrochim.Acta*, **73** (1992) 2277.

[27] Z. Werner, A. Jaskiewicz, M. Pisarek, M. Janik-Czachor, M. Barlak, and *Z. Phys. Chem.*, **219** (2005) 1461-1479.

[28] A. Jaśkiewicz, *Stability of the passive state and the composition of anodic oxide films on Al-Ta and Al-Nb amorphous alloys*, thesis, Warsaw University of Technology, 2007.

[29] T. Valand and K. E. Heusler, *J. Electroanal. Chem.*, **149** (1983) 71.

[30] H. Habazaki, P. Skeldon, G. E. Thompson, and G. C. Wood, *Philosophical Mag. B*, **71** (1995) 1497.

[31] H. Habazaki, K. Shimizu, G. E. Thompson, G.C. Wood, and X. Zhou, *Corros. Sci.*, **39** (1997) 731.

[32] A. Crossland, G. E. Thompson, J. Wan, H. Habazaki, K. Shimizu, P. Skeldon, and G.C. Wood, *J.Electrochem.Soc.*, **144** (1997) 847.

[33] M. Janik-Czachor, *Corr. Sci.*, **33** (1992) 1327.

[34] H. Habazaki, K. Schimizu, P. Skeldon, G. E. Thompson, and G. C. Wood, *Proc. R. Soc. A* **453** (1997) 15.

[35] G. Alcala, S. Mato, P. Skeldon, G. E. Thompson, P. Bailey, T. C. Q. Noakes, H. Habazaki, and K. Shimizu; *Corr. Sci.*, **45** (2003) 1803-1813.

[36] M. Janik-Czachor, A. Jaśkiewicz, P. Kędzierzawski, and Z. Werner, *Mater. Sci. Eng. A*, **358** (2003) 171.

[37] Avantage Data System version 3.83 - Thermo Scientific software.

[38] L. C. Feldman and J. W. Mayer; Fundamentals of surface and thin film analysis, North-Holland, New York – Amsterdam – London, 1986.

[39] R. Z. Valiev and I. V. Alexandrov, *Nanostructured Mater.* **12** (1999) 35.

[40] M. Furukawa, Z. Horita, M. Nemoto, and T. G. Langdon, *Mater. Sci. Eng.* A (2002) 82.

[41] M. Suś-Ryszkowska, T. Wejrzanowski, Z. Pakieła, and K. J. Kurzydłowski, *Mat. Sci. Eng. A* **369** (2004)151.

[42] K.J. Kurzydłowski, *Bulletin of the Polish Academy of Sciences, Technical Sciences* **52 (4)** (2004)301.

[43] J-Ch. Hung and Ch. Hung, *J. Mater. Process. Tech.* **104** (2000) 226.

[44] M. Lewandowska, *J. Microsc.* **224** (2006) 34.

[45] H. Garbacz, M. Lewandowska, W. Pachla, and K.J. Kurzydłowski, *J. Microsc.* **223** (2006)272.

[46] D. Klassek, T. Suter, P. Schmutz, W. Pachla, M. Lewandowska, K. J. Kurzydłowski, and O. von Trzebiatowski, *Solid State Phenom* **114** (2006)189.

[47] Z. Zhang, E. Akiyama,Y. Watanabe, Y. Katada, and K. Tsuzaki, *Corros. Sci.* **49** (2007) 2962.

[48] K. Krasilnikov, W. Lojkowski, Z. Pakiela, and R. Valiev, *Mat. Sci. Eng.* **397** (2005)330.

[49] *Stainless Steels and Ferrous Alloys*, in: R.P. Frankenthal and J. Kruger, editors. Passivity of Metals. New Jersey: The Electrochemical Society, INC., Princeton, 1978, p. 646-770.

[50] N. Sato and K. Hashimoto, editors. Passivation of Metals and Semiconductors, Oxford, Pergamon Press, 1990. The Sixth International Symposium on Passivity, Sapporo, Japan, 1989.

[51] K.E. Heusler, editor. Passivation of Metals and Semiconductors, Special issue of Mater. Sci. Forum 185–188, (1995): p. 221–364.

[52] P. M.Natishan, H.S. Isaacs, M. Janik-Czachor, V.A. Macagano, P. Marcus, and M. Seo, editors, *Passivity and its Breakdown*, The Electrochemical Society, NJ, Pennington, 1998.

[53] L.M. Bastidas, C.L. Torres, E. Cano, and J.L. Polo, *Corros. Sci.* **44** (2002) 625.

[54] Ch.Ch. Shih, Ch.-M. Shih, Y.-Y. Su, L.H.J. Su, M.-S. Chang, and S.-J. Lin, *Corros. Sci.* **46** (2004) 427.

[55] K. Hashimoto and K. Asami, Factors determining corrosion resistance of chromium-bearing alloys, in: R.P.Frankenthal and J.Kruger, editors, *Passivity of Metals*, New Jersey, Princeton, The Electrochemical Society, INC., 1978, p.749.

[56] M. Janik-Czachor, *J. Electrochem. Soc.* **513C**-519C (1981).

[57] M. Janik-Czachor, G.C. Wood, G.E. Thompson., *Brit. Cor. J.* **15** (1980) 154.

[58] H.J. Engell and R.A. Oriani, *Corros. Sci.* **29** (1989) 119.

[59] J.B. Lumsden and R.W. Staehle, *Scripta Mat.* **6** (1972) 1205.

[60] M. Pisarek, P. Kędzierzawski, T. Płociński, M. Janik-Czachor, and K.J. Kurzydłowski, *Mater.Charac. 59 (2008) 1292.*

[61] S. Yang and D.D. Mcdonald, *Electrochim. Acta* **52** (2007) 1871.

[62] M. Pisarek, P. Kędzierzawski, M. Janik-Czachor, and K. J. Kurzydłowski, *Corrosion NACE* **64 (2)** (2008) 131.

[63] M. Pisarek, P. Kędzierzawski, M. Janik-Czachor, and K.J. Kurzydłowski, *Electrochem. Commun.* **9** (2007) 2463.

[64] M. Pisarek, P. Kędzierzawski, M. Janik-Czachor, and K. J. Kurzydłowski, *J. Solid State Electrochem.* **13** (2009) 283.

[65] G. S. Eklund, *J. Electrochem. Soc.* **121** (1974) 121.

[66] J. Scotto, G. Ventura, and E. Traverso, *Corros. Sci.* **19** (1979) 237.

[67] B. Baroux and G. Gorse, The effect of pH and potentiostatic polarisation on the pitting resistance of stainless steels: relation to non-metallic inclusions or passive film modifications. Proc. European Symp. on Modification of Passive Films, EFC-CPS. London, Institute of Materials 1994; 12:300.

[68] E.F.M. Jansen, W.G. Sloof, and J.H. W. de Wit, Inclusions in stainless steels – their role in pitting initiation. Proc. European Symp. on Modification of Passive Films, EFC-CPS. London, Institute of Materials 1994; 12:290.

[69] M. Janik-Czachor, A. Szummer, *Corrosion Review*, M. Schorr, editor, **11** (1993)118.

[70] A. Rossi, B. Elsner, G. Hahner, M. Textor, and N.D. Spencer, *Surf. Interface Anal.*, **29** (2000) 460-467.

[71] M. Janik-Czachor, *Mater. Sci. Forum* **1** (1995) 185-188.

[72] P. Schmuki, H. Hildebrand, A. Freidrich, and S. Virtanen, *Corros. Sci.* **47** (2005) 1239-1250.

[73] A. Szummer, M. Janik-Czachor, and S. Hofmann, Proc European Symp on Modification of Passive Films, EFC-CPS. London, Institute of Materials, **12** (1994) 280-288.

[74] G. Okamoto, T. Shibata, Passivity and the breakdown of passivity of stainless steel, in R. P. Frankenthal, and J. Kruger, editors, *Passivity of Metals*, New Jersey, Princeton, The Electrochemical Society, INC., 1978; p. 646.

[75] H. H. Kung, *Catal. Rev. Sci. Eng.* **22** (1980) 235.

[76] K. Klier, *Adv. Catal.* **31** (1982) 243.

[77] J. C. J. Bart and R. P. A Sneeden, *Catal. Today* **2** (1987) 1.

[78] A. Baiker, in: G. Ertl, H. Knozinger, and J. Weitkamp (Eds.), Handbook of Heterogeneous Catalysis, vol. 2, Wiley-VCH, Weinheim, 1997, p.803.

[79] A. Molnar, G.V. Smith, and M. Bartok, *Adv. Catal.* **36** (1989) 329.

[80] K. Hashimoto, *Mater. Sci. Eng.* **A226–228** (1999) 891.

[81] A. Baiker, R. Schlogl, E. Armbruster, and H.-J. Guntherodt, *J. Catal.* **107** (1987) 221.

[82] K. Hashimoto, H. Habazaki, M. Yamasaki, S. Meguro, T. Sasaki, H. Katagiri, T. Matsui, K. Fujimura, K. Izumiya, N. Kumagai, and E. Akiyama, *Mater. Sci. Eng. A* **304–306** (1999) 88.

[83] A. Szummer, M. Pisarek, M. Dolata, A. Molnar, M. Janik-Czachor, M. Varga, and K. Sikorski, *Mater. Sci. Forum* **377** (2000) 15.

[84] M. Janik-Czachor, A. Szummer, J. Bukowska, A. Molnar, P. Mack, S.M. Filipek, P. Kedzierzawski, A. Kudelski, M. Pisarek, M. Dolata, and M. Varga, *Appl. Catal. A* **235** (2002) 157.

[85] M. Pisarek and M. Janik-Czachor, *Microsc.Microanal.*, **12** (2006) 228-237.

[86] G. Ertl and J. Kueppers, *Low Energy Electrons and Surface Chemistry*, pp. 17–64, Weinheim: VCH (1985).

[87] M. Pisarek, M. Janik-Czachor, P. Kedzierzawski, A. Molnar, B. Rac, and A. Szummer, *Pol. J. Chem.* **78** (2004) 1379.

[88] M. Pisarek, M. Janik-Czachor, A. Gebert, A. Molnar, P. Kedzierzawski, and B. Rac, *Appl. Catal.* **A267** (2004) 1.

[89] M. Pisarek, M. Janik-Czachor, and A. Molnar, B. Rac, *Electrochim. Acta*, **50** (2005) 5111.

[90] A. Molnar, L. Domokos, T. Martinek, T. Katona, G. Mulas, G. Cocco, I. Bertoti, and J. Szepvolgyi, *Mater. Sci. Eng. A* **226–228** (1997) 1074.

[91] R. J. Madix, *Adv. Catal.* **29** (1980) 1.

[92] J. Cunningham, G.H Sayyed, J.A. Cronin, J.L.G. Fierro, C. Healy, W. Hirschwald, M. Ilyas, and J. P. Tobin, *J. Catal.* **102** (1986) 160.

[93] F. Pepe, R. Polini, and L. Stoppa, *Catal. Lett.* **14** (1992) 15.

[94] Y. Han, J. Shen, and Y. Chen, *Appl. Catal.* **A205** (2001) 79.

[95] R.M. Rioux and M.A. Vannice, *J. Catal.* **216** (2003) 362.

[96] A. Guerrero-Ruiz, I. Rodrigez-Ramos, and J. L. G. Fierro, *Appl. Catal.* **72** (1991) 119.

[97] N. Kanoun, M. P. Astier, and G.M. Pajonk, *J. Mol. Catal.* **79** (1993)217.

[98] A.J. Marchi, J. L .G. Fierro, J. Santamaria, and A. Monzon, *Appl. Catal.* **A142** (1996) 375.

[99] A. Aboukais, R. Bechara, C. F. Aissl, J. P. Bonnelle, A. Ouqour, M. Loukah, G. Coudurier, and J. C. Vedrine, *J. Chem. Soc., Faraday Trans.* **89** (1993) 2545.

[100] I. E. Wachs, R. J. Madix, *J. Catal.* **53** (1978) 208.

[101] I. E. Wachs, R. J. Madix, *Appl. Surf. Sci.* **1** (1978) 303.

[102] M. A. Chester, E. M. Mccash, *Spectrochim. Acta* **43A** (1987) 1625.

[103] B. A. Sexton, *Surf. Sci.* **88** (1979) 299.

[104] W. R. Patterson, J. A. Roth, and R. L. Burwell, *J. Am. Chem. Soc.* **93** (1971) 839.

[105] D. A. Chen and C. M. Friend, *Langmuir* **14** (1998) 1451.

[106] M. K. Weldon and C. M. Friend, Chem. Rev. **96** (1996) 1391.

[107] M. Kraus, G. in, H. Ertl, J. Weitkamp Knozinger editors, Handbook of Heterogeneous Catalysis, vol. 5, Wiley-VCH, Weinheim, 1997, p. 2159.

[108] L. Nondek and J. Sedlacek, *J. Catal.* **40** (1975) 34.

[109] N.Ismail, M.Uhlemann, A. Gebert and J.Eckert, *J. Alloy Compd.* **298** (2000) 146.

[110] N.Ismail, A.Gebert, M. Uhlemann, J. Eckert, and L. Schultz, *J. Alloy Compd.* **314** (2001) 170.

[111] M. Pisarek, M. Janik-Czachor, A. Molnar, and K. Hughes, *Appl. Catal. A* **283** *(2005) 177.*

[112] A. Molnar, I. Bartok, J. Szepvolgyi, G. Mulas, and G. Cocco, *J. Phys. Chem.* B**102** (1998) 9258.

[113] S. Majorowski and B. Baranowski, *J. Phys. Chem. Solids* **43** (1982) 1119.

[114] E. Fromm and E. Gebhardt, Gase und Kohlenstoff in Metallen. pp. 406–416. Berlin, Heidelberg, New York: Springer-Verlag (1976).

[115] S. Mroz, *Acta Phys. Pol.*, **89** (1996) 183–194.

[116] D. Briggs and J. T Grant, Surface Analysis by Auger and X-Ray Photoelectron Spectroscopy, Chichester, UK: IM Publications (2005).

[117] F. Reniers and C. Tewell, *J Electron Spectrosc,* **142** (2005) 1-25.

[118] M.-J. Chung, S.-H. Han, K.-Y. Park, and S.-K. Ihm, *J. Mol. Catal.* **79** (1993) 335.

[119] M. Pisarek, *Chemical methods of surface modification of Cu-based amorphous alloys for catalytic applications*, thesis, Warsaw University of Technology, Warsaw, 2004.

*4*

# Physicochemical Characterization of Passive Films and Corrosion Layers by Differential Admittance and Photocurrent Spectroscopy


Francesco Di Quarto, Fabio La Mantia, and Monica Santamaria

*Dipartimento di Ingegneria Chimica dei Processi e dei Materiali, Universita' degli studi di Palermo, Viale delle Scienze, 90128 Palermo, Italy*


## I.   INTRODUCTION

The stabilization of metallic surfaces against corrosion processes in natural and industrial environment rests on the onset of *passivity* condition with a subsequent drastic reduction of the corrosion rate of the underlying metallic substrate. In spite of a longstanding controversy it is now universally accepted that a passive metal is usually covered by a thin or thick external layer the whose physicochemical properties control the evolution of corrosion process as well as the possible reactions occurring at metal/oxide and oxide/electrolyte interfaces.[1,2] In many cases of practical importance, passivity of metals is reached in presence of very thin (few nm thick) layer which makes the complete physico-chemical characterization a very complex task requiring the use of different


*Modern Aspects of Electrochemistry,* Number 46, edited by S. –I. Pyun and J. -W. Lee, Springer, New York, 2009.



S.-I. Pyun and J.-W. Lee (eds.), *Modern Aspects of Electrochemistry 46:*          231
*Progress in Corrosion Science and Engineering I,*
DOI 10.1007/978-0-387-92263-8_4, © Springer Science+Business Media, LLC 2009

powerful in situ and/or ex situ techniques. This is particularly true if we want to get information pertaining to the chemical composition, morphology, crystalline or disordered nature and solid-state properties of the passive layers.

Many useful information on the film composition can be gathered from ex situ techniques (Auger, ESCA, XPS, SIMS, RBS, GDOES) although they suffer some limitations and drawback specially when investigating very thin films, owing to the risk of changing structure and composition of the passive film on going from the potentiostatic control in solution to the vacuum. According to this a large agreement exists on the advantages of passive film characterization based on in situ techniques or controlled transfer under inert atmosphere on going from electrochemical cell to analytical equipment.[3–11]

Besides more traditional (and mainly optical) techniques (differential and potential modulated reflectance, ellipsometry, interferometry, Fourier transform infrared spectroscopy, Raman and Mössbauer spectroscopy), new in situ techniques have been introduced in the last years, which are capable of providing useful information on structure, composition, morphology and thickness of passive films, such as: EXAFS, XANES, EQCM, STM and AFM. The use of these analytical techniques has improved our understanding of the structure and composition of passive films grown on metal and alloys.[1–14]

As for the study of solid state and electronic properties of passive films and corrosion layers differential admittance (DA) and photocurrent spectroscopy (PCS) techniques have been largely used by electrochemists and corrosion scientists during the years. Both techniques take advantage from the pioneering works of Garrett and Brattain in the mid 1950s and from the advent of Gerischer's theory and its systematic use for interpreting the kinetics of electron and ion transfer reactions (ETR and ITR) at the semiconductor/electrolyte interface.[15–25] We have to mention that both ITR and ETR are involved in determining the kinetics of growth and the breakdown processes of passive films.[1,2,26–33] In many cases the combined use of both DA and PCS techniques is able to provide the information necessary to locate the characteristic energy level (flat band potential, $U_{fb}$, conduction and valence band edge $E_C$, $E_V$) of the passive film/electrolyte junction.

The location of characteristics energy levels of the metal/passive film and passive film/electrolyte junctions is a preliminary task for understanding the kinetics of ion and electron transfer reactions at passive film electrolyte-interface. In fact, both the ITR and ETR are controlled by the:

- energetics of the metal/film and film/electrolyte interfaces
- electronic properties of the passive film.

In the case of differential admittance technique apart the seminal papers of Mott[34] and Schottky[35] on the rectifying properties of metal/semiconductor contacts (copper/cuprous oxide) there is no doubt that the classical papers of Dewald on the ZnO/electrolyte interface,[36] showing the validity of Mott-Schottky analysis for the location of characteristic levels of the semiconductor/electrolyte junction, opened up a new route for a deeper understanding of the structure of SC/electrolyte interface.

If we take into account that in the case of passive films the extreme thinness and their disordered or amorphous structure adds further complications to the possible interpretative models of DA data, it is not surprising that in spite of the numerous studies on passive film/electrolyte interfaces an unambiguous picture or generalized acceptance of interpretative models is still lacking.

Together with differential admittance studies and among other optical methods, PCS has gained a large consideration in the last decades as in situ technique for the characterization of photoconductive passive film able to provide information not only on the location of characteristic energy level of passive film/electrolyte junction and internal photoemission threshold ($E_{th}$) at the metal/passive film interface, but also as a possible analytical tool for identifying the nature of passive film and corrosion layers.[37–40] The attractive features of PCS are due to the fact that it is a non-destructive technique based on the analysis of the electrochemical response (photocurrent or photopotential) of the passive electrode/electrolyte interface under irradiation with photons of suitable energy and intensity. Often the choice of a potentiostatic control is preferred, by taking into account the prominent role of the electrode potential in the establishment of electrochemical equilibrium involving different metal oxidation states and reactivity of the passive films.[41]

It is in the aims of this work to stress the limitations of a traditional approach, based on the theory of ideal crystalline semiconductor, to the study of semiconducting behavior of passive films and to provide, possibly, a more general interpretative frame for a deeper understanding of the semiconducting properties of passive films. At this aim after an initial introduction to the theory of Mott-Schottky barrier valid for ideal single crystal SC/electrolyte junction a more extended introduction to the theory of amorphous semiconductor Schottky barrier will be presented. In this frame some inconsistencies usually encountered in literature of thin passive films will be discussed and possible alternative explanations suggested in agreement with the more realistic model of the solid state properties in disordered and amorphous materials.

In the second part of the work a short introduction to the optical properties of crystalline semiconductor/electrolyte interface will be presented as preliminary to a more extended discussion on the influence of lattice disorder on the optical properties of semiconductor and insulating materials. New features related to the extremely thin thickness of passive films will be presented and discussed. Finally the use of PCS as an analytical tool for identifying the possible nature of corrosion and passive film will be discussed on the basis of a semi-empirical correlation between optical band gap of oxides and composition of passive films grown on metal and alloys.

## II.   IMPEDANCE MEASUREMETS FOR PASSIVE FILMS

### 1.   Semiconductor/Electrolyte Interface

With very few exceptions, most of passive layers grown on metals and alloys behave like semiconducting or insulating materials. Accordingly, widespread use of the theory of crystalline semiconductors has been made to discuss the behavior of passive film/electrolyte junction. Such an extension is not always critically checked by taking into account whether the hypothesis underlying the theory of bulk crystalline materials are also valid in the case of the thin passive films investigated. Moreover, in spite of the exist-

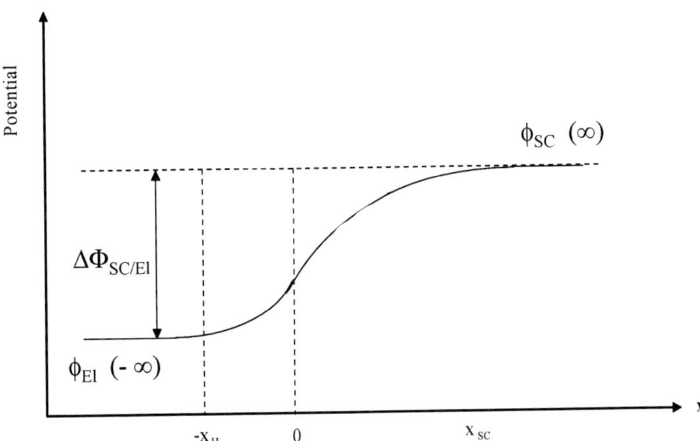


Figure 1. Galvani potential distribution across an n-type SC/El interface in absence of specific adsorption of ions in solution and surface states.

ing evidence that in many cases the passive films are amorphous or strongly disordered such an information is not usually taken into account to explain the admittance behavior of the junction neither to modify the traditional approach to the study of the interface by using the more pertinent theory of amorphous semiconductor Schottky barrier. For a better understanding of the influence of amorphous nature on the differential admittance and photocurrent spectroscopy measurements a brief introduction on the structure of semiconductor/electrolyte interface will be given to derive some equations usually employed to investigate the structure of such interface.

With respect to the metal/electrolyte interface in the case of SC/electrolyte interface the main differences stem out from the different electronic structure of the electrode.[16–25] In fact, differently than in a metal, owing to a much lower density of free carriers in the conduction band of semiconductor a much larger screening length in SC electrode is necessary to neutralize the excess of charge lying in the solution side.

In Fig. 1 we report the schematic Galvani potential distribution across an n-type SC/El interface in absence of specific absorption of ions in solution. According to Fig. 1 the total potential drop across the SC/El interface is given by:

$$\Delta\Phi_{SC/El} = \left\{\phi_{SC}(\infty) - \phi_{SC}(0)\right\} + \left\{\phi_{SC}(0) - \phi_{El}(-\infty)\right\} \quad (1)$$

where the first term in the bracket of Eq. (1) represents the Galvani potential drop from the bulk (zero electric field) to the surface of SC/El junction, whilst the term in the second bracket is the Galvani potential drop occurring into the compact and diffuse double layer (if any) of the electrolytic solution. In presence of concentrated electrolyte the diffuse double layer is missing and we can rearrange the previous equation as:

$$\Delta\Phi_{SC/El} = \Delta\Phi_{SC} + \Delta\Phi_H \quad (2)$$

where $\Delta\Phi_{SC}$ and $\Delta\Phi_H$ represent the potential drop into SC and the Helmholtz double layer respectively. In equilibrium conditions the potential drop inside the semiconductor can be calculated by solving the Poisson equation (see below) under the same conditions used for an ideal semiconductor/metal Schottky barrier and by taking into account that the potential drop within the semiconductor is only a part of the total potential difference measured with respect to a reference electrode. Moreover, by assuming that in presence of a sufficiently concentrated (> 0.1 M) electrolytic solution the potential drop in the diffuse double layer is negligible, the measured differential capacitance of the interface, $C_m$, in electrochemical equilibrium can be defined as:[36]

$$C_m = \left|\frac{\partial Q_{SC}}{\partial\Delta\Phi_{SC/El}}\right| = \left|\frac{\partial Q_{SC}}{\partial\Delta\Phi_{SC}}\frac{\partial\Delta\Phi_{SC}}{\partial\Delta\Phi_{SC/El}}\right| = C_{SC}\left(1 - \frac{C_m}{C_H}\right) \quad (3)$$

where $C_{SC}$ is the capacitance of the space charge region in the semiconductor, and $C_H$ is the capacitance of the Helmholtz compact double layer. In terms of equivalent electrical circuit the interface can be represented by two capacitors in series:

$$\frac{1}{C_m} = \frac{1}{C_{SC}} + \frac{1}{C_H} \quad (4)$$

The value of $C_{SC}$ changes with the width of space charge region within the semiconductor, $X_{SC}$, and it is a function of the potential drop, $\Delta\Phi_{SC}$, within the SC whilst a constant value (10–30 $\mu F\ cm^{-2}$) is usually assumed for $C_H$.

The region of SC necessary for screening the potential drop $\Delta\Phi_{SC}$ defines the space charge region. The width of the space-charge region in crystalline semiconductor (c-SC) changes with the potential drop according to the following equation:

$$X_{SC} = X_{SC}^0 \left( \left| \Delta\Phi_{SC} \right| - \frac{kT}{e} \right)^{0.5} \tag{5}$$

where $X^0{}_{SC}$ represents the space-charge width into the SC electrode at 1 V of band bending ($\Delta\Phi_{SC} = \psi_{SC} = 1$ V) and its value depends on the concentration of mobile carriers into the SC. In the hypothesis of completely ionized donors (n-type SC) or acceptor (p-type SC) the expression for such a characteristics length is given by:

$$X_{SC}^0 = \left( \frac{2\varepsilon_{ox}\varepsilon_0}{eN_{D/A}} \right)^{0.5} \tag{6}$$

where $N_{D/A}$ are the donor (or acceptor) concentration in $cm^{-3}$, $\varepsilon_{ox}$ and $\varepsilon_0$ are the SC dielectric constant and the vacuum permittivity, respectively. By assuming for the passive film dielectric constant the typical value of ~ 10 and a donor concentration of $10^{19}\ cm^{-3}$ we get for $X^0{}_{SC}$ a value around 100 Å $V^{-0.5}$.

In Figs. 2a and 2b we report the schematic diagram of an n-type SC/electrolyte junction in energy-distance coordinates at flat band conditions ($X_{SC} = 0$) and under slight depletion ($X_{SC} > 0$). $\Delta\Phi_{SC} = 0$ (no potential drop within the SC) corresponds to the special flat band condition reported in Fig. 2a whilst an anodic $\Delta\Phi_{SC} > 0$ (n-type SC) polarization corresponds to the conditions depicted in Fig. 2b. In the case of p-type material a space charge layer develops inside the SC for $\Delta\Phi_{SC} < 0$.

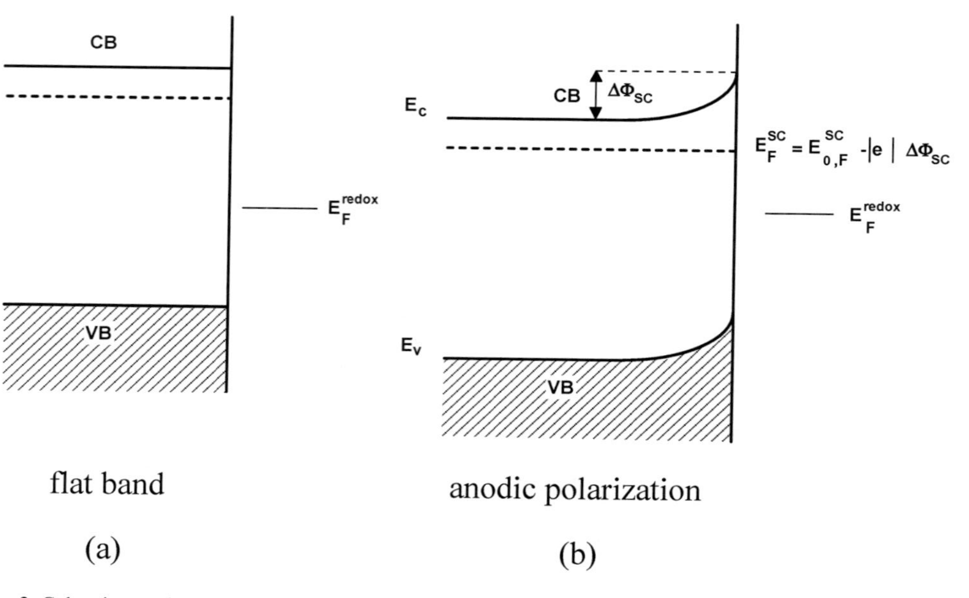



Figure 2. Galvani potential distribution across an n-type SC/El interface at flat band conditions (a) and under slight depletion (b).

In terms of electrode potential, $U_E$, for not heavily doped SC and in absence of an appreciable density of electronic surface states (SS), we can write:

$$\Delta\Phi_{SC} = U_E - U_{fb}(ref) \tag{7}$$

where $U_{fb}(ref)$ represents the flat band potential measured with respect to a reference electrode in the electrochemical scale.

### 2.  Location of Characteristic Energy Level in c-Semiconductor/Electrolyte Interface

The first task in the location of the energetics of a semiconductor/ electrolyte interface is to derive the flat band potential of the junction, corresponding to the condition of zero Galvani potential drop inside the SC. A common practice in electrochemistry is to get such a parameter by measuring the differential capacitance of the junction defined as above:

$$\frac{1}{C_m} \cong \frac{1}{C_{SC}} \tag{8}$$

by assuming $1/C_H \ll 1/C_{sc}$. In order to get the expression of the differential capacitance $C_{SC}$ it is first necessary to solve the Poisson-Boltzmann equation for the band-bending in the SC and then calculate the charges from Gauss's law.[36 and refs therein] Under the simplifying assumptions of:

- crystalline semiconductor electrode homogeneously doped under depletion regime;
- fully ionized single donor (or acceptor, for p-type SC) level;
- absence of deep lying donor (acceptor) levels in the forbidden gap of the SC;
- negligible contribution of surface states and minority carriers to the measured capacitance;
- absence of Faradaic processes at the SC/El interface;
- width of space-charge layer much lower than the semiconductor thickness;

it is possible to get a simplified expression for the space charge capacitance, which coincides with the well-known Mott-Schottky equation:

$$\left(\frac{1}{C_{SC}}\right)^2 = \left(\frac{2}{\varepsilon_{ox}\varepsilon_0|e|N_D}\right)\left(\Delta\Phi_{SC} - \frac{kT}{|e|}\right) \tag{8}$$

where $\Delta\Phi_{SC} = (U_E - U_{fb})$ and $N_D$ is the density of ionized donors in the n-type SC. The previous equation holds for a p-type material, with the minus sign in front of $\Delta\Phi_{sc}$ and $N_A$ instead of $N_D$. The extrapolation to zero of the first term in M-S equation provides the value of the intercept with voltage axis $U_0$, whilst from the slope of Eq. (8) it is possible to derive the concentration of donor (or acceptor) in the SC. A more general equation taking into account the degree of dissociation of donors (acceptors) has been reported[36] but rarely employed in the study of passive film.[28]

The assumption of neglecting the contribution of Helmholtz capacitance to $C_m$ can seriously affect the value of flat band potential. In fact we recall that the following relationship exists between the intersection of the M-S plot with potential axis, $U_0$, and the flat potential value:[42–44]

$$U_{fb} = U_0 - \frac{kT}{e} + \frac{\varepsilon_{ox}\varepsilon_0 eN_D}{2C_H^2} \qquad \text{for n-type SC} \tag{9a}$$

$$U_{fb} = U_0 + \frac{kT}{e} - \frac{\varepsilon_{ox}\varepsilon_0 eN_A}{2C_H^2} \qquad \text{for n-type SC} \tag{9b}$$

The required correction in the $U_{fb}$ value remains negligible as long as the dielectric constant value and donor density are small but it becomes important as $\varepsilon_{ox}$ or $N_D$ (or both) increase close to values as those sometimes reported for passive films (see below). The location of the Fermi level of the c-SC, in the electrochemical scale, is carried out by means of the equation:[39]

$$E_f^0(El) = -|e|U_{fb}(ref) \tag{10}$$

According to the theory at any other electrode potential the Fermi level of the SC, in the hypothesis of electronic equilibrium trough the space charge region, will be located by assuming $E_f = E^0_f - |e|\Delta\Phi_{sc}$.

The location of the remaining energy levels of the junction is performed by means of the usual relationships for n- and p-type semiconductors:

$$E_C = E^0_F + kT \ln\left(\frac{N_C}{N_D}\right) \qquad \text{for n-type SC} \quad (11a)$$

$$E_V = E^0_F - kT \ln\left(\frac{N_V}{N_A}\right) \qquad \text{for p-type SC} \quad (11b)$$

$$E_g = E_C - E_V \qquad\qquad\qquad\qquad\qquad (11c)$$

where $N_C$ (and $N_V$) is the effective density of states (DOS) at the bottom (top) of the SC conduction (valence) band, $E_C$ and $E_V$ the conduction and valence band edges, respectively, and $E_g$ is the band gap of the SC. The simplified method reported above and based on the use M-S theory for locating the energy levels at the SC/El interface is very popular among the corrosion scientists. However, we have to mention that, although the validity of Eq. (8) has been tested rigorously for several SC/El interfaces,[43–50] since the seminal work of Dewald[36] on single crystal ZnO electrodes, in many cases a misuse has been made of such equation.

In order to highlight this aspect we come back to the limitations included in the use of the above reported equations. As for the effective DOS at conduction (valence) band edge for n-type (p-type) SC it is given by:

$$N_C = 2M_C\left(\frac{2\pi m^* kT}{h^2}\right)^{3/2} \qquad\qquad (12)$$

where $M_C$ is the number of equivalent minima (maxima) in the conduction (valence) band, $m^*$ is the effective mass for the DOS in the corresponding band and the other symbols have their usual meaning. By assuming $M_C = 1$ and $m^*$ equal to the mass of free electron, $m_0$, we get $N_C = 2.5 \times 10^{19}$ cm$^{-3}$ at room temperature. By taking into account that Eq. (11) on the hypothesis that the following relationship:

$$(E_v + 3\ kT) \leq E_F \leq (E_c - 3\ kT) \tag{13}$$

holds, we can estimate for an hypothetical donor (acceptor) density equal to $10^{19} cm^{-3}$ an effective mass of electron (holes) in the conduction (valence) band $m^* \geq 7$ . Such a value of $m^*$ is not unusual for transition metal oxides so that we can also estimate the corresponding space charge thickness of the semiconducting oxide by means of Eq. (6) reported above. According to the previous calculations and by assuming $\Delta\Phi_{SC} = 0.5$ V a value of space charge length of ~ 9 nm for iron oxide ($\varepsilon_{ox} = 15$) and ~ 8 nm for nickel oxide ($\varepsilon_{ox} = 12$) are estimated. Both these values are larger than the oxide film thickness usually reported for passive iron (3 ~ 5 nm), nickel (2 ~ 3 nm), chromium and stainless steels (2 ~ 3 nm) in different solutions before the onset of the transpassive region,[51–56] so that the hypothesis of SC space charge width much lesser than the thickness of SC could be untenable and the electrical equivalent circuit of the junction should be modified by adding a further capacitance in parallel with the $C_{SC}$ term which accounts for the metal contribution to the total stored charge.[57]

On the other hand donor or acceptor concentration larger than $10^{20} \sim 10^{21}$ $cm^{-3}$ for passive films on iron, nickel and stainless steel have been reported[58–66] and up to $10^{22}$ $cm^{-3}$ in some chromium carbon steel.[67] At such high level of donor or acceptor concentration the applicability of classical M-S theory to the study of a degenerate (or *strong* impurity metal[68]) semiconductor/electrolyte interface is open to serious doubts. By considering that in this case the Fermi level should be located well above (below) the conduction (valence) band edge, the passive film/electrolyte interface becomes now more similar to a semimetal/electrolyte interface for which a different theoretical approach has been suggested.[31,69–70]

In presence of such large density of donor (acceptor) concentration the experimental data should be taken with caution and the electrical equivalent circuit, employed to extract the space-charge capacitance data, carefully scrutinized in order to verify if the restrictions underlying the simple M-S analysis are satisfied specially in absence of ideally polarizable interface. Moreover at such high donor (acceptor) concentration, neglecting the contribution of the capacitance of the Helmholtz double layer to the measured capacitance is no more acceptable whilst the assumption of a potential independent $C_H$ value, at negative potentials with respect to the flat band condition, should be tested.[31] For highly doped materials, as previously mentioned, the values of $U_{fb}$ can differ considerably from the intercept value $U_0$. By substituting the value of dielectric constant usually reported for passive iron ($\varepsilon_{ox} = 15$) and an average $N_D$ value of $10^{21}$ cm$^{-3}$ we obtain (see Eqs. 9a,b) a difference between the flat band potential and the intersection voltage $U_0$ ranging from 0.26 V up to 1.06 V by assuming for $C_H$ values ranging between 20 and 10 μF cm$^{-2}$.

Further limitations in the application of classical M-S analysis to passive films come out from the strong frequency dependence usually observed in the differential capacitance values of the junctions affecting considerably both the slope ($N_D$, $N_A$) and the flat band potential values. As possible explanations of such a behavior the presence of deep lying donor (acceptor) level and/or the presence of surface states has been frequently invoked in the case of crystalline semiconductors and semiconducting passive film. However, the apparent measured density of donor (acceptor) carried out trough the M-S equation is meaningless whilst the location of characteristics energy levels of the junction performed by means of Eqs. (11) may be misleading as discussed in the specialized literature several years ago.[71,72] In the case of c-SC the possible physical cause of such frequency dependence has been also attributed to the presence of an external (disordered or amorphous) layer, which after chemical etching of the surface of c-SC could be removed to restore the expected behavior.[73]

In the case of thin passive film the presence of a frequency dependent differential capacitance is intrinsic to the formation of a layer having an amorphous or strongly disordered nature requiring the use of interpretative models accounting for such a specific feature. According to this it appears preferable to afford such a common aspect in the study of passive film/electrolyte interface by using a more general theoretical approach based on the theory of amorphous semiconductor Schottky barrier which is able to provide a better physical description also for the behavior of crystalline semiconductor junction containing a distribution of donor (acceptor) level in the forbidden energy gap of the SC.

On the other hand semiconducting passive films in a large range of thickness (2 ~ 200 nm) can be grown on different valve metals (Ti, Ta, Nb, W) with donor density values of $10^{17} \sim 10^{20}$ cm$^{-3}$, measured according to M-S equation. For these systems a thickness dependent, as $(d_{ox})^{-2}$, donor concentration changing with the growth conditions, initial surface treatment, post-anodizing annealing and nature of investigated metal has been reported by different groups.[28,74–75] Also for these systems a systematic dependence of the measured capacitance values from the frequency of ac signal has been observed. Usually at constant frequency a transition from semiconductor-like to insulating behavior was reported with increasing film thickness at not too low estimated donor concentration ($N_D > 10^{18}$ cm$^{-3}$),[74] whilst at large film thickness the same transition may occur with increasing ac signal frequency.[75–78]

In this frame the theory of amorphous semiconductor Schottky barrier provides a unified approach to the theory of electrical admittance of electrode/electrolyte junction including as limiting case the ideal SC/electrolyte junction on which the traditional M-S approach is based. The theory and the results of such studies will be presented and discussed in the next Section.

## 3. Differential Admittance in Semiconductor/Electrolyte Junction

In order to understand the main differences in the behavior of a-SC Schottky barrier with respect to the case of crystalline semiconductor it may be helpful to compare preliminarily the density of states distribution in both materials. We stress that usually amorphous materials maintain the same short-range order than their crystalline counterparts and that the main differences come out from the absence of the long-range order, typical of crystalline phases.[78–84] It is now generally accepted that the band structure model retains its validity also in absence of the long-range lattice periodicity. This means that the long-range disorder perturbs but does not annihilate the band structure: its main effect is the presence of a finite DOS within the so-called *mobility gap*, $E_C - E_V$, of the amorphous semiconductor (a-SC) or insulator.

In Fig. 3 we report the model of DOS distribution vs. energy for generic crystalline and amorphous material. Although the gen-

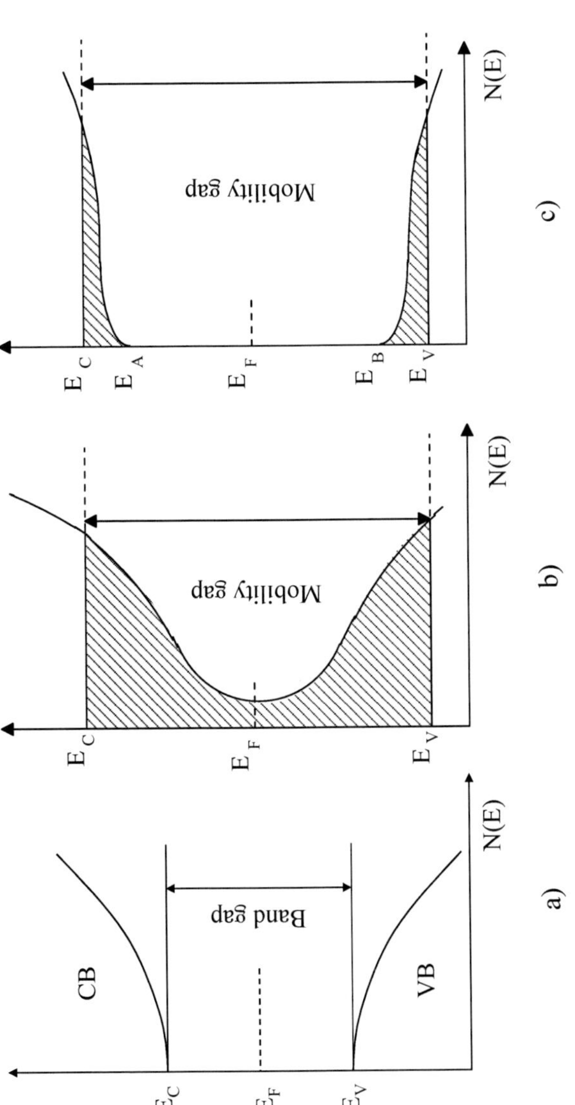


Figure 3. Model of the electronic structure for a crystalline semiconductor (a), amorphous semiconductor following the Cohen-Fritzsche-Ovishinsky model (b), and amorphous semiconductor following the Mott-Davis model (c).

eral features of DOS distribution are preserved, especially if we compare Fig. 3a with Fig. 3c, some differences are evident and they strongly affect the response of a-SC/El junction under ac electrical as well under light stimulus (see below). The DOS distribution of Fig. 3b, initially proposed by Cohen-Fritzsche-Ovishinsky (CFO model),[79] takes into account the possible existence of lattice defects inside the semiconductor which originate a continuous distribution of electronic states within the mobility gap (like in a-Si:H). On the other hand the DOS distribution of Fig. 3c, due to Mott and Davis,[81] has been proposed for an ideal amorphous material in which only the long-range lattice disorder is taken into account. Other models have been suggested for explaining the behavior of different classes of amorphous materials, but they involve only minor modifications to those of Fig. 3, when the existence of specific defects in the investigated material is considered.

As evidenced in the Fig. 3, the general features of the DOS of crystalline materials are preserved also for disordered phases, but some differences are now evident with respect to the crystalline SC and these can be summarized as follows:

(a) Existence of a finite DOS within the mobility gap, defined by two mobility edges, $E_C$ and $E_V$, in the conduction band (CB) and in the valence band (VB), respectively.

(b) For energy levels below $E_C$ or above $E_V$, the free electron-like DOS, $N(E) \propto E^{1/2}$ is no more generally valid. In these energy regions the presence of a tail of states, exponentially[82] or linearly[81] decreasing, has been suggested by different authors for explaining the optical properties of different amorphous materials.

(c) Different mechanisms of charge carriers transport are invoked in extended (below $E_C$ or above $E_V$) or localized (within the mobility gap) electronic states. A free carrier-like mechanism of transport is involved in the first case, whilst a transport by hopping (thermally activated) is assumed in localized states.

These differences between the distributions of electronic states in crystalline and disordered materials have noticeable influence on both the impedance and the photoelectrochemical behavior of the a-SC/El junction. In order to highlight such differences in the impedance behavior of the a-SC/El junction, we will derive some of the most important results of the theory of Schottky barriers which are relevant for a better understanding of the difference in admittance behavior of c-SC/electrolyte and passive film/ electrolyte junctions.[76,78,85–90]

It was shown by different authors that the existence of deep electronic states in the mobility gap of the material influences both the shape of the space-charge region and the frequency response of the barrier to the modulating *ac* signal.[85–90] As for the first aspect, the potential distribution inside the SC can be obtained by solving the one dimensional Poisson equation with the suitable boundary conditions:

$$\frac{\partial^2 \psi(x)}{\partial x^2} = \frac{\rho(x)}{\varepsilon_{ox}\varepsilon_0} \tag{14}$$

where $\psi(x)$ represents the band bending at a point $x$ within the barrier (the galvani potential has opposite sign), and $\rho(x)$ the volumetric density of charge at $x$. $\rho(x)$ is strictly related to the local potential so that Eq. (14) can be integrated by changing the variable $x$ with $\psi$:

$$d\left(\frac{\partial \psi}{\partial x}\right)^2 = \frac{2}{\varepsilon_{ox}\varepsilon_0}\rho(\psi)d\psi \tag{15}$$

If the space charge region is very small with respect to the total thickness of the semiconductor, the usual boundary conditions for infinite length semiconductor holds:

$$\psi = 0 \quad \text{and} \quad d\psi/dx = 0 \qquad \text{for } x \rightarrow \infty;$$
$$\psi = \psi_{SC} \qquad\qquad\qquad \text{for } x = 0.$$

The first integral of Poisson Eq. (15) gives:

$$\left.\frac{\partial \psi}{\partial x}\right|_\psi = \pm\left(\int_0^\psi \frac{2}{\varepsilon_{ox}\varepsilon_0}\rho d\psi'\right)^{\frac{1}{2}} = \pm\sqrt{H(\psi)} \tag{16}$$

where $+$ is used if $\psi$ is negative and $-$ if $\psi$ is positive. By further integration we can find the value of $x$ for different values of band bending in the semiconductor $\psi_0$:

$$x(\psi_{SC}, \psi_0) = \mp \int_{\psi_0}^{\psi_{SC}} \frac{d\psi}{\left(\int_0^{\psi} \frac{2}{\varepsilon_{ox}\varepsilon_0}\rho d\psi'\right)^{\frac{1}{2}}} \tag{17}$$

The integration of Eq. (17) in the hypothesis of a constant DOS distribution, $N$, has been carried out in reference[85] and it provides the dependence of band bending on the distance from the semiconductor/electrolyte interface as:

$$\psi(x) = \psi_{SC} e^{-x/x_0} \tag{18}$$

where $x_0 = (\varepsilon_{ox}\varepsilon_0/e^2 N)$ and $N$ is the DOS in $cm^{-3}\,eV^{-1}$.

The charge per unit area stored in the semiconductor, $Q_{SC}$, is equal to:

$$Q_{SC} = \int_0^{\infty} \rho dx = \varepsilon_{ox}\varepsilon_0 \int_0^{\infty} d\frac{\partial\psi}{\partial x} = -\varepsilon_{ox}\varepsilon_0 \left.\frac{\partial\psi}{\partial x}\right|_0 \tag{19}$$

The general expression of the steady state space charge capacitance of the junction is given by:

$$C_{SC}(\psi_{SC},0) = -\varepsilon_{ox}\varepsilon_0 \frac{d}{d\psi_S}\left(\left.\frac{\partial\psi}{\partial x}\right|_0\right) = \mp\left(\frac{\varepsilon_{ox}\varepsilon_0}{2}\right)^{\frac{1}{2}} \frac{\rho(\psi_S)}{\left(\int_0^{\psi_S}\rho(\psi')d\psi'\right)^{\frac{1}{2}}} \tag{20}$$

As expected, the response of the semiconductor is dependent on $\rho$ which is a function of the band bending and charge distribution into localized states within the forbidden gap (mobility gap) of the crystalline (amorphous) semiconductor.

The dependence of the density of charge on the band bending for a n-type c-SC with deep traps, behaving as donor-like levels, can be written as:

$$\rho(\psi) = |e| \sum_{i=1}^{m_T} N_{T,i} \left[ \frac{\exp\left(\dfrac{E_{T,i} + |e|\psi - E_F}{k_B T}\right)}{1 + \exp\left(\dfrac{E_{T,i} + |e|\psi - E_F}{k_B T}\right)} - \frac{\exp\left(\dfrac{E_{T,i} - |e|\psi - E_F}{k_B T}\right)}{1 + \exp\left(\dfrac{E_{T,i} - E_F}{k_B T}\right)} \right] \quad (21)$$

where $E_{T,i}$ is the energy level of the i-th deep trap and $N_{T,i}$ the concentration of deep traps at $E_{T,i}$ in $cm^{-3}$. It can be easily shown that the previous equations give, as particular case, the M-S equation once the expression of the volumetric density of charge for crystalline SC as a function of the band bending is substituted in Eq. (20). In fact for a single donor-like level completely ionized ($E_{T,i} - E_F \gg 0$) Eq. (21) reduces to the Mott-Schottky case:

$$\rho(\psi) = |e| N_D \left[ 1 - \exp\left( -\frac{|e|\psi}{k_B T} \right) \right] \quad (22)$$

By using Eq. (22) in Eq. (20), after simple algebraic manipulation, the M-S equation, previously reported (Eq. 8), is obtained. In this last case the frequency dependence in Eq. (20) is absent owing to the fact that there is no delay in following the *ac* signal by the electrons staying in extended states of the conduction (or valence band), whichever is the chosen frequency of the modulating *ac* signal.

On the other hand by using Eq. (21) together with Eq. (20) the static space charge capacitance for an n-type semiconductor with multiple deep donor levels is obtained as reported in a previous work.[91] Such an expression does not take into account the finite time of answer of electrons lying in deep trap[71,72] so that a new model which accounts for the frequency dependence of the a-SC Schottky barrier containing deep donor levels and/or a distributed density of states within the mobility gap of a-SC was successively put forward by the same authors in agreement with the theory of a-SC Schottky barrier.[92]

## 4.  Static Differential Capacitance in a-Semiconductor with Constant DOS

In Fig. 4 the energetics of n-type a-SC/El interface is reported, under the simplifying hypothesis of a spatially homogeneous a-SC having a constant DOS distribution in energy into the mobility gap. Both the assumptions (homogeneity and constancy in energy) are necessary to get an analytical solution for the general expression of $C_{SC}$ in a-SC Schottky barrier.[85–90] Both these assumptions will be relaxed in discussing the admittance behavior of passive film/electrolyte junction.

Owing to a large density of localized states within the mobility gap we can assume as negligible the contribution of the free electrons in the conduction band ($E_C - E_F > 3kT$) to the density of charge $\rho(\psi)$. Moreover an abrupt change in the electron occupancy of the localized states at the Fermi level is assumed.[86] The electric charge density for a-SC can be written, within the given assumption, as:

$$\rho(\psi) = |e| \int_{E_F - |e|\psi}^{E_F} N(E)dE = |e|N\psi \tag{23}$$

The contribution of minority carriers to the density of charge is assumed negligible, whilst the electron quasi Fermi level is assumed flat throughout the semiconductor as long as the band bend-

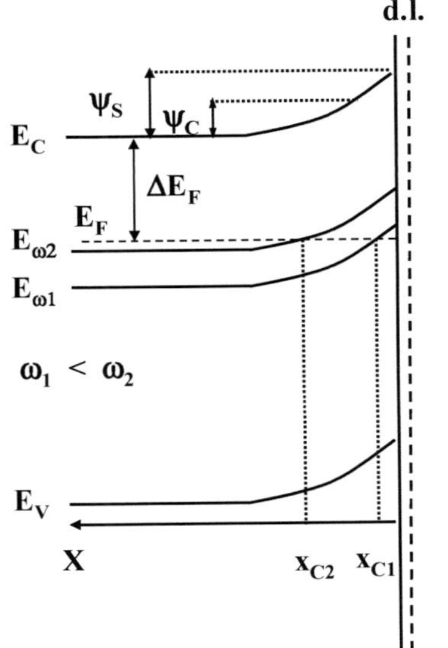


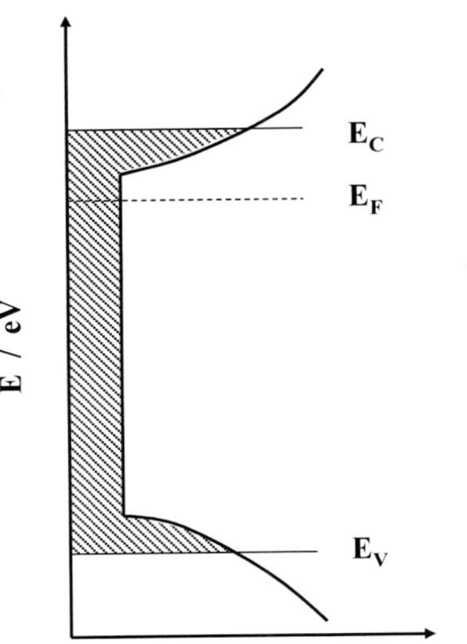


Figure 4. Electronic structure (a) and band bending distribution (b) across an n-type a-SC/El interface.

ing is not too high (see below). Under these conditions the static capacitance (see Eq. 20) can be calculated as:

$$C_{SC}(\psi_{SC},0) = \left( \varepsilon_{ox}\varepsilon_0 |e|^2 N \right)^{1/2} \tag{24}$$

It may be useful to compare the results for $C_{SC}$ obtained for a-SC with those reported by Gerischer[69] for graphite as well as the extension of space charge layer in amorphous semiconductor with constant DOS: $N(E) = N$. By assuming the space charge region as the distance inside the a-SC where the band bending is in the order of $kT/e$ we can derive from Eq. (17) the extension of $X_{SC}$ as:

$$X_{SC} = X_0 \ln\left( \frac{|e|\psi_{SC}}{kT} \right) \qquad \text{for } |e|\psi_{SC} > 3kT \tag{25}$$

By assuming a DOS at $E_F$ around $10^{20}$ eV$^{-1}$ cm$^{-3}$ we get for $C_{SC}$ a constant value of 3.8 $\mu$F cm$^{-2}$ and a space charge length $X_{SC}$ = 5.4 nm for a band bending around 0.25 V with $\varepsilon_{ox}$ = 10. We like to stress that in the case of base metals like: Ni, Fe, and Cr the value of $X_{SC}$ at 0.25 V of band bending results higher than the thickness of passive film reported in literature for such metals. If we take into account that the value of $C_{SC}$ is of the same order of the Helmholtz double layer and that small changes in $C_H$ as well as in the value of dielectric constant of the passive film can occur with changing potential during the capacitance measurements it is, once more, necessary to be aware that the rather unusual donor (acceptor) density measured on these systems could be affected by different pitfalls in the assumed physical model.

In order to evidence the main differences in the admittance behavior of a-SC with respect the crystalline counterpart it is necessary to derive, however, the expression of differential admittance in a-SC-Schottky barrier as a function of the frequency of $ac$ signal.

## 5.  AC Response of a-Semiconductor Schottky Barrier

According to the schematic DOS distribution reported in Fig. 4 the most part of electronic charge is now located below the Fermi level in localized states quite distant in energy from the conduction band mobility edge separating the extended states region from localized ones. According to the theory,[85–90] at variance with the case of crystalline SC, the filled electronic states into the gap (see Fig. 4b) do not follow instantaneously the imposed *ac* signal, but they need a finite response time. This response time depends on their energy position with respect to the Fermi level and it can be much longer than the period of the *ac* signal having angular frequency $\omega$. In fact the relaxation time, $\tau$, for the capture/emission of electrons from electronic states $E$ below $E_F$ is assumed to follow the relationship:

$$\tau = \tau_0 \exp\left(\frac{E_C - E}{kT}\right) \tag{26}$$

where, at constant temperature, $\tau_0$ is a constant characteristic of each material usually ranging between $10^{-14} \sim 10^{-10}$ s. According to Eq. (26), by decreasing the energy of the localized state in the gap, $\tau$ increases sharply so that deep states (for which $\omega\tau \gg 1$) do not respond to the *ac* signal.

By assuming a full response for states satisfying the condition $\omega\tau \ll 1$ and a null response for states having $\omega\tau \gg 1$, a sharp cut-off energy level, $E_\omega$, separating states responding from those not responding to the signal, can be defined from the condition: $\omega\tau = 1$.

The location of the cut-off level $E_\omega$ is found by imposing in Eq. (26) $\omega\tau = 1$ for $E = E_\omega$, which gives:

$$E_C - E_\omega = -kT \ln(\omega\tau_0) \tag{27}$$

This condition occurs at some position within the barrier ($X = X_C$) which is a function of band bending and *ac* frequency (see Fig. 4b). The intersection of the cut-off level, $E_\omega$, with the Fermi level allows to locate the point within the barrier, $X_C$, which separates two regions of the a-SC (see Fig. 4b). The first region (for $X > X_C$) represents the region where all electronic states fully respond to the *ac* signal ($\omega\tau \ll 1$), the second one (for $X < X_C$) where they do not respond at all ($\omega\tau \gg 1$). The corresponding band bending at $X_C$ is given according to Fig. 4b by:

$$|e|\psi_C = -kT\ln(\omega\tau_0) - \Delta E_F \tag{28}$$

with $\Delta E_F = (E_C - E_F)_{bulk}$. We like to stress that $X_C$ is now a distance in the barrier which changes with changing frequency, $\omega$, and band bending $\psi_S$. In particular $X_C$ increases with increasing frequency, at constant polarization, or with increasing polarization at constant frequency. From the theory it comes out that the total capacitance is sum of two series contribution coming from the $X < X_C$ and $X > X_C$ regions of the a-SC. The contribution to the conductance comes mainly from the region around $X = X_C$ dividing the total response from null response regions. In the hypothesis of a constant DOS the total capacitance is given by the sum of the two contributions[77,78,93,94] (see Fig. 5):

$$\frac{1}{C_P(\psi_{SC},\omega)} = \frac{1}{C(\psi_C,0)} + \frac{X_C}{\varepsilon\varepsilon_0} \tag{29}$$

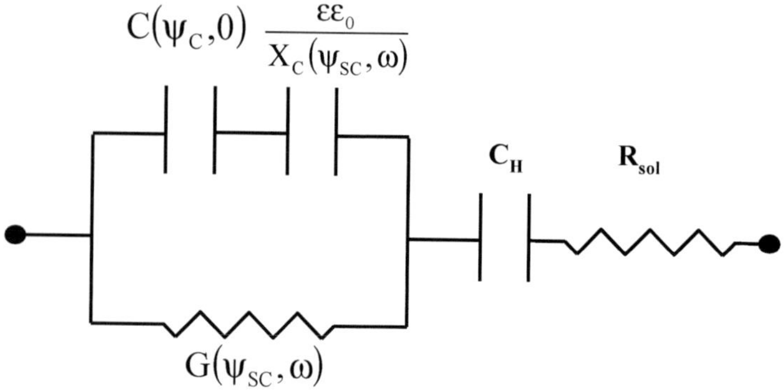


Figure 5. Equivalent electrical circuit of an ideally polarizable a-SC/electrolyte interface in absence of surface state contribution.

where $C(\psi_C,0)=\sqrt{\varepsilon\varepsilon_0 e^2 N}$ coincides with the static space charge capacitance of a-SC previously derived and it is frequency independent whilst $X_C = \sqrt{\dfrac{\varepsilon\varepsilon_0}{e^2 N}}\ln\dfrac{\psi_{SC}}{\psi_C}$ is the distance from the surface of the barrier at which the Fermi level crosses the cut-off energy level $E_\omega$ inside the space charge region. After substitution the following relationship is obtained for the total capacitance:

$$C_{LBB}(\psi_{SC},\omega)=\sqrt{\varepsilon\varepsilon_0 e^2 N}\left(1+\ln\frac{\psi_{SC}}{\psi_C}\right)^{-1} \qquad (30)$$

while the parallel conductance of the junction is given by:

$$G_{LBB}(\psi_{SC},\omega)=\left(\frac{\omega\pi kT}{2}|e|\psi_C\right)\sqrt{\varepsilon\varepsilon_0 e^2 N}\left(1+\ln\frac{\psi_{SC}}{\psi_C}\right)^{-2} \qquad (31)$$

The analytical solutions for the admittance components of the junction have been derived under conditions that $\psi_{SC} > \psi_C > 3kT/|e|$ and at not too high band bending (Low Band Bending regime).[76–78]

It has been shown that $G(\psi_{SC},\omega)$ has a spectroscopic character with respect to the distribution of electronic states within the gap, whilst variations in DOS cause only minor changes in the $C(\psi_{SC},\omega)$ vs. potential plots provided that the DOS varies little over an energy range of $kT$. At variance with the M-S analysis in the case of a-SC a fitting procedure is required for getting flat band potential and DOS distribution around the Fermi level. As reported in previous works[77–78] the fitting must be carried out on both components of the admittance of the junction. For an ideally polarizable a-SC/electrolyte interface and in absence of surface state contribution to the measured admittance the electrical equivalent circuit of such an interface is shown in Fig. 5. Such an equivalent circuit has been used by the authors for fitting both the components of space charge admittance, $Y_{SC,}$, at high frequency when the contribution of surface state admittance, $Y_{SS}$, is negligible with respect to $Y_{SC}$. The fitting procedure must be performed under condition that at any employed *ac* frequency both the $1/C_{SC}$ vs. $\psi_{SC}$ and the $G_{SC}$ vs. $\psi_{SC}$ plots give the same $U_{fb}$ value, within an assigned uncertainty (in our case, 0.025 V). Moreover an additional constraint arises from Eq. (28) requiring for $\psi_C$ a variation of 59 mV for decade of frequency at room temperature.[76–78,93–98]

As reported previously[76–78] a very meaningful difference with respect to the use of M-S analysis is the procedure for locating the mobility edge, $E_C$ (n-type) or $E_V$ (p-type), by considering that now the most part of electronic charge is located into the localized states below the Fermi level. At this aim instead of Eq. (11) valid for c-SC we can make use of Eq. (28) once the $\psi_C$ parameter and flat band potential $U_{fb}$ have been obtained. In order to perform such a task we need to know for each material the constant $\tau_0$. For our purposes an average $\tau_0$ value of $10^{-12}$ s can be assumed in absence of further information. This choice could affect the location of the mobility edge ($E_C$ or $E_V$) by a quantity equal to 0.12 eV which is not too bad if we take into account the absence of a sharp boundary between localized and extended states region.

The ability of previous admittance equations, in low band banding approximation, to fit the experimental curves of different amorphous passive film SC/electrolyte junctions has been tested in previous works.[76–78] It is in our opinion that such an approximation could be generally employed for analysing the admittance behavior of thin passive films on base metal (Fe, Ni, Cr) where it is expected that the space charge region reaches the metal interface at rather low values of band bending (see above for estimates of $X_{SC}$) for reasonable values of DOS ($10^{20}$ cm$^{-3}$ eV$^{-1}$) near the Fermi level.

In the case of thicker films like those grown on valve metals (Ti, Nb, W, Ta, etc.) may be of interest for a better understanding of the mechanism of growth of the film and for deriving information on the DOS distribution inside the a-SC to investigate the admittance behavior of the passive film/electrolyte junction in a larger range of band bending values. At high band bending a parabolic potential distribution will appear in the deep depletion region at the surface of a-SC/El junctions where the quasi-Fermi level of the a-SC is now pinned, due to the equilibrium of the emission rate for electrons and holes from localized states to the respective band

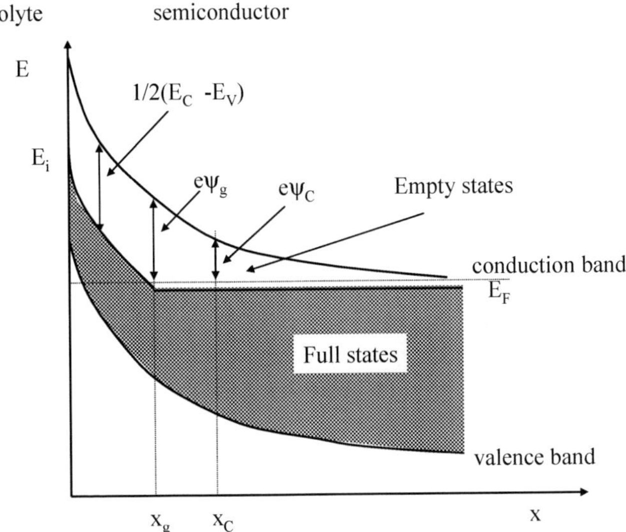



Figure 6. Band bending distribution across an n-type a-SC/El interface under the conditions of high band bending (HBB).

edge. It is possible to define a band banding $\psi_g$ (see Fig. 6) corresponding to the equilibrium of emission/capture process:

$$|e|\psi_g = \frac{E_g}{2} + \frac{k_B T}{2}\ln\left(\frac{v_n}{v_p}\right) - \Delta E_F \tag{32}$$

The point at which the band bending of the semiconductor is equal to $\psi_g$ is $x_g$ and it can be obtained from Eq. (17) by imposing $\psi_0 = \psi_g$. In order to include the deep depletion region the previous equations have been modified[95-98] as follows:

$$\frac{1}{C_{HBB}(\psi_{SC},\omega)} = \frac{1}{\sqrt{\varepsilon\varepsilon_0 e^2 N}}\left(\ln\frac{\psi_g}{\psi_C} + \sqrt{1 + \frac{2}{\psi_g}(\psi_{SC} - \psi_g)}\right) \tag{33}$$

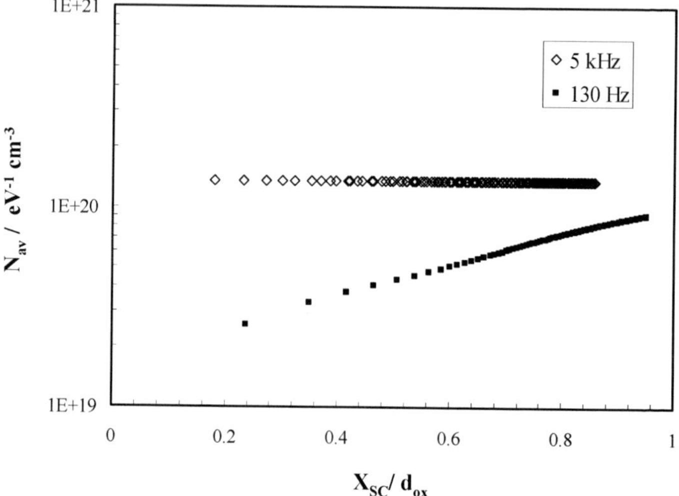


Figure 7. DOS distribution for a-WO$_3$ obtained from Eq. (33) as a function of the position inside the passive film at different frequencies.

$$G_{HBB}(\psi_{SC}, \omega) = \pi^2 f \frac{kT}{|e|\psi_C} \sqrt{\varepsilon\varepsilon_0 e^2 N}$$

$$\times \left( \ln \frac{\psi_g}{\psi_C} + \sqrt{1 + \frac{2}{\psi_g}(\psi_{SC} - \psi_g)} \right)^{-2} \quad (34)$$

Equations (33) and (34) have been derived under the same conditions valid for the low band bending (LBB) expression and they coincide with the previous ones for $\psi_{SC} = \psi_g$. The first test on the ability of Eqs. (33) and (34) to fit the two components of the differential admittance in a relatively large range of band bending values was carried out for passive film on WO$_3$ and Nb$_2$O$_5$ about 25 nm thick.[94] It was noticeable, in the case of WO$_3$, that from Eq. (33) an almost constant average DOS equal to about $10^{20}$ eV$^{-1}$ cm$^{-3}$ was derived for a-WO$_3$ in agreement with previous results, obtained in the low band bending approximation.[78] However a more close inspection of the admittance behavior at different scan rates and *ac* frequencies suggested the possible presence of a more complex dependence of localized DOS both from energy as well as from the spatial coordinate (metal/oxide interface distance) (see Fig. 7).

In more recent papers[95–97] a study aimed to get information on the possible spatial and energy dependence of DOS near the Fermi level of a-SC has been carried out trough a detailed investigation of a-$Nb_2O_5$ passive films anodically grown. The choice of a-$Nb_2O_5$ was dictated by its long term stability in aqueous solution of variable pH and by the fact that anodic film of $Nb_2O_5$ are amorphous and easily prepared in a wide range of film thickness, from few to hundredths of nanometers. These aspects make a-$Nb_2O_5$ an ideal candidate for testing any proposed model of a-SC/electrolyte junction. From such studies it comes out the importance of the choice of the equivalent circuit to be used to subtract from the measured admittance the components of the admittance depending on the electrochemical reactions at the a-SC/electrolyte interface by including also the presence of intrinsic surface states which could result from the amorphous nature of the material (see Fig. 8).

Moreover it came out that in order to fit the EIS spectra of the junction in the overall investigated range of frequency (0.1 Hz–

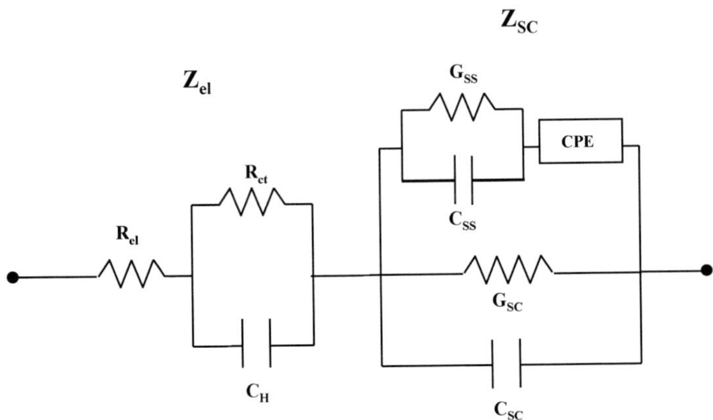


Figure 8. Equivalent electrical circuit of an a-SC/electrolyte interface, considering the contribution of an electrochemical reaction occurring through electrons/holes exchange with the surface states.

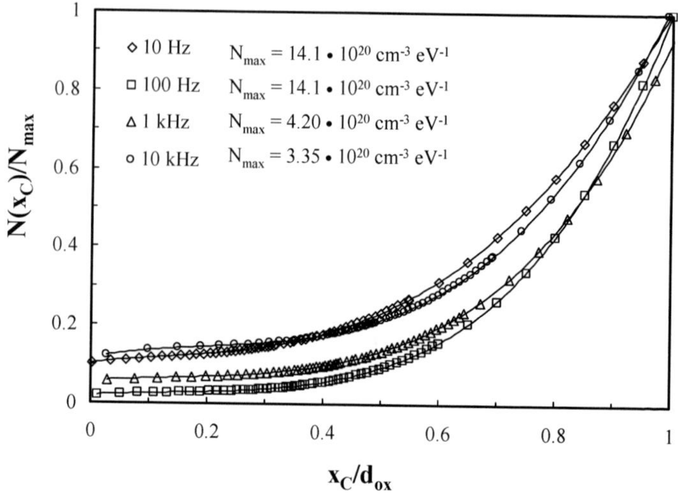


Figure 9. DOS distribution for a-$Nb_2O_5$ obtained from Eq. (36) as function of the position inside the passive film at different frequencies. Reprinted from F. Di Quarto, F. La Mantia, and M. Santamaria, "Physicochemical characterization of passive films on niobium by admittance and electrochemical impedance spectroscopy studies." *Electrochim. Acta* **50** (2005) 5090, Copyright (2005) with permission from Elsevier.

100 kHz) as well the two components—$C_p(\omega,\psi)$, $G_p(\omega,\psi)$—of the admittance curves in a wide range of frequencies (10 Hz–10 kHz) and electrode potentials ($\sim 5$ V) a DOS distribution slightly changing with energy and along the spatial coordinate was necessary (see Fig. 9 and 10).

As for the spatial dependence of DOS a fitting of the admittance curves, $C_{SC}(\omega, \psi_S)$ and $G_{SC}(\omega, \psi_S)$, at different frequencies (10 Hz $\leq f \leq$ 10 kHz), was carried out[95–97] by using Eqs. (33) and (34) modified as:

$$\frac{1}{C_{SC}(x,\psi_{SC},\omega)} = \frac{1}{C(\psi_{SC},\omega)} f_\omega(x(\psi_{SC})) \tag{35}$$

$$G_{SC}(x,\psi_{SC},\omega) = G(\psi_{SC},\omega) g_\omega(x(\psi_{SC})) \tag{36}$$

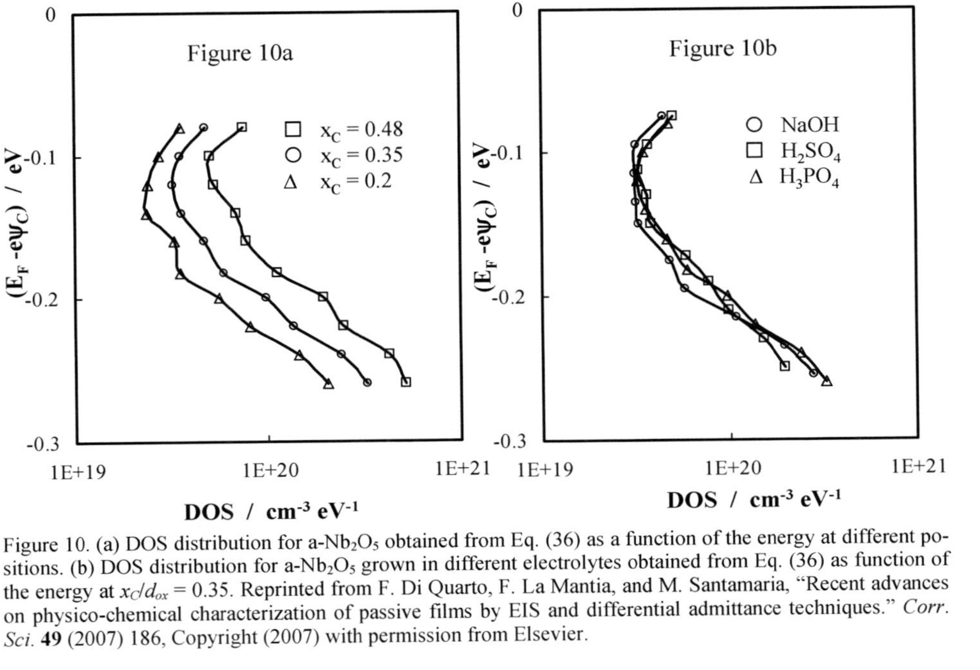


Figure 10. (a) DOS distribution for a-Nb$_2$O$_5$ obtained from Eq. (36) as a function of the energy at different positions. (b) DOS distribution for a-Nb$_2$O$_5$ grown in different electrolytes obtained from Eq. (36) as function of the energy at $x_C/d_{ox}$ = 0.35. Reprinted from F. Di Quarto, F. La Mantia, and M. Santamaria, "Recent advances on physico-chemical characterization of passive films by EIS and differential admittance techniques." *Corr. Sci.* **49** (2007) 186, Copyright (2007) with permission from Elsevier.

where $f_\omega(x(\psi_S))$ and $g_\omega(x(\psi_S))$ are respectively two different trial functions depending only on the electrode potential but changing with employed frequency. The term multiplying the two trial functions can be considered as coincident with the expression of $(C_{SC})^{-1}$ and $G_{SC}$ in absence of spatial variation in DOS (homogeneous film) but averaged in energy ($N(E) = N_{av}$).

By using the spectroscopic character of the conductance in a-SC Schottky barrier,[87,88] the DOS at $x_C$, $N(E_F - e\psi_C)$, was obtained by using the following equation:[95]

$$N\left(E_F - |e|\psi_C\right) = \frac{2}{\pi kT}\frac{G_{SC}\left(\psi_{SC},\omega\right)/\omega}{C_{SC}^2\left(\psi_{SC},\omega\right)}f_\omega\left(x\left(\psi_{SC}\right)\right)C\left(\psi_C,0\right)$$

$$\times \int_{E_F - |e|\psi_C}^{E_F} N(E)\,dE \quad (37)$$

and by substituting to the integral the term $e\psi_C N_{av}$. As first approximation for $N_{av}$ that one derived from the fitting of $C_{SC}$ at the same $\omega$ and $x_C$ have been chosen. As mentioned in previous work,[94,97] the choice of the other fitting parameters ($U_{fb}$, $\psi_g$, $\psi_C$) could be carried out through an educated guess procedure by taking into account that:

- as for the $\psi_C$ parameter the constraint of a decrease of 0.06 eV for decade of frequency has to be introduced in the fitting procedure. A check of the acceptability of the $\psi_C$ choice is the prediction of flattening of capacitance curves at a frequency corresponding to $\psi_C = 0$;
- in absence of further information the $\psi_g$ parameter can be calculated by assuming an equal emission-capture time constant for electrons and hole in the conduction and in valence band (Eq. 32) and by using the measured mobility gap for passive film (a mobility gap equal to 3.40 eV was used for a-$Nb_2O_5$ film);
- the $U_{fb}$ value derived from fitting the $C_{SC}(x, \omega, \psi_S)$ and $G_{SC}(x, \omega, \psi_S)$ curves should have to differ no more than $k_B T/e$ volt.

The DOS distribution in a-$Nb_2O_5$ grown in different electrolytic solutions as a function of energy and distance from the film/solution interface are reported in Fig. 10a. The results suggest that the DOS in anodic film is changing with distance in energy from the Fermi level but such a distribution of DOS is largely independent from the nature of the electrolytic solution[96,97] (see Fig. 10b). It was worth to note that almost coincident values of DOS, as a function of the energy distance from the Fermi level, were derived from EIS analysis and DA measurements thus confirming the spectroscopic character of the admittance measurements as a function of the ac signal frequency. We like to stress that, at variance with the simple M-S approach, in the frame of the theory of amorphous Schottky barrier the frequency dependence of the differential admittance curves is explicitly taken into account and the study of such dependence is able to provide further information on the DOS distribution in energy. As for the spatial dependence it seems that a DOS distribution steadily increasing on going from the oxide/electrolyte interface toward the metal/oxide interface is compatible with the experimental results. Such a finding seems quite reasonable if it is assumed that the DOS distribution is related to the anodizing process and in particular to some specific mobile defects becoming frozen when the anodizing process is stopped. An exact quantitative analysis of these aspects needs to solve the problem of taking into account the effects of a finite film thickness on the measured values of admittance particularly in the case of thin films where the space charge region could reach the total film thickness. However if carefully used the DA measurements can provide useful information, also in a quantitative way, on the electronic structure of passive film beside the ordinary estimate of the flat band potential which in dubious cases should be confirmed by other technique like photocurrent spectroscopy (see below). This last technique in presence of insulating passive film is a possible route to the estimation of the flat band potential of the passive film/electrolyte junction and a complementary way to get information on the solid-state properties of passive film as will be discussed in the next Section.

## III.    PHOTOCURRENT SPECTROSCOPY IN PASSIVITY STUDIES

The first experimental report on the interaction between light and passive films on metal can be traced to the Becquerel's study of photoeffects at metal-electrolyte interface dating back to the first half of 19th century.[99] Earlier studies on photoelectrochemical behavior of oxidized metals were summarized in a review of the initial 1940s[100] but the beginning of photoelectrochemical science is marked by the work of Brattain and Garret in the mid-1950.[15] However, the first attempt to use photoelectrochemical technique in passivity studies dates more recently to the end of 1960s when Oshe and Rozenfeld[101] proposed to characterize the nature of passive films on metals and alloys by using a photopotential method initially proposed by Williams[102] for bulk semiconductors. Some applications of such a method in passivity studies can be find in Refs. [103, 104] and in references therein. The inadequacy of Oshe and Rozenfeld's method in characterising complex metal/oxide/electrolyte interfaces was initially evidenced by Hackerman et al.[105] This fact and the onset of more refined theories of photocurrent generation at illuminated semiconductor/electrolyte interface in the mid-1970s[106–107] made obsolete the Oshe and Rozenfeld's method of characterisation of passive metal/electrolyte interface. In fact, as a result of an intensive research effort on the photoelectrochemical behavior of semiconductor electrodes, aimed to harvest solar energy by photoelectrochemical solar cells, different electrochemical techniques started to be exploited as analytical tools for in situ characterisation of semiconductor/electrolyte (SC/El) interface.[108–110] In many cases the investigated photoelectrodes were oxides so that it was evident to electrochemists that passive films and corrosion layers having semiconducting or insulating behavior could be scrutinised by using the same techniques used for studying SC/El interface.[37–40]

It is worth to note that photoelectrochemical techniques were practically absent in the passivity meeting held in Airlie[3] but they gained importance since the fifth Symposium on Passivity of Metals and Semiconductors held in 1983.[4] Since this last symposium PCS technique became a constant presence at passivity meetings and workshops around the world. This is due to the fact that PCS is able to provide information on:

(1) the energetics at the metal/passive-film/electrolyte interfaces (flat band potential determination, conduction and valence band edges location, internal photo-emission thresholds);

(2) the electronic structure and indirectly (through the optical band gap values) chemical composition of passive films in situ and under controlled potential in long lasting experiments.

With respect to other optical techniques PCS offers the further advantage that the photocurrent response of the passive film is directly related to the amount of absorbed photons. This means that the technique is not demanding in terms of surface finishing so allowing the monitoring of long lasting corrosion processes, where changes of surface reflectivity are expected owing to possible roughening of metal surfaces covered by corrosion products. As for the risk of electrode modifications under illumination, it can be minimised by improving the sensitivity of the signal detection by using a lock-in amplifier, coupled to a mechanical light chopper, which allows scrutinizing very thin films (1–2 nm thick) also under relatively low intensity photon irradiation. More details on the use of Lock-in technique to measure photocurrent signal can be found in Refs. [39, 40] and references therein. PCS presents limitations owing to the following aspects:

(1) the technique is able to scrutinize only photoactive passive films and corrosion layers;

(2) the investigation of surface layers having optical band gap lower than 1.0 eV or larger than 5.5 eV requires special set-up or they are experimentally not accessible in aqueous solutions;

(3) structural information and chemical composition of the layers are not accessible directly and complementary investigation based on other techniques can be required.

The first two limitations are rather apparent than real. With the exception of noble metals (Ir, Ru, etc.) which are covered by conducting oxides only at high electrode potentials, most of metals are thermodynamically unstable by immersion in aqueous solution, and they become covered by oxide or hydroxide films having often insulating or semiconducting properties. Moreover, with the exception of very few oxides grown on metals of lower electronegativity, the most common base metal oxides have band gap values largely lying within the optical window experimentally accessible by PCS.[39–40]

The third limitation is the principal one but it has been shown by different authors in the last years that PCS can provide indirectly compositional information once some interpretative model of the photo-electrochemical behavior of passive film/electrolyte interface is introduced which accounts also for the complex electronic structure of amorphous materials.[35–39] Complementary information accessible by other in situ and ex situ techniques can help this task and may be unavoidable for very complex systems (see below). The aim of this Section is to:

(a) provide a general interpretative framework of the photoelectrochemical behavior of passive metal electrodes by discussing some features which are related to the extreme thinness and/or amorphous nature of the passivating layers;

(b) highlight the more recent quantitative use of PCS for characterising the chemical nature of passive films and corrosion layers.

A very short theoretical background on the photoelectrochemistry of crystalline semiconductors will be provided for readers not acquainted to the subject, in order to evidence the differences between the photoelectrochemical behavior of passive films and bulk crystalline semiconductors. More extensive and detailed introduction to the principles of photoelectrochemistry of semiconductors can be found in classical books and workshop discussions published on the subject.[14,21,22,38,108,109] Theoretical interpretations of the experimental results will be presented on the basis of simple models developed initially for passive films grown on valve-metals (Al, Ta, Zr, Nb, Ti, W) and their alloys. In order to show the ability of PCS to scrutinise also complex systems we will discuss some results pertaining to passive films grown on base metals and alloys (Fe, Cr, Ni, stainless steels, Mg). Moreover recent results of a quantitative use of PCS for the chemical characterization of passive films on metallic alloys and conversion coatings will be presented.

## 1. Semiconductor/Electrolyte Junctions under Illumination: the Gärtner-Butler Model

The modelling of photocurrent vs. potential curves at fixed irradiating wavelength (photocharacteristics) for a crystalline SC/El junction has been carried out by several authors.[107,111–114] starting from the seminal paper of Gärtner[115] on the behavior of illuminated solid state Schottky barrier.

In Fig. 11 the absorption process of incident light in the bulk of a SC is sketched: $\Phi_0$ (in cm$^{-2}$ s$^{-1}$) is the photon flux entering the

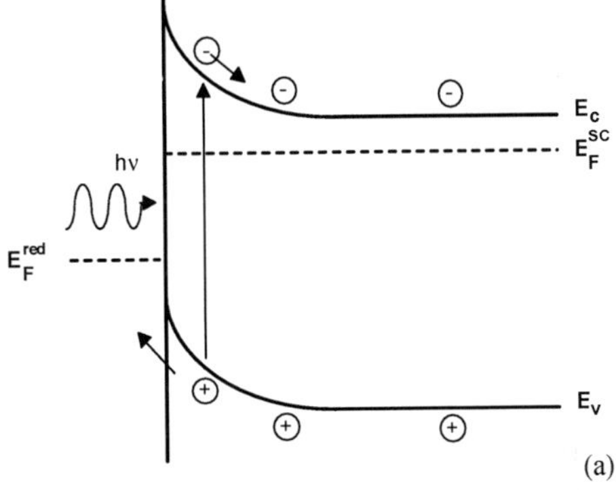


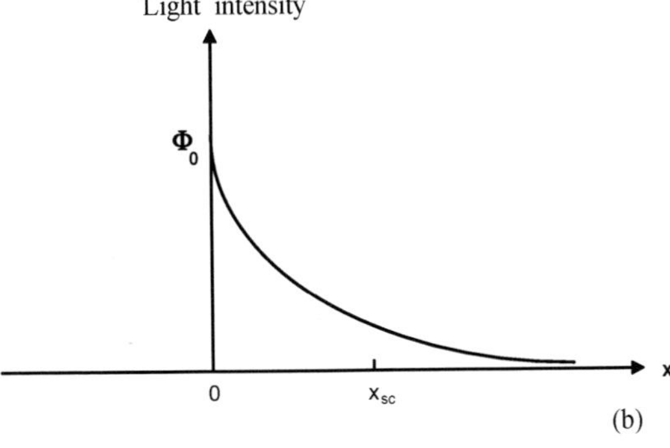


Figure 11. Schematic representation of a crystalline n-type SC/El interface under illumination, showing the electron-hole pair generation (a) and the change of light intensity due to the absorption process within the semiconductor (b).

semiconductor (corrected for the reflections losses at the SC/El interface), which is absorbed following the Lambert-Beer law. The number of electron-hole pairs generated per second and unit volume at any distance from the SC surface, $g(x)$, is given by:

$$g(x) = \Phi_0\, \alpha\, e^{-\alpha x} \tag{38}$$

where $\alpha$ (in cm$^{-1}$), the light absorption coefficient of the semiconductor, is a function of the impinging wavelength. It is assumed that each absorbed photon, having energy $h\nu \geq E_g$ originates a free electron-hole couple. In the Gärtner-Butler model the total photocurrent collected in the external circuit is calculated as sum of two terms: a migration term, $I_{drift}$, and a diffusion term, $I_{diff}$. The first one takes into account the contribution of the minority carriers generated into the space-charge region; the second one accounts for the minority carriers entering the edge of the space-charge region from the bulk field free region ($x > X_{SC}$) of SC. No light reflection at the rear interface is assumed, so that all the entering light is absorbed within the SC. Moreover, it is assumed that minority carriers generated in the space- charge region of the SC do not recombine at all, owing to the presence of an electric field which separates efficiently the photogenerated carriers. The same assumption is made for the minority carriers arriving at the depletion edge from the bulk region of the SC. In order to calculate $I_{diff}$, Gärtner solved the transport equation for minority carriers, which for a n-type SC is:

$$\frac{dp(x)}{dt} = D_p \frac{d^2 p(x)}{dx^2} - \frac{p - p_0}{t} + g(x) \tag{39}$$

with the boundary conditions: $p = p_0$ for $x \to \infty$ and $p = 0$ for $x = X_{SC}$. In Eq. (39), $p$ is the hole concentration under illumination, $p_0$ the equilibrium concentration of hole in the bulk of the (not illuminated) SC and $D_p$ the diffusion coefficient of the holes. The zero value for $p$ at the boundary of the depletion region comes out from the previous assumption that all the holes generated into the space-charge region are swept away without recombining. According to Gärtner, for the total photocurrent we can write:

$$I_{ph} = I_{drift} + I_{diff} = e \int_0^{X_{SC}} g(x)dx + eD_p \left[ \frac{dp(x)}{dx} \right]_{x=X_{SC}} \tag{40}$$

where $e$ is the absolute value of the electronic charge. By solving Eq. (39) in the steady-state approximation for getting out the distribution of holes in the field free region, and by substituting Eq. (38) for $g(x)$ in the integral of the drift term, we get finally the Gärtner equation for a n-type semiconductor:[115]

$$I_{ph} = e\, \Phi_0 \left[ 1 - \frac{\exp(-\alpha X_{SC})}{1+\alpha L_p} \right] + ep_0 \frac{D_p}{L_p} \tag{41}$$

where $L_p$ is the hole diffusion length. The same equation holds for p-type SCs, with $D_n$ and $L_n$ (electron diffusion coefficient and diffusion length, respectively) instead of $D_p$ and $L_p$ and $n_0$ (electrons equilibrium concentration) instead of $p_0$.

## 2. PCS Location of $U_{fb}$ and $E_g$ Determination in c-Semiconductor/Electrolyte Junction

For wide band gap SCs, where the concentration of minority carriers into the bulk is very small, Eq. (41) can be further simplified by neglecting the last term. In this case, by using also Eq. (5) for $X_{SC}$, Butler derived the following expression for the photocurrent at a crystalline SC/El junction:[107]

$$I_{ph} = e\, \Phi_0 \left[ 1 - \frac{\exp\left(-\alpha X_{SC}^0 \sqrt{\Delta\Phi_{SC} - \dfrac{kT}{e}}\,\right)}{1+\alpha L_p} \right] \tag{42}$$

In this equation, $X_{SC}^0$ represents the space-charge width into the SC electrode at 1 V of band bending, and $\Delta\Phi_{SC} = (U_e - U_{fb})$. It is easy to show[107] that if $\alpha X_{SC} \ll 1$ (slightly absorbed light) and $\alpha L_p \ll 1$ (small diffusion length for minority carriers), the photocurrent crossing the n-type SC/El interface can be rewritten as:

$$I_{ph} = e\,\Phi_0 \alpha X_{SC}^0 \left( U_e - U_{fb} - \frac{kT}{e} \right)^{1/2} \tag{43}$$

Equation (43) foresees a quadratic dependence of the photo-current on the electrode potential, which can be used for getting out the flat band potential of the junction. In fact, by neglecting the term $kT/e$, a plot of $(I_{ph})^2$ vs. $U_E$ should intercepts the voltage axis at the flat band potential, $U_{fb}$, regardless the employed $\lambda$ as long as $\alpha X_{SC} \ll 1$ condition is obeyed. Apart from the initial assumption of Gärtner of an ideal Schottky barrier, we have to mention that several hypotheses underlie to the use of Eqs. (41)–(43) for interpreting photoelectrochemical data. Other authors have introduced the possible existence of a kinetic control at the SC surface or within the space charge region[111–114,116–117] showing that:

(a) in presence of strong surface recombination effects the on-set photocurrent may occur at much higher band bending than that foreseen by the Gärtner-Butler equation;[111]

(b) a square root dependence of the photocurrent on the electrode potential is still compatible with the existence of some mechanism of recombination (first order kinetics) within the space charge region,[113] so that the determination of the flat band potential from the square of the photocurrent vs. electrode potential plot must be taken with some caution.

According to this it is possible to write a generalized Gartner-Butler equation for the photocurrent at illuminated crystalline SC/El junction as :

$$I_{ph} = \frac{S_t}{S_t + S_r}\, e\,\Phi_0 \alpha X_{sc}^0 \left( U_E - U_{fb} - \frac{kT}{e} \right)^{1/2} \tag{44}$$

In the previous equation the $S_t /(S_t + S_r)$ term accounts for the ratio between the minority carriers transfer reaction rate, $S_t$, and total recombination rate $S_r$. At high band bending such a ratio goes to 1 so that in presence of low recombination rate the onset photo-current potential, $U^*$, could provide a close estimate of the flat band potential[118]. This last value is expected to be coincident or more anodic (cathodic) than $U_{fb}$ of n-type (p-type) SC/El junction.

These aspects must be carefully considered when the photocurrent vs. potential curves are used for deriving the flat band potential of SC/El junctions, especially in the case of corrosion layers which are far from the ideal behavior of crystalline semiconductors previously assumed. Finally we have to remark that all previous equations pertain to the steady-state values of the $dc$ photocurrent. The equations derived for steady-state remain valid also for chopped conditions provided that the lock-in measured signal remains proportional to the steady-state chopped value.[39]

A second important aspect embodied in Eq. (44) is the direct proportionality between the measured photocurrent and the light absorption coefficient. By considering that in the vicinity of the optical absorption threshold of the SC,[21] the relationship between the absorption coefficient and optical band gap of material, $E_g$ can be written as:

$$\alpha = A \frac{\left(hv - E_g^{opt}\right)^{n/2}}{hv} \tag{45}$$

it is possible to derive the following expression:

$$Qhv = \left(hv - E_g^{opt}\right)^{n/2}\left(X_{sc}^0 \sqrt{\left|U_e - U_{fb}\right|}\right) \tag{46}$$

$Q = (I_{ph})/e\Phi_0$ represents the photocurrent collection efficiency and $E_g^{opt}$ the optical threshold for the onset of photocurrent at the illuminated electrode. Eq. (46) shows that, at constant electrode potential, it is possible to get the optical band gap of the material from the dependence of the photocurrent on the wavelength (shortly referred as the photocurrent spectrum of the junction) of incident light at constant photon flux. In fact by plotting $(Qhv)^{2/n}$ vs. hv at constant electrode potential ($U_E - U_{fb}$ = const.) we get a characteristic photon energy $hv_0 = E_g^{opt}$. For an ideal SC/El junction $E_g^{opt}$ coincides with the minimum distance in energy between the filled states of VB and empty states of CB (band gap, $E_g$) and n can assume different values depending on the nature of the optical transitions between states of the VB and states of the CB.

Optical transitions at energies near the band gap of a crystalline material may be direct or indirect. In the first case no intervention of other particles is required, apart the incident photon and the electron of the VB; in the second case the optical transition is assisted by the intervention of lattice vibrations. By assuming a parabolic electronic density of states distribution, DOS, ($N(E) \propto E^{1/2}$) near the band edges, in the case of direct transitions n assumes values equal to 1 or 3, depending on whether the optical transitions are allowed or forbidden in the quantum mechanical sense.[21] In the case of indirect optical transitions the value of n in Eq. (45) is equal to 4. It will be shown in the case of amorphous materials that the measured optical band gap does not necessarily coincides with the band gap of crystalline material but it can be still related to the material composition and morphology.

### 3.  Passive Film/Electrolyte Interface under Illumination

On going from crystalline thick SC electrodes to very thin insulating or semiconducting corrosion films on metals, new experimental features are observed which require the extension of previous interpretative models and the introduction of new theoretical concepts in order to account for novel results not observed for crystalline bulk materials. In the following we discuss the main differences in the optical properties of amorphous and crystalline materials[79–82] and how they can affect the photoelectrochemical behavior and band gap values of passive films.

### (i)  *Electronic Properties of Disordered Passive Films*

The electronic properties of disordered passive films described in previous Sections affect the generation and transport process of photocarriers and then the photoelectrochemical answer of the a-SC/El junction. The main differences in the photocurrent response of disordered thin films with respect to the case of bulk crystalline semiconductors arise from the following facts:

(a)  the optical band gap of an amorphous material may coincide or not with that of the crystalline counterpart, depending on the presence of different types of defects which can modify the DOS distribution;

(b) at variance with crystalline materials, the generation process of free carriers by the absorption of photons having energy equal or higher than the optical band gap of the film may depend on the electric field, owing to the presence of initial (geminate) recombination effects;

(c) the small thickness of passive film makes possible the optical excitation at the inner metal/film interface. This allows injecting photocarriers from the underlying metal into the VB or CB of thick film (internal photoemission), or directly into the electrolyte (external photoemission) in the case of very thin films (1–2 nm thick);

(d) the presence of reflecting metal/film and film/electrolyte interfaces makes possible the onset of multiple reflections, even for photons having energy higher than the optical absorption threshold. This fact originates interference effects in the photocurrent vs. film thickness curves. This last aspect will be not treated here for brevity; but interested readers can make use of previous published works.[39–40] In the following we derive an equation for the photocurrent in amorphous SC and insulators by taking into account the influence of the amorphous nature on the electronic properties of materials.

### (ii) Amorphous Film/Electrolyte Junction under Illumination

Due to the low mobility of carriers in amorphous materials it is reasonable to assume that a negligible contribution to the measured photocurrent arises from the field free region of the semiconductor. In this case it is quite easy to derive an expression for the migration term in the space-charge region of the a-SC, in a quite similar way to that followed by Gärtner but introducing also a recombination probability for the photocarriers generated in the space charge region of the a-SC.

Like in the Butler model, we will assume the absence of kinetic control in the solution and a negligible recombination rate at surface of the semiconductor. The limits of validity of such assumption have been discussed previously for the case of crystalline SC/El junctions[112] and they will not be repeated here. We will assume also an efficiency of free carrier generation, $\eta_g$, position independent under illumination with light having energy higher than the SC mobility gap.

Under steady-state conditions the recombination of photogenerated carriers in the space-charge region can be taken into account by assuming that the probability of any carrier photogenerated at a position x to leave the space-charge region is given by:[119]

$$P(x, \overline{F}) = \exp\left(-\frac{x}{L_d}\right) \tag{47}$$

where $\overline{F}$ is the mean electric field in the space-charge region of the a-SC and $L_d$ is the drift length of the photocarriers ensemble in the average field approximation, given by:

$$L_d = \mu\tau\overline{F} \tag{48}$$

$\mu$ and $\tau$ being the drift mobility and the lifetime of the photocarriers, respectively. According to these equations and to the assumptions made, we can write:

$$I_{ph} = I_{drift} = \eta_g e \Phi_0 \int_0^{X_{sc}} \alpha \exp(-\alpha x) P(x, \overline{F}) dx \tag{49}$$

where $\Phi_0$ is the photon flux corrected for the reflection at the electrolyte/film interface, having assumed negligible reflections at the film/metal interface, and $\eta_g$, the efficiency of free carriers generation in presence of geminate recombination effects, is a function of the thermalization distance, $r_0$, and of $\overline{F}$. By integration of Eq. (49) we get:[119]

$$I_{ph} = \eta_g e \Phi_0 \frac{\alpha L_d}{1+\alpha L_d} \left[1 - \exp\left(-X_{sc}\frac{1+\alpha L_d}{L_d}\right)\right] \tag{50}$$

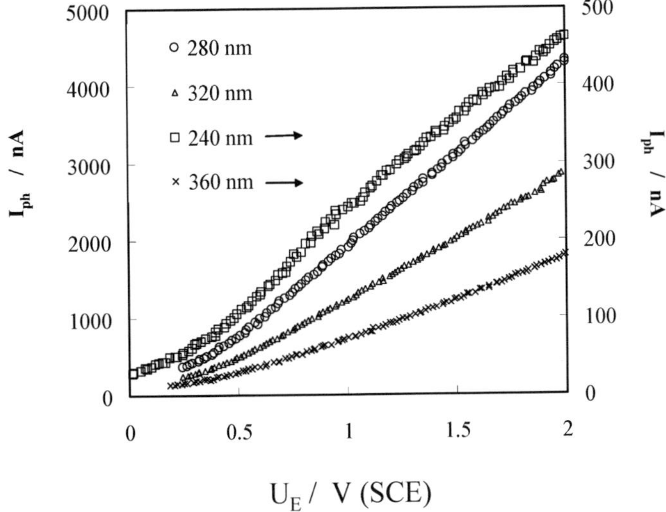



$$U_E / V (SCE)$$

Figure 12. Photocurrent vs. potential curves at different λ relating to an anodic films grown to 4 V(SCE) on sputter-deposited Ti-10at.%Zr in 0.5 M $H_2SO_4$. $v_{scan}$ = 10 mV s$^{-1}$ and solution: 0.5 M $H_2SO_4$.

An expression for the efficiency of free carriers generation, $\eta_g(r_0, \overline{F})$, for amorphous materials was given by Pai and Enck.[120] This last expression shows that very low efficiency of free carriers generation are expected at low electrical fields and thermalization lengths $r_0$ i.e., at photon energy near the mobility gap.[121] This finding could affect in some extent the measured optical band gap value of amorphous anodic films.

According to Eq. (50) any dependence of $I_{ph}$ vs. $U_E$ curves (photocharacteristics) from the energy of incident photons must be traced out to the dependence on energy and electric fields, and it has been frequently observed in amorphous SC (refs), as shown in Fig. 12 where we report the experimental photocurrent vs. potential curves at different λ relating to an anodic films grown to 4 V(SCE) on sputter-deposited 90Ti-10Zr.[122]

We like to stress two aspects in the expression of the photocurrent in a-SC/El junction. The first one is that from Eq. (50) it follows a direct proportionality between the photocurrent and the

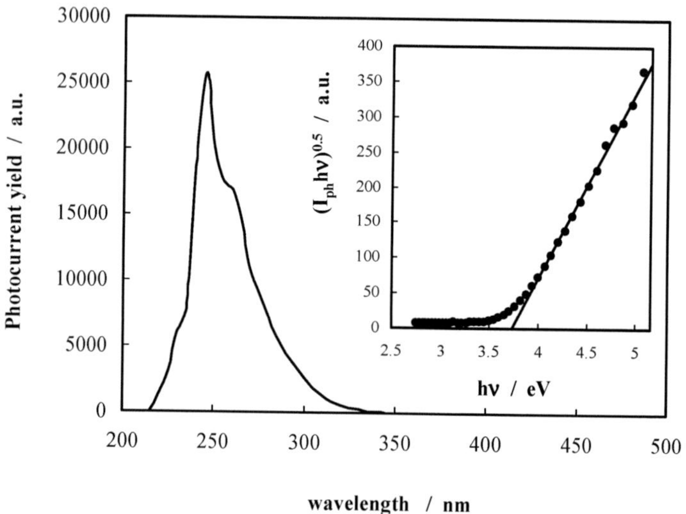


Figure 13. Photocurrent action spectrum recorded at +2 V(SCE) for a film grown up to 4 V(SCE) on sputter-deposited Ti-42at.%Zr in 0.1 M ammonium pentaborate electrolyte. Inset: determination of the optical band gap by assuming indirect transitions. Reprinted from M. Santamaria, F. Di Quarto, and H. Habazaki, "Photocurrent spectroscopy applied to the characterization of passive films on sputter-deposited Ti–Zr alloys." *Corr. Sci.* **50** (2008) 2012, Copyright (2008) with permission from Elsevier.

absorption light coefficient for $\alpha L_d \gg 1$ (no recombination) and $\alpha X_{SC} \ll 1$ (slightly absorbed light), as previously derived for crystalline materials (see Eq. 43). On the other hand for $\alpha L_d \ll 1$ still a direct proportionality between $I_{ph}$ and $\alpha$ is assured by the fractional term $\alpha L_d/(1 + \alpha L_d)$. According to these considerations, we can still assume for amorphous SCs a direct proportionality between the photocurrent yield, $Q = I_{ph}/e\Phi_0$, and the light absorption coefficient, $\alpha$, in the vicinity of the absorption edge under constant electrode potential. Like for crystalline materials, this allows to replace $\alpha$ with the photocurrent yield in deriving the optical band gap of amorphous semiconducting films from the photocurrent spectra (see Fig. 13).

A second aspect we like to stress is that Eq. (50) contains as a particular case the expression of the photocurrent for an amor-

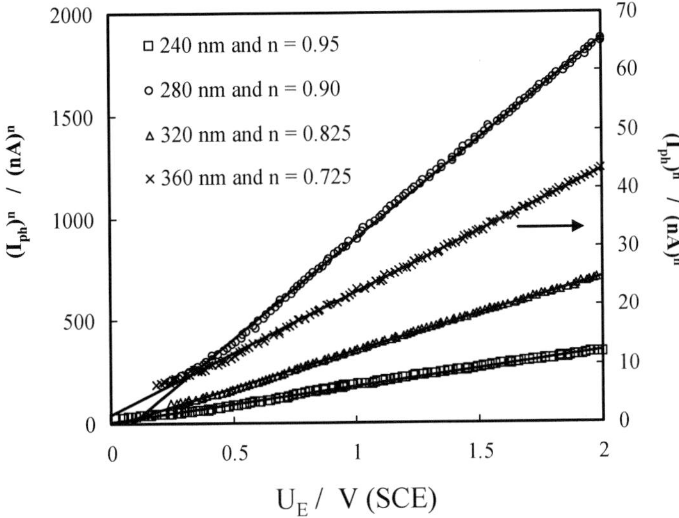



Figure 14. Fitting of the photocurrent vs. potential curves of Figure 12.

phous insulator/electrolyte junction after substitution of the film length, $d_f$, to the space charge region $X_{SC}$. According to this we can write for insulating film/electrolyte junction the relationship:

$$I_{ph} = e\Phi_0\,\eta_g\,\frac{\alpha\,L_d}{1+\alpha\,L_d}\left[1\;-\;\exp\left(-d_f\,\frac{1+\alpha\,L_d}{L_d}\right)\right] \qquad (51)$$

The usual expression for the mean electric field in insulator $F_{av} = (U_E - U_{fb})/d_f$ holds in absence of trapping phenomena.[39,123] In both cases the variation of the efficiency of generation with the electric field and photons energy, through $r_0$, can account for changes in $E_g^{opt}$ values measured at different electrode potentials as well as for the dependence from the incident photon energy of the photocurrent vs. potential curves. This last aspect has been deeply discussed both for semiconducting (a-WO$_3$ and a-TiO$_2$) and insulating (a-Ta$_2$O$_5$) anodic films in previous works.[39,121,124] In Fig. 14 we report the fitting curves of experimental data for films grown to 4 V(SCE) on 90Ti-10Zr. The details of fitting procedure can be found in Ref. [122]; here we have to mention that the flat band potential can now be derived as a fitting parameter of the photo characteristics.

### (iii)    Optical Gap in Amorphous Materials

It was previously mentioned that optical transitions at energies near the band gap of a crystalline material may be direct (allowed or forbidden) or indirect. In the case of amorphous materials, owing to the relaxation of the k-conservation selection rule:

> "no intervention of phonons is invoked to conserve momentum and all energy required is provided by the incident photons."[81]

By assuming again a parabolic DOS distribution in the vicinity of the mobility edges of both the conduction and valence band (above $E_C$ and below $E_V$, with reference to Fig. 3) it has been shown[82] that for amorphous materials the following relationship holds:

$$\alpha h v = const \left( h v - E_g^m \right)^2$$

(52)

where $E_g^m = E_C - E_V$ is now the mobility gap of the a-SC (see Fig. 3). The exponent 2 is reminiscent of the indirect optical transitions in crystalline material but now photons interact with the solid as a whole: this type of transition in amorphous materials is termed non-direct. Because some tailing of states is theoretically foreseen for a-SC by any proposed model of DOS, $E_g^m$ represents an extrapolated rather then a real zero in the density of states. On the other hand in presence of a DOS distribution varying linearly with energy in the ranges $E_C - E_A$ and $E_B - E_V$ of Fig. 3b, it is possible to get for the absorption coefficient[81] of amorphous material the following relationship:

$$\alpha h v = const \left( h v - E_g^{opt} \right)^2$$

(53)

where $E_g^{opt}$ now represents the difference of energy $(E_A - E_V)$ or $(E_C - E_B)$ in Fig. 3b, whichever is smaller, whilst the constant assumes values close to $10^5$ $eV^{-1}cm^{-1}$. The range of energy in which Eq. (53) should be valid is in the order of 0.4 eV or less.[81] In order to distinguish between these two different models of optical transitions, both giving a similar dependence of absorption coefficient on the photon energy, we will refer to the first one as the Tauc's approximation and to the second one as the Mott-Davis approximation. From the first one we derive an estimation of the mobility gap and from the second one the optical gap of amorphous materials. If $(\alpha h\nu)^{0.5}$ vs. hv plots display a linear region larger than 0.4 eV it seems more correct to interpret the data on the basis of the Tauc's model of optical transitions. The coexistence of both types of transitions has been reported for thin anodic films grown on niobium,[125] with the presence of a mobility gap in the order of 3.5 eV in the high photon energy range extending around 1 eV and an optical gap (in the Mott's sense) of about 3.05 eV.

In the case of anodic films on valve-metals, an exponential decrease in the photocurrent yield (Urbach tail) as a function of photon energy is frequently observed at photon energies lower than the mobility gap. A possible origin of such dependence can be attributed to a variation of the light absorption coefficient according to the following law:

$$\alpha = \alpha_0 \exp\left(-\gamma\frac{E_0 - h\nu}{kt}\right) \tag{54}$$

with $\gamma$ and $\alpha_0$ constant. This relationship, which has been found to hold also for crystalline materials, has been rationalized in the case of a-SCs by assuming an exponential distribution of localized states in the band edge tails.[126] In this case $E_0$ marks the energy where $\ln\alpha$ vs hv (Urbach plot) ceases to be linear. This value frequently coincides with the mobility gap value determined according to Eq. (52).

A typical example is reported in Fig. 15 for the anodic film of Fig. 13: the value of $E_g^m$, equal to about 3.75 eV, is in good agreement with the value of $E_0$ ($\cong$ 3.85 eV) derived form the Ur-

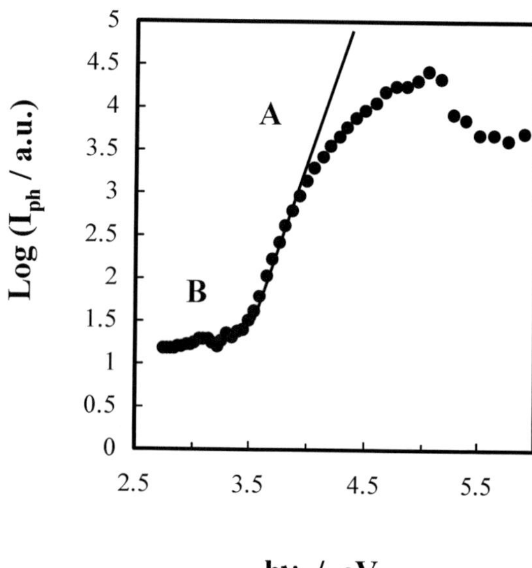


**hv / eV**

Figure 15. Urbach plot relating to the anodic film of Figure 13. Reprinted from M. Santamaria, F. Di Quarto, and H. Habazaki, "Photocurrent spectroscopy applied to the characterization of passive films on sputter-deposited Ti–Zr alloys." *Corr. Sci.* **50** (2008) 2012, Copyright (2008) with permission from Elsevier.

bach plot. Other explanations have been suggested for this behavior in the case of crystalline materials.[81] At very low absorption levels, a second exponential part in the Urbach plot (weak absorption tail, B in Fig. 15c) can be frequently observed. Such behavior has been interpreted by assuming a dependence of the light absorption coefficient on energy according to the following relationship:

$$\alpha \propto \exp\left(\frac{h\nu}{E_t}\right) \tag{55}$$

where the energy parameter $E_t$ is always larger than $E_0$.[127] This part of the log $\alpha$ vs. *hv* plot is not as well reproducible as that at higher energy due the high structure sensitivity of the light absorption coefficient in this energy range.[127]

In agreement with a general statement reported in Ref. [81], it has been suggested[39–40] that, in absence of appreciable differences in short range order of amorphous and crystalline counterparts, the mobility gap of amorphous anodic films should be equal or larger than the band gap of the crystalline counterpart. Such a difference in optical band gap value can be assumed as a measure of the influence of lattice disorder on optical gap of the films.

In Table 1 we report the mobility gap and band gap values of some passive films grown on valve metals. The difference

$$\Delta E_{am} = (E_g^m - E_g^{cryst})$$ in the range of $0.1 - 0.35$ eV, is in agree-

ment with the expected extension of the localized states regions near the band edges due to the lattice disorder.[81] Values of $E_0$ (see Eq. 54) nearly coincident with the mobility gap, $E_g^m$, have been

## Table 1

**Measured Optical Gap, $E_g^m$ for Passive Films on Pure Metals Compared with the Band Gap of the Crystalline Counterpart, $E_g^{cryst}$. $\Delta E_{am}$ is the Difference Between $E_g^m$ and $E_g^{cryst}$.**

| Phase | $E_g^m$ (eV) | $E_g^{cryst}$ (eV) | $\Delta E_{am}$ (eV) |
|---|---|---|---|
| $ZrO_2$ | 4.70~4.80[137] | 4.50 [a] | 0.20~0.30 |
| $Ta_2O_5$ | 3.95~4.05[137] | 3.85 [a] | 0.10~0.20 |
| $Nb_2O_5$ | 3.30~3.40[137] | 3.15 [a] | 0.15~0.25 |
| $TiO_2$ | 3.20~3.35 [a] | 3.05 (rutile) [a] <br> 3.20 (anatase) [a] | 0.15~0.20 |
| $WO_3$ | 2.95~3.15[137] | 2.75[137] | 0.20~0.40 |
| $MoO_3$ | 2.95~3.10[137] | 2.90[144] | 0.05~0.20 |
| $Cr_2O_3$ | 3.30~3.55[137] | 3.30[137] | 0.0~0.25 |
| $NiO$ | 3.43[137] | 3.45~3.55[137] | 0 |
| $Cu_2O$ | 1.86[137] | 1.86[137] | 0 |
| $Fe_2O_3$ | 1.90~1.95[137] | 1.90[137] | 0~0.05 |
| $Fe_{0.25}Ti_{0.75}O_{1.875}$ | 2.95[137] | 2.80[a] | 0.15 |
| $Fe_{0.1}Ti_{0.9}O_{1.95}$ | 3.15[145] | 3.00[a] | 0.15 |

[a] Estimated from the corresponding crystalline phases according to Eqs. (60a) and (61).

frequently derived for passive films on valve metals. It seems quite reasonable to suggest, for such a class of amorphous materials, a band model similar to that shown in Fig. 3b with an exponentially decreasing DOS in the mobility gap of the films at energies lying below $E_C$ and/or above $E_V$. A mobility gap of passive film lower than the band gap of the crystalline counterpart must be interpreted as an indication that differences are present in the short-range order of the two phases. A different short-range order can imply the formation of a defective structure, with a high density of localized states within the mobility gap as well as changes in the density of the passive film, which is known to affect also the value of the optical gap in amorphous materials.[79–81] The experimental findings on passive films and corrosion layers suggest that large differences in optical gap values, between amorphous and crystalline counterparts, should be traced out to a different chemical environment around the metallic cation or to the presence of large amount of defects within the passive films, originating electronic states within the mobility gap. A remarkable case is reported in Ref. [128], where the incorporation of organic species into anodic films, grown on electropolished Al samples in tartrate containing solution, originated a DOS distribution within the band gap of a-$Al_2O_3$ so allowing the onset of anodic photocurrent at photon energies ($h\nu \cong 3.0$ eV) well below the band gap of $Al_2O_3$ ($E_g \geq 6.30$ eV).

### (iv) *Photoemission Phenomena at the Metal/Passive Film Interface*

In this Section we discuss the role of the inner metal/film interface in the generation processes of photocarriers for thin and thick passive films. In presence of thin passive films it is possible that under illumination a large fraction of photons impinging the film/solution interface arrive at the metal/film interface, by exciting metal electrons to higher energy levels and leaving vacant states below the Fermi level of the metal. The fate of the excited states into the metal depends on the occurrence of different physical deactivation processes at this interface. Apart the thermal deactivation by scattering of excited electrons with the lattice vibrations, photoemission phenomena of excited photocarriers can be observed. In the case of very thin passive films ($d_{ox} \leq 2$ nm) external (into the electrolytic solution) photoemission processes become possible by tunnelling of excited electrons or holes at the metal surface throughout the film. A hole photoemission process has been suggested in the case of a gold electrode covered with a very thin oxide.[129] The photoemission of electrons directly from the metal to the ground state of liquid water has been observed more frequently through very thin oxide films covering metals.[130–132] When such an external photoemission process occurs, in absence of diffuse double layer effects or large adsorbed molecules, it is possible to write for the emission photocurrent the so-called *five half (5/2) power law*, which gives the dependence of photocurrent from photon energy and electrode potential as:[39,40,133,134]

$$I_{ph} = \text{const.}(h\nu - h\nu_0 - |e|U_E)^{5/2} \tag{56a}$$

where $U_E$ is the electrode potential measured with respect to a reference electrode, $h\nu_0$ is the photoelectric threshold at zero electrode potential (changing with reference electrode) and $h\nu$ is the photon energy in eV. At constant potential the photocurrent yield can be expressed as:

$$Q^{0.4} = \text{const.}(h\nu - E_{th}) \tag{56b}$$

where $E_{th} = h\nu_0 + |e|U_E$ is the measured photoemission threshold, dependent on the imposed electrode potential. It follows that a change in the photoemission threshold vs. potential of 1 eV/V is expected. By assuming that the electron photoemission process occurs from the metal Fermi level to the bottom of the conduction band[134] of solvent, $E_{solvent}^C$, this last level can be estimated (with respect to the vacuum) according to the equation:

$$E_{solvent}^C = E_{th} - |e|U_E(ref) - |e|U_{E,ref}(vac) \tag{57}$$

where $E_{th}$ is the photocurrent emission threshold calculated according to Eq. (56b), $U_E$ is the electrode potential with respect to a reference electrode, and $U_{E,ref}(vac)$ is the reference electrode potential with respect to the vacuum. A value of 4.60 V has been assumed on the vacuum scale for the NHE according to different authors.[23]

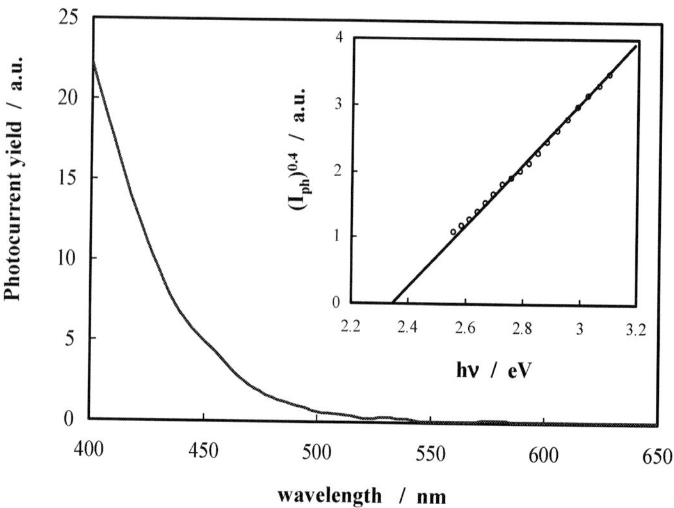



Figure 16. Photocurrent spectrum relating to Mg after mechanical treatment recorded by polarizing the electrode at - 1.94 V (MSE) in 0.1 M $Na_2SO_4$ (pH = 6.15). Inset: estimate of the external photoemission threshold. Reprinted from M. Santamaria, F. Di Quarto, S. Zanna, and P. Marcus, "Initial surface film on magnesium metal: A characterization by X-ray photoelectron spectroscopy (XPS) and photocurrent spectroscopy (PCS)." *Electrochim. Acta* **53** (2008) 1314, Copyright (2007) with permission from Elsevier.

Figure 16 shows the photocurrent spectrum, in the long wavelengths region, recorded at constant potential $U_E = -1.94$ V(MSE) (more cathodic than the respective $U_{OC}$ potential), for a freshly prepared mechanically cleaned Mg electrode (MTE) at pH = 6.15, from which an external photoemission threshold of 2.35 eV was estimated (see inset)[135]. At very close electrode potential ($-1.97 \leq U_E \leq -1.84$ V(MSE)) an almost identical photoemission threshold was measured for MTE Mg electrodes in a wide range of solution pH (6 < pH < 14), allowing to estimate an almost constant $E^C_{H2O} \cong$ $-1$ eV (see Fig. 17). This value agrees quite well with previously reported value ($-1 \pm 0.1$ eV).[39,40,133,136]

The same photoemission processes can operate in organic solvent, as shown in Fig. 18, where we report the cathodic photocurrent spectra of MTE Mg electrode recorded at - 2.4 V(SCE) in 0.1

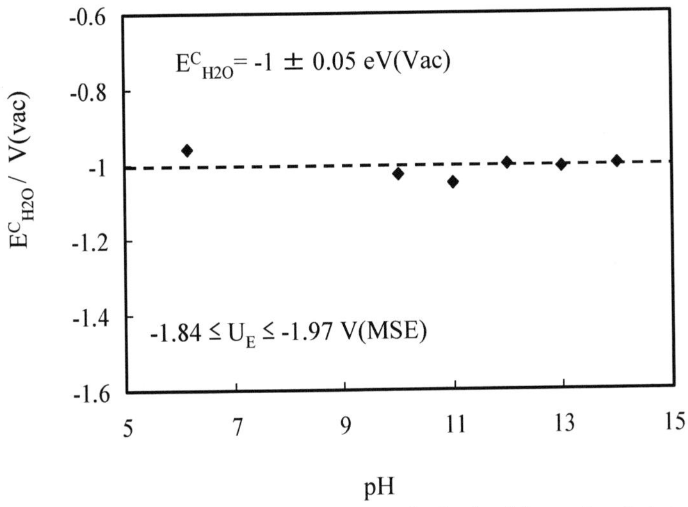



Figure 17. Estimated energy levels of water conduction band from external photo-emission data relating to Mg after mechanical treatment as a function of solution pH. Reprinted from M. Santamaria, F. Di Quarto, S. Zanna, and P. Marcus, "Initial surface film on magnesium metal: A characterization by X-ray photoelectron spectroscopy (XPS) and photocurrent spectroscopy (PCS)." *Electrochim. Acta* **53** (2008) 1314, Copyright (2007) with permission from Elsevier.

M $LiClO_4$ in propylene carbonate electrolyte. As evidenced in the figures (see inset) the photocurrent yield plotted according to the 5/2 law (see Eq. 56) gives an onset photocurrent threshold value of ~ 1.8 eV. The threshold values are reported as a function of $U_E$ in Fig. 19. A slope of ~ 1 eV/V was derived, in agreement with the theoretical expectation.[134] It is interesting to mention that the $(I_{ph})^{0.4}$ vs. $U_E$ plots at different wavelengths ($\lambda = 450, 400$ nm) (see Fig. 20) showed a zero photocurrent potential value which shifted with photon energy (about 1 V/eV) as foreseen from the theory (see Eq. 56).[134]

The data of Fig. 19 and 20 allow to locate the bottom of the conduction band of the organic solvent (or the energy level of photoemitted electron in propylene carbonate) by subtracting to the photoemission threshold (4.15 eV), at zero electrode potential, the Fermi energy level of SCE measured with respect to the vacuum (–4.84 eV). It came out that the bottom of the conduction band of

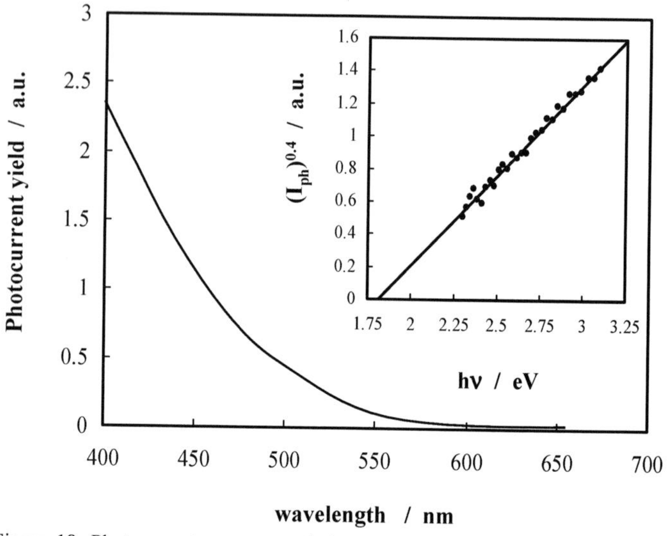



Figure 18. Photocurrent spectrum relating to Mg after mechanical treatment recorded by polarizing the electrode at - 2.40 V(SCE) in 0.1 M LiClO$_4$ propylenc carbonate electrolyte. Inset: estimate of the external photoemission threshold. Reprinted from M. Santamaria, F. Di Quarto, S. Zanna, and P. Marcus, "Initial surface film on magnesium metal: A characterization by X-ray photoelectron spectroscopy (XPS) and photocurrent spectroscopy (PCS)." *Electrochim. Acta* **53** (2008) 1314, Copyright (2007) with permission from Elsevier.

propylene carbonate is located at about –0.7 eV with respect to the vacuum. Such a value incorporates an uncertainty on the potential drop at the organic solvent/SCE junction. However the value derived for the conduction band bottom edge of propylene carbonate compares quite well with corresponding values reported in Ref. [134] for other organic solvents and it allows to get an estimate for the Volta potential difference ($\approx$ –0.3 V) between Mg and propylene carbonate at zero charge.

In the case of thicker films ($d_{ox} \geq 5$ nm), where the external photoemission processes are forbidden, the possibility of an internal photoemission process due to the injection of photoexcited electrons (or holes) from the metal into the CB (or VB) of the passive film must be considered. In such a case the internal photocur-

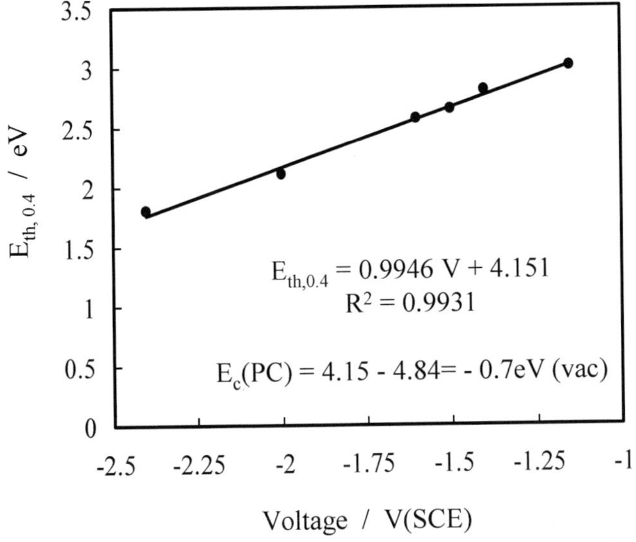



Figure 19. External photoemission threshold as a function of electrode potential relating to Mg after mechanical treatment. Sol: 0.1 M $LiClO_4$ in propylene carbonate. Reprinted from M. Santamaria, F. Di Quarto, S. Zanna, and P. Marcus, "Initial surface film on magnesium metal: A characterization by X-ray photoelectron spectroscopy (XPS) and photocurrent spectroscopy (PCS)." *Electrochim. Acta* **53** (2008) 1314, Copyright (2007) with permission from Elsevier.

rent emission varies with the photon energy according to the so-called Fowler photoemission law:[134]

$$Q = const \left(h\nu - E_{th}\right)^2 \tag{58}$$

where $E_{th}$ is the internal photoemission threshold energy, which can be obtained from a plot of $QQ^{0.5}$ vs. the photon energy, $h\nu$, at constant photon flux. This threshold is a measure of the distance in energy between the Fermi level of the metal and the CB (electron photoemission) or VB (hole photoemission) edge of the film. The

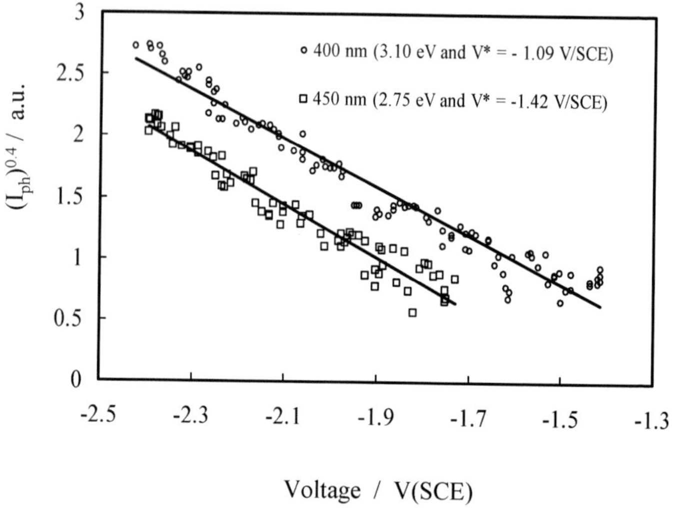



Figure 20. $(I_{ph})^{0.4}$ vs. $U_E$ plots for mechanical treated Mg, recorded by scanning the electrode potential at 10 mV s$^{-1}$. Sol: 0.1 M LiClO$_4$ in propylene carbonate. Reprinted from M. Santamaria, F. Di Quarto, S. Zanna, and P. Marcus, "Initial surface film on magnesium metal: A characterization by X-ray photoelectron spectroscopy (XPS) and photocurrent spectroscopy (PCS)." *Electrochim. Acta* **53** (2008) 1314, Copyright (2007) with permission from Elsevier.

occurrence of electron or hole internal photoemission in the case of insulating films is established by the direction of the electric field, and in turn by the electrode potential value with respect to the photocurrent sign inversion potential. In absence of trapping effects, the inversion photocurrent potential can be used to determine the flat band potential of insulating passive films. In the case of insulating anodic films on valve-metals internal electron photoemission processes are usually observed under cathodic polarization and under illumination with photons having energy lower than the optical band gap of the film.[130,132] In Fig. 21 we report the determination of the internal photoemission threshold for a passive film grown on Mg metal after mechanical treatment and immersion in 0.1 M NaOH at 80°C for 1 h, recorded by polarizing the electrode at −1.53 V(MSE) in 0.1 M NaOH at room temperature.

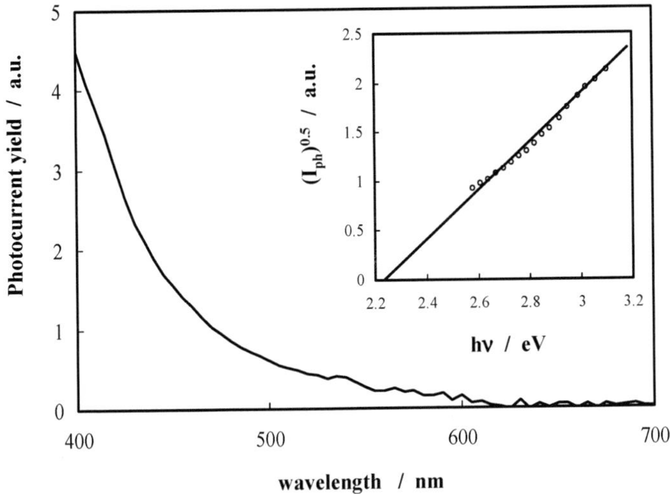

Figure 21. Photocurrent spectrum relating to Mg after mechanical treatment and immersion in 0.1 M NaOH at 80°C for 1 h, recorded by polarizing the electrode at - 1.93 V(MSE) in 0.1 M NaOH at room temperature. Inset: estimate of the Fowler photoemission threshold. Reprinted from M. Santamaria, F. Di Quarto, S. Zanna, and P. Marcus, "Initial surface film on magnesium metal: A characterization by X-ray photoelectron spectroscopy (XPS) and photocurrent spectroscopy (PCS)." *Electrochim. Acta* **53** (2008) 1314, Copyright (2007) with permission from Elsevier.

It is interesting to stress that the thickening of the film grown in alkaline solution at high temperature hinders the direct photo-emission process of electrons into the electrolyte observed for the same metal after scraping. In the case of semiconducting films no evidence of internal photoemission process is expected, owing to the absence of any electric field at the metal/film interface, as long as the space charge region of the SC is less than the film thickness. The knowledge of the internal photoemission thresholds allow to locate the energy level of the conduction band of the oxide films with respect to the Fermi level of the underlying metal, once the work function of the metal is known. In Table 2 the internal thresholds for cathodic photoemission of a series of insulating ox-ide films grown on different valve-metals have been reported.

     

## Table 2
### Threshold Energy Values Derived from the Fowler Plots for the Internal Photoemission Process at the Metal/Passive Film Interface. The Uncertainty is of ± 0.1 eV.

| Interface | $E_{th}^{F}$ (eV) |
|---|---|
| Y/Y$_2$O$_3$ | 2.10[39] |
| Bi/Bi$_2$O$_3$ | 1.40[39] |
| Pb/PbO | 1.00[39] |
| Zr/ZrO$_2$ | 1.80[39,146] |
| Hf/HfO$_2$ | 2.15[39,147] |
| Al/Al$_2$O$_3$ | 2.00[39] |
| Ta/Ta$_2$O$_5$ | 1.50[39] |
| Mg/Mg(OH)$_2$ | 2.15[135] |
| anodic oxide on Mo-79at.%Ta | 1.75[39,144] |
| anodic oxide on Al-34at.%Ta | 1.53[39,143] |
| anodic oxide on Al- 25at.%Nb | 2.25[a] |
| anodic oxide on Al- 44at.%Hf | 2.40[a] |
| anodic oxide on Al- 34at.%Hf | 2.48[a] |

[a]Unpublished data.

## 4. Band Gap and Oxide Film Composition

### (i) Binary Crystalline Oxides

In spite of the relevant information on the structure of the passive film/electrolyte junctions obtainable by PCS a more widespread use of this technique in corrosion studies has been hampered by the lack of a viable theory which relates the measured E$_g$ values to the passive film composition. In general terms such a task is a very challenge one, also for advanced theoretical methods based on quantum mechanical calculations. It is within the aims of this chapter to show that it is possible to relate the band gap values of numerous oxides to their composition by means of a semi-empirical approach. In a previous work[137] it has been shown that such a task could be accomplished by proposing the following general correlation between the band gap of crystalline oxides and the difference of electronegativity of their constituents:

$$E_g = 2[E_I(\chi_M - \chi_O)^2 + \Xi] \tag{59}$$

where according to Phillips[138] $E_I$ is the extra-ionic energy unit orbitally dependent, assumed "to vary with hybridisation configuration, i.e., with different atomic co-ordination in different crystal structures." $\chi_M$ and $\chi_O$ are the electronegativities of metal and oxygen respectively. By plotting the band gap values of numerous oxides as a function of $(\chi_M - \chi_O)^2$ it comes out that the proposed correlation was able to provide:

(a) apart few exceptions (see below), two clearly separated interpolating lines were found for sp-metal and d-metal oxides, according to the following equations:

d-metal oxide: $\quad E_g = 1.35(\chi_M - \chi_O)^2 - 1.49 \quad$ in eV   (60a)

sp-metal oxide: $\quad E_g = 2.17(\chi_M - \chi_O)^2 - 2.71 \quad$ in eV   (60b)

(b) a better fitting of the experimental data with respect to previous proposed correlation as evidenced by the higher correlation coefficient values.

From the d-metal correlation it follows that metallic oxides $(E_g \cong 0)$ are expected to form on metals having Pauling's electronegativity value around 2.45 in agreement with the common experience that noble metal oxides at higher oxidation states ($RuO_2$, $IrO_2$ notably) usually display metallic conductivity. From a practical point of view two more aspects are interesting for corrosion studies:

(a) in this correlation NiO stays neatly on the sp-metal oxides interpolating line, whilst $Cr_2O_3$, FeO, $Fe_2O_3$, $Cu_2O$ and CuO are well interpolated as d-metal oxides;
(b) three non transitional-metal oxides (PbO, $In_2O_3$, $Tl_2O_3$) are better interpolated like d-metal oxides. According to this an intriguing d-/sp-metal oxides dividing line along the diagonal Zn, In, Pb/Ga, Sn, Bi appears, with some of sp-metals (In, Tl, Pb) of higher atomic number showing a d-like behavior in terms of optical band gap vs. $(\chi_M - \chi_O)^2$ correlation. Like all semiempirical approaches the correlation cannot account for such a different behavior, moreover a further complication comes out from the fact that in some ma-

terials the nature of optical transition (direct or indirect) can affect sensibly the measured band gap value.

As for the electronegativity values in all calculations the Pauling's scale of electronegativity,[139] integrated with the Gordy-Thomas values[140] has been used with the exception of Tl(III) for which the value given by Allred[141] has been preferred. The electronegativity value of different elements, calculated by using the experimental band gap values of the corresponding oxides and according to the best fitting straight lines, differ from those reported in Refs. [139-141] by a quantity of about 0.05, which is more or less the uncertainty given by the authors. According to this procedure in Table 3 the estimated electronegativity for a group of lanthanides metals, obtained by using the d-metal correlation and the band gap values reported in the same table, is reported. The electronegativity values for such a group of metals stay within the limits 1.1–1.3 usually reported for f-block elements and in fair agreement with those reported by Allred[141] for the same elements at different oxidation states. Although the proposed correlation seems to work nicely also for f-block elements, some uncertainty still remains as for the parameters of the sp-metal oxides correlation owing to the limited numbers of oxides band gap values used to derive it as well as to the difficulty to get reliable optical band gap values for sp-metals having very low electronegativity parameters.

**Table 3**

**Experimental Band Gap Values and Electronegativity Parameter Estimated by Using the d-Metal Correlation for a Group of Lanthanides Metals.[40]**

| Phase | $E_g$ / eV | $\chi_M$ |
|---|---|---|
| $Sm_2O_3$ | 5.0 | 1.31 |
| $Dy_2O_3$ | 5.0 | 1.31 |
| $Yb_2O_3$ | 5.2 | 1.27 |
| $Gd_2O_3$ | 5.3 | 1.26 |
| $La_2O_3$ | 5.4 | 1.24 |
| $CeO_2$ | 5.5 | 1.22 |

## (*ii*) *Ternary Crystalline Oxides*

The most interesting aspect embodied in the proposed correlation is the possibility to use such relationships for predicting the band gap of mixed oxides so opening a new route to the quantitative characterization of corrosion layers on metallic alloys. It was suggested that Eq. (60) could be extended to ternary oxides, $A_aB_bO_o$,[137] containing only d,d-metal or sp,sp-metal oxides, by substituting to the metal electronegativity, $\chi_M$, the arithmetic mean for the cationic group, $\chi_c$, defined as:

$$\chi_c = \frac{a\,\chi_A + b\,\chi_B}{a+b} \tag{61}$$

where $a$ and $b$ are the stoichiometric coefficients of the cations in the ternary oxide, and $\chi_A$ and $\chi_B$ their electronegativities.

In Table 4 the experimental band gap values for a number of binary and ternary d,d-metal oxides covering a quite large range of

## Table 4
**Experimental Optical Band Gaps Values for d-Metal and d-d-Metals Alloys Oxides and Comparison Between Metal Electronegativity Estimated According to Eqs. (60a) and (61), $\chi_{exp}$, and Pauling Electronegativity.[40]**

| Phase | $E_g^{exp}$ (eV) | $\chi_{exp}$ | $\chi_{Pauling}$ |
|---|---|---|---|
| $Y_2O_3$ | 5.50 | 1.22 | 1.20 |
| $Sc_2O_3$ | 5.40 | 1.24 | 1.30 |
| $CuYO_2$ | 3.50 | 1.55 | 1.55 |
| $MnTiO_3$ | 3.10 | 1.65 | 1.60 |
| $La_2Ti_2O_7$ | 4.00 | 1.48 | 1.445 |
| $FeTiO_3$ | 2.85 | 1.71 | 1.725 |
| $Y_3Fe_5O_{12}$ | 3.00 | 1.68 | 1.64 |
| $CuScO_2$ | 3.30 | 1.62 | 1.60 |
| $Fe_{18}Ti_2O_{31}$ | 2.09 | 1.87 | 1.88 |
| $Fe_8Ti_2O_{16}$ | 2.17 | 1.85 | 1.85 |
| $Fe_6Ti_4O_{17}$ | 2.35 | 1.81 | 1.80 |
| $Fe_{10}Ti_{10}O_{35}$ | 2.50 | 1.78 | 1.775 |
| $Fe_{18}Ti_{22}O_{71}$ | 2.60 | 1.76 | 1.76 |

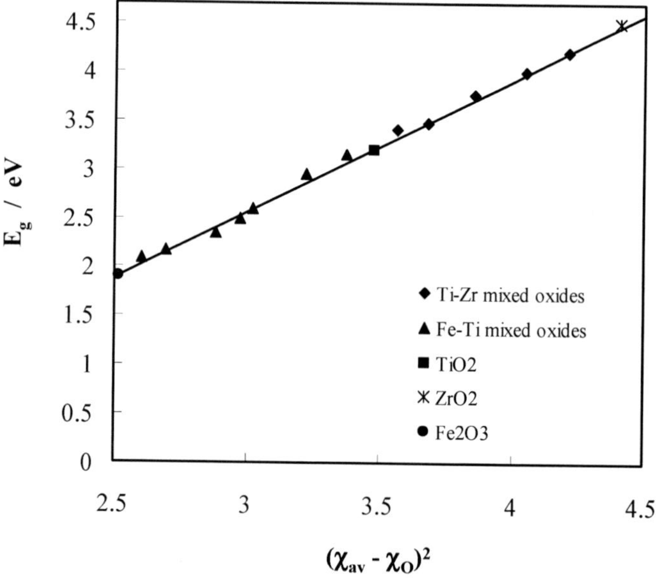


Figure 22. Experimental band gap values relative to Fe-Ti mixed oxides[145] and Ti-Zr mixed oxides[122] as a function of $(\chi_{av} - \chi_O)^2$. The continuous line represents the theoretical prediction according to Eq. (60)a.

band gap values (1.90–5.50 eV) and d-metal electronegativity ($\Delta\chi \cong 0.7$) are reported. A very good agreement is observed between the $\chi$ experimental value obtained by PCS and the Pauling's ones. The difference between the two values stays always within the experimental uncertainty given in the Pauling's book.[139] A further confirmation of the validity of the proposed correlation for d,d-metal ternary oxide comes from the data of Fig. 22 showing the fitting of the experimental $E_g$ values of ternary Ti-Fe and Ti-Zr mixed oxides as a function of different cationic ratio into the films. It is interesting to stress that for amorphous or disordered mixed oxides slightly larger $E_g$ values are experimentally observed  By taking into consideration the results at our disposal for d-metal and d,d-metal mixed oxides we can reasonably state that Eq. (60a) together with Eq. (61) are able to predict the band gap of transition metal oxides within the experimental uncertainty.

Unfortunately we are not aware of a similar large range of band gap values for sp-metal mixed oxides. The few available data seem to confirm the validity of sp-metal correlation also for ternary oxides with differences in the experimental and calculated $E_g$ values in the order of 10%.

As for ternary sp,d-metal oxides it was found that the d-metal correlation was able to fit quantitatively their band gap values provided that the difference in the electronegativity between the metallic cations is less than $0.5^{137}$ (see Table 5). Due to the limited number of systems investigated it remains unsolved the limits of applicability of such correlation to the sp,d-metal mixed oxides. Very recent results both for crystalline bulk Mg-Zn[142] oxides and amorphous anodic oxide on Al-W alloys[143] alloys seem to suggest that when the atomic fraction of d-metal in the ternary oxides reaches values lower than 20% the experimental $E_g$ data are better interpolated by the sp-metal oxide correlation. These aspects need further investigations aimed to better define the limits of validity of the proposed semiempirical correlation.

**Table 5**

**Experimental Optical Band Gaps Values for sp-d-Metal Alloys Oxides and Comparison Between Metal Electronegativity Estimated According to Eqs. (60) and (61), $\chi_{exp}$ , and Pauling Electronegativity.[40]**

| Phase | $E_g^{exp}$ (eV) | $\chi_{exp}$ | $\chi_{Pauling}$ |
|---|---|---|---|
| $Bi_{0.7}Y_{0.3}O_{1.5}$ | 3.0 | 1.68 | 1.69 |
| $SrZrO_3$ | 5.40 | 1.24 | 1.20 |
| $MgTiO_3$ | 3.70 | 1.54 | 1.425 |
| $La_2NiO_4$ | 4.0 | 1.48 | 1.43 |
| $Mg_{0.19}Zn_{0.81}O$ | 3.76 | 1.53 | 1.52 |
| $Mg_{0.27}Zn_{0.73}O$ | 3.92 | 1.50 | 1.49 |
| $Mg_{0.36}Zn_{0.67}O$ | 4.19 | 1.45 | 1.46 |

### (*iii*)   *Amorphous Oxide Films*

The previous correlation was extended to amorphous anodic films and corrosion layers by taking into account the influence of the amorphous nature on the optical band gap of passive films as previously discussed. Numerous investigations on anodic oxide films of valve metals have shown that amorphous oxides usually displayed optical band gap values larger than crystalline counterpart in absence of other specific defects. According to these results it has been proposed to take into account the influence of amorphous nature of passive films on their optical band gap values by modifying the correlation, used for crystalline oxides, as follows:[144]

$$E_{gf} - \Delta E_{am} = A(\chi_c - \chi_O)^2 + B \tag{62}$$

where $E_{gf}$ is the optical band gap of a passive film and $\Delta E_{am}$ represents the difference between $E_{gf}$ and the optical band gap of the crystalline counterpart in eV. The choice of $A$, $B$ and $\Delta E_{am}$ depends on the nature (sp or d) of the metal cations as well as on their relative atomic fraction in the case of mixed oxides.

As for the influence of the lattice disorder on $E_g$ a value of $\Delta E_{am} \leq 0.35 \sim 0.40$ eV seems able to account for the experimental data (see Table 1). $\Delta E_{am}$ values in the order of $0.35 \sim 0.40$ eV are typical of truly amorphous oxides ($MoO_3$, $WO_3$, $Al_2O_3$,), whilst lower values are expected for anodic film oxides having a tendency to grow in microcrystalline forms ($ZrO_2$, $TiO_2$, $Ta_2O_5$). In Fig. 22 it is reported a fitting, according to Eq. (62), of the band gap values as a function of the compositional parameter $\chi_c$ obtained according to Eq. (61) for two different ternary oxides systems containing only d-metal cations and forming both amorphous and crystalline phases as a function of cationic ratio into the film. The oxide films were obtained by anodizing metallic alloys (Ti-Zr magnetron sputtered alloys) or by metal-organic chemical vapour deposition.[122,145] As previously mentioned in both cases a $\Delta E_{am}$ value near to zero, in agreement with the theoretical expectation, is derived for crystalline phases, whilst from the fitting of experimental data a $\Delta E_{am}$ value different from zero is obtained for amorphous phases.

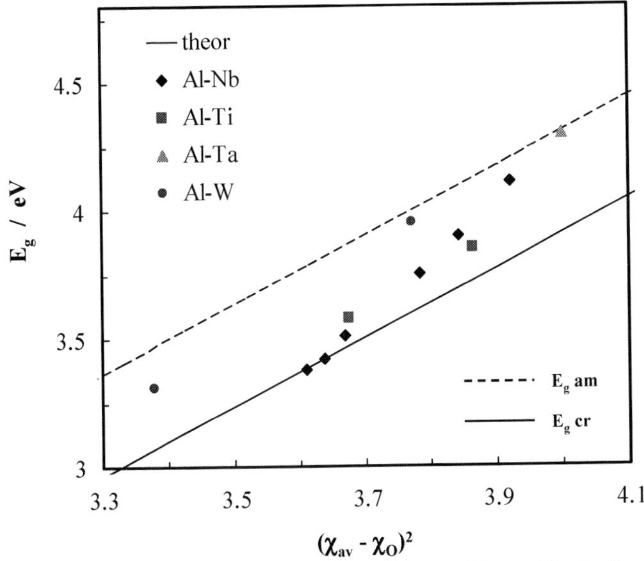


Figure 23. Experimental band gap values relative to Al-d metal mixed oxides[143] as a function of $(\chi_{av} - \chi_O)^2$. The continuous line represents the theoretical prediction according to Eq. (60)a. The dotted line represents the theoretical prediction for amorphous oxides with $\Delta E_{am} = 0.4$ eV to Eq. (62).

The value of $\Delta E_{am}$, as previously mentioned, seems to be a measure of the degree of long range disorder provided that the oxide stoichiometry remains almost the same in the two phases. According to this systems showing a tendency to form microcrystalline phase display band gap values very near to the crystalline counterpart. This is evidenced in Fig. 23 where it is reported the fitting of the band gap values as a function of the compositional parameter $\chi_c$ obtained for different amorphous ternary oxides, containing both sp-metal (i. e. $Al^{3+}$) and d-metal cations, grown by anodizing sputter-deposited or physical vapour deposited metal alloys.[143] In both cases a $\Delta E_{am}$ value in agreement with the theoretical expectation was derived from the fitting of experimental data and, significantly, larger $\Delta E_{am}$ value were measured for anodic films made by oxides having a tendency to grow in truly amorphous state. Further evidence in favour of the proposed correlation for anodic films grown on Mo-Ta, Ti-Zr, Hf-W metallic alloys can be found in Refs. [144,146-148]. We have to mention that apart the nature of the oxide other experimental parameters can affect the lattice disorder degree of passive films and then the $\Delta E_{am}$ value in anodic films.[127,143,146]

## (iv)   *Correlation for Hydroxides and Oxi-hydroxide Films*

For a further improvement toward the possibility to use PCS in a quantitative way for the investigation of corrosion layers grown in different conditions, we need to correlate the optical band gap of hydroxides and oxy-hydroxides to their composition. The experimental data collected on a number of systems suggest that hydroxides films display lower optical band gap values with respect to the corresponding anhydrous oxides. This finding can be rationalized on the basis of the correlation between the band gap of the films and the electronegativity of their constituents. In the case of hydroxide phases we can postulate that the band gap depends on the difference between the electronegativity of metal cation, $\chi_M$, and hydroxyl group, $\chi_{OH}$. The latter can be calculated as the arithmetic mean of those pertaining to oxygen (3.5) and hydrogen (2.2): accordingly a value of 2.85 is obtained.

In order to interpret some experimental data pertaining to passive layers on metals, for which the formation of hydroxides was inferred on the basis of PCS study and supported by thermodynamic considerations or confirmed by surface analytical investigations (XPS and XANES), we assumed[137] that, in agreement with the findings on pure and mixed oxides, also for hydroxides the optical band gap value depends on the square of the difference between the electronegativity of the metallic cation and the average anionic electronegativity ($\chi_M$ - $\chi_{OH}$). In the electrochemical literature there were few, but experimentally well defined, investigated systems which seemed to support our assumption. Further studies, carried out in our laboratory on selected systems, have now provided a number of experimental data sufficient to derive numerical correlation regarding the optical band gap of hydroxide films on both sp- and d-metals.[135,149,150] Figure 24 shows the photocurrent spectrum relating to an $Mg(OH)_2$ film on mechanically polished magnesium formed by anodizing the metal in 0.1 M

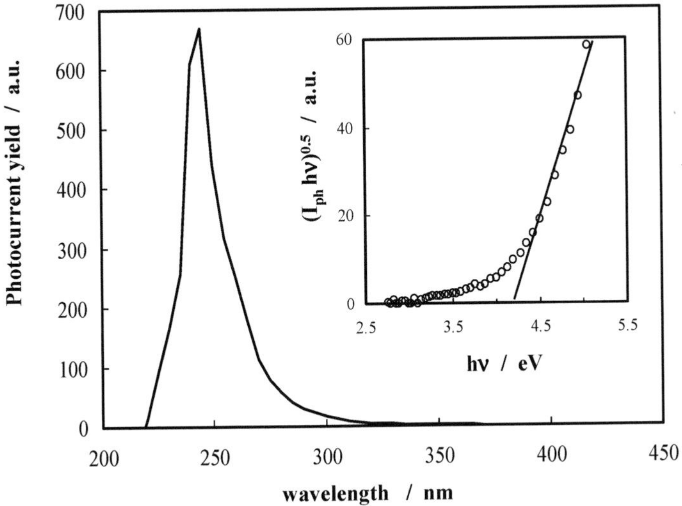

Figure 24. Photocurrent spectrum relating to Mg after mechanical treatment and anodizing in 0.1 M NaOH to 0.6 V(MSE), recorded by polarizing the electrode at the formation potential in the same electrolyte. Inset: $(I_{ph}h\nu)^{0.5}$ vs. $h\nu$ plot. Reprinted from M. Santamaria, F. Di Quarto, S. Zanna, and P. Marcus, "Initial surface film on magnesium metal: A characterization by X-ray photoelectron spectroscopy (XPS) and photocurrent spectroscopy (PCS)." *Electrochim. Acta* **53** (2008) 1314, Copyright (2007) with permission from Elsevier.

NaOH to 0.6 V(MSE): an optical band gap of 4.25 eV is determined, by assuming indirect optical transitions (see inset of the figure), a value well below the value of 7.8 eV, now widely accepted, for MgO. On the basis these experimental data,[149] the following correlation has been proposed:

$$E_g = 1.21(\chi_M - \chi_{OH})^2 + 0.90 \tag{63a}$$

for hydroxides grown on sp-metal, and:

$$E_g = 0.65(\chi_M - \chi_{OH})^2 + 1.38 \tag{63b}$$

for hydroxides grown on d-metals. It is noteworthy that, as previously reported for metal oxides, different relationships hold for sp-metal and d-metal hydroxides. With respect to the analogous correlation obtained for sp-metal oxides (Eq. 60), the most relevant difference in Eq. (63a) comes out from the second term of the right side, suggesting that sp-metal hydroxides present always a finite optical band gap, in the order of 1.70 eV for Tl, the most electronegative sp-metals ($\chi = 2.03$). Although this last correlation suffers some limitations owing to the rather limited number of investigated systems it is able to rationalise some results reported in the literature. In fact from Eq. (63a) an optical band gap around 2.25 eV is derived for $Ni(OH)_2$ by assuming $\chi_{Ni} = 1.80$ according to the Pauling scale and under the hypothesis that also for Ni-hydroxides holds the sp-metal correlation valid for NiO. The optical gap estimated according to Eq. (63a) is in good agreement with the value estimated from photocurrent spectra reported in the literature[151] for passive films on Ni ($E_g = 2.20$ eV) and for $Ni(OH)_2$ electrochemically deposited and soaked in KOH solution.[152] These findings confirm once more that in the proposed correlation Ni(II) oxide and hydroxide conforms to the sp-metal behavior.

Further support to the proposed hydroxides correlations comes from more recent data obtained from the PCS study of conversion coating grown on Mg metal in stannate bath, where the formation of a mixed $MgSn(OH)_6$ layer has been reported,[153] as well as from PCS characterisation of $Co(OH)_2$ and $La(OH)_3$ films grown in alkaline solution.[150,154] Owing to the lack of information on the band gap values of a larger number of hydroxides, the correlation between the optical band gap and the difference of electronegativity of the hydroxides constituents are based on a restrict number of systems and, thus, it must be taken with some caution. Nevertheless, it is possible to make some general considerations on the behavior of some investigated systems lending a further support to their validity.

By comparing the correlation relative to transition metals, it comes out that larger band gap values are expected for the hydroxide phase than for the corresponding anhydrous oxide, when the cation electronegativity is higher than 1.95. Such a finding can help to rationalize the experimental results reported for gold electrode as well as for other noble metals. In the case of gold an optical band gap of about 1.50 eV is expected for the hydroxide phase in comparison with the near to zero band gap value foreseen for the anhydrous oxide. Analogously, for platinum metal ($\chi_M = 2.1$), an increase of the optical gap is calculated from our correlations on going from the oxide ($E_g = 1.16$ eV) to the hydroxide phase (1.75 eV). Although the nature and the photoelectrochemical behavior of the oxide films grown on these noble metals are still under debate,[39] this finding could help to explain the complex behavior observed experimentally.

The last point we like to discuss is the possibility to relate the band gap values of passive films to their hydroxylation degree. At this aim it has been suggested[137] on a purely heuristic basis, a connection formula between the band gap of oxides and the band gap of the corresponding oxi-hydroxide, $MO_{(y-m)}OH_{(2m)}$. According to this suggestion it has been proposed that the band gap of anhydrous and hydroxilated oxide phase can be related trough the relationship:

$$E_g^{hyd} = \frac{E_g^{anh}}{1 + k_m x_{OH}} \tag{64}$$

where $k_m$ is a constant that can be calculated for each system once the band gap values of the anhydrous oxide and hydroxide are known. $x_{OH}$ is the fraction of hydroxyl group into the oxi-hydroxide phase defined as:

$$x_{OH} = \frac{2m}{y + m} \tag{65}$$

It is interesting to note that Eqs. (64) and (65) have been successfully used to get the composition of Chromate Conversion Coating (CCC) formed on aluminium from their band gap values as a function of the hydration degree and as a function of the Cr(III)/Cr(VI) ratio.[155] The photoelectrochemical characterization of CCC grown for different conversion times and in different electrochemical baths data showed that a quite reproducible optical band gap value, around 2.55 eV, is derived for freshly prepared CCCs. Such a value is different from that measured for passive chromium[155] and assigned to Cr(OH)$_3$ anodic film (2.40 eV), and has been attributed to the formation of an amorphous Cr(III)-Cr(VI) mixed oxy-hydroxide of formula $Cr_{0.667}Cr_{0.333}O_{1.2}(OH)_{1.6}$, having a stoichiometric ratio O/Cr equal to 2.80, in fair agreement with data obtained by XANES and Auger spectroscopy.[155]

### (v) *PCS Analysis of Passive Films and Corrosion Layers on Base Metals and Alloys*

One of the most challenging task for any experimental analytical technique is to be able to provide useful information on the electronic properties and chemical composition of passive films on base metals (like: Cu, Fe, Cr, Ni) and their alloys (stainless steels (SS), Fe-Cr alloys, etc.). The previous correlation has been used to relate quantitatively PCS data to passive films composition formed on some of these metals.[131,137,156–159] In the last years an increasing use of such correlation has been registered in PCS studies aimed to characterize passive film on base metals and alloys (Fe-Cr, SS).[160–166] In these complex systems further information on the oxidation state of each metallic cation and their atomic fraction can be necessary in order to use quantitatively the previous correlation. Moreover in some cases the expected optical band gap values of oxides and hydroxides of the same metal, at different oxidation states, are very close or almost coincident so that it becomes difficult to get quantitative information from PCS measurements alone.

In order to highlight this point, in Table 6a we report the band gap values of crystalline oxides and hydroxides of a group of base metals (Cu, Cr, Fe, Ni) experimentally measured by PCS technique or derived from the literature. In the same table we report the Pauling electronegativity parameter for the different oxidation states. Apart from the electronegativity value of Cr(VI) for which we are not aware of other reliable values in literature, we like to stress that the electronegativity data obtained by Eq. (60a) are in very good agreement with those reported in literature[139–141] so that we suggest to use these last values as preferred PCS electronegativity parameters for characterisation of passive films on pure base metals and on their alloys. According to this we reported in Table 6b the estimated band gap values of the corresponding hydroxides as derived from the proposed correlation and by using the PCS preferred $\chi_M$ values.

It comes out from data of Table 6 that oxides and hydroxides of d-metal having electronegativity around 1.90 display very close band gap values, so making very difficult to distinguish by PCS

**Table 6**

**(a)Band Gap Values of Crystalline Oxides and Hydroxides of a
Group of Base Metals. Comparison Between Pauling
Electronegativity and PSC Estimated Electronegativity.[40]**

| Phase | $E_g$ (eV) | $\chi_{Pauling}$ | $\chi_{PCS}$ |
|---|---|---|---|
| $Cr_2O_3$ | 3.30 | 1.60 | 1.62 |
| $Cr(OH)_3$ | 2.43 | 1.60 | 1.62 |
| $CrO_3$ | 2.0 | – | 1.89 |
| $ZnO$ | 3.34 | 1.60 | 1.60 |
| $Cu_2O$ | 1.86 | 1.90 | 1.90 |
| $CuO$ | 1.40 | 2.00 | 2.04 |
| $FeO$ | 2.40 | 1.80 | 1.80 |
| $Fe_2O_3$ | 1.90 | 1.90 | 1.91 |
| $NiO$ | 3.80 (direct) | 1.80 | 1.80 |
|  | 3.58 (indirect) |  | 1.77 |
| $Ni(OH)_2$ | 2.31 | 1.80 | 1.77 |

**(b) Band gap values of some base metal hydroxides estimated
according to Eq. (63)b.**

| Phase | $\chi_{PCS}$ | $E_g$ (eV) |
|---|---|---|
| $Zn(OH)_2$ | 1.60 | 2.39 |
| $CuOH$ | 1.90 | 1.97 |
| $Cu(OH)_2$ | 2.04 | 1.80 |
| $Fe(OH)_2$ | 1.80 | 2.10 |
| $Fe(OH)_3$ | 1.91 | 1.95 |

alone which type of passive films is formed. This is the case of
copper and iron passive films for which at least three different
phases at different hydroxylation degree exist with very close optical band gap values. In such a case further information gathered by
other (possibly in situ) analytical techniques are necessary to rationalise PCS data of passive films.

Other authors[167–169] have proposed a different route for the
identification of passive film. According to these authors a comparison of PCS data (as well as of the general semiconducting behavior) between passive films anodically grown and sputtered oxides could help to identify the nature of passive films. This approach may be useful provided that any difference in the defect

structure and hydroxylation degree for oxide phases grown in such different ways is negligible.

The data of Table 6 and the previous considerations could help to rationalise some of the experimental findings on PCS analysis concerning passive films on iron, for which band gap values ranging between 1.90 eV and 2.10 eV are reported in the literature.[169] A band gap value of 1.90 eV has been estimated for passive films formed on iron in borate buffer solution (pH = 8.4) at high potential ($U_E$ = 0.6 ~ 0.8 V(SCE)).[169] This value matches well the one reported in the same work for evaporated $Fe_2O_3$ and the one reported in Table 6a for crystalline $Fe_2O_3$, thus allowing identifying the passive film composition as Fe(III) oxide. This conclusion is in agreement with recent XANES data suggesting the formation of micro-crystalline iron oxide structure (LAMM phase)[170] on passive iron anodized under similar conditions. The higher $E_g$ measured at lower polarizing voltage ($U_E$ < 0.4 V(SCE)) have been be ascribed to the presence of crystallographic disorder, as in amorphous passive films, or to the partial reduction of Fe(III) to Fe(II) with a subsequent variation of film composition.[171] According to Eqs. (60a) and (61) a band gap of 2.0 eV is expected for a 10% Fe(II) containing oxide in agreement with the value reported in Ref. [169] for oxidised magnetite, while $E_g$ = 2.1 eV can be estimated for $Fe_3O_4$ in contrast with the value reported in the same work. The $E_g$ variation with $U_E$, reported in Ref. [169], is predicted by the above mentioned equations as a function of the Fe(II) content into the film, and it should agree with the description of passive film on iron, described as a Fe(II) deficient magnetite of formula $Fe(II)_{1-x}Fe(III)_2O_{4-x}$, with Fe(II) content in the range of 10-25% depending on the passivation potential.[169]

By considering that a band gap value around of 1.90 ± 0,05 eV has been estimated for Fe(III) oxide and hydroxide, almost coincident with that of $Fe_2O_3$ and $Fe(OH)_3$ is estimated in Table 6, we can conclude that $E_g$ = 1.85 ~ 1.95 eV is expected for iron passive films having composition equal to FeOOH, and thus PCS is not sensitive to variation of the hydration degree of Fe passive films due to the small dependence of the band gap values on the OH content. A detailed discussion of the influence of anodizing solution, spanning a large range of pH (1 ≤ pH ≤ 15), and electrode potential values on the measured optical band gap of passive film on metallic iron can be found in Ref. [171].

The use of PCS in the identification of passive films on iron-chromium-nickel alloys and stainless steel appears even more difficult, despite the appreciable differences in the band gap values of the corresponding pure crystalline oxides. The possible formation of a passive film with an unknown hydration degree and with different metals at different oxidation state makes the PCS analysis very complex. By the way, different photoelectrochemical data on such systems are reported in the literature and the interested reader can access to the published works.[104,160–166,172–183]

Owing to space limitations we will discuss shortly and qualitatively some selected PCS data pertaining to such systems, which can be compared in some extent to more recent quantitative analysis based on in situ and ex situ techniques (XANES, XPS, EQCM).[184–187]

The formation of $Cr(OH)_3$ on passive Cr at low potentials ($U_E \leq -0,6V(MSE)$) and in a wide range of pH ($0 < pH < 9$) has been suggested on the basis of the measured optical band gap value ($E_g = 2.45 \pm 0.1$ eV, see Refs. [159.160-162,188]). Such a hypothesis is in agreement with XPS and XANES data reported in Refs. [184–187] for films grown on Cr and on Fe-Cr alloys in acidic solutions as well as in slightly alkaline solutions.[189]

As for PCS data of passive films grown on Fe-Cr alloys (Cr content $\leq$ 30at.%) in borax buffer solution ($8.4 \leq pH \leq 9.2$) band gap values ranging from 1.90 eV to 2.20 eV with increasing Cr content have been reported in Refs. [177,181]. In a more recent work,[190] by using steady-state PCS data band gap around 3.0 eV have been reported for passive film on Fe-Cr alloys (Cr content $\leq$ 20at.%) at variance with the previous ones. Such difference in $E_g$ values could be attributable, in principle, to a loss of sensitivity in the last method, which could miss to detect the lower photocurrent values usually measured for photon energies near the band gap of passive films. This last hypothesis could also explain higher $E_g$ values measured, for passive films grown on experimental ferritic SS[174] in chloride containing solution, at low anodic potentials ($E_g = 2.80$ eV). On the other hand $E_g$ values around 2.60 eV were measured, at higher electrode potentials ($U_E = 0.6V(SCE)$), for passive films formed on analogous ferritic SS containing small amount ($<$ 4% in weight) of Mo and Ni.[172]

Due to the large range of $E_g$ values collected in the literature for passive films on Fe-Cr alloys, the loss of sensitivity of PCS for photon energies near the band gap of passive films does not seems a satisfactory explanation. The data on passive films composition obtained by in situ and ex situ analytical techniques reported in Refs. [184-187] for Fe-Cr passive films grown at pH values $\leq 4.5$ can be useful to justify these results.

XANES data on various Fe-Cr alloys[184] strongly suggest that at pH = 4.5 (acetate buffer solution) films formed at low potentials ($U_e = -0.3$V(MSE)) are enriched, with respect to the base alloy, of Cr(III) usually present in the films as hydrated species, whilst at higher potentials ($U_E = 0.2$ V(MSE)) the passive films display both a minor Cr enrichment as well as a decrease in hydroxides content. These findings suggest that passive films on Fe-Cr alloys are oxides containing both the metals with a Fe/Cr ratio and a hydration degree dependent on experimental conditions. In a more recent paper[187] the formation of $Cr(OH)_3$ in acidic solution of 0.5 M $H_2SO_4$ has been reported on the basis of a XPS analysis of passive film on Fe20Cr grown at various potentials. More interestingly in the same paper the formation of a chromite ($Fe(II)Cr_2O_4$) compound at anodic potential near to +0.4 V(SHE) was suggested. At still higher potential $U_E > 0.8$V(SHE) the co-presence of Fe(III) and Cr(III) oxide species was suggested in presence of a decreasing amount of Fe(II) and $Cr(OH)_3$ species. A band gap value near to 2.5 eV, very near to that estimated for $Cr(OH)_3$, has been reported in Ref. [160] for a film grown in acidic solution on similar alloy, whilst both the presence of an optical band gap near to 2.5 eV as well as a second optical band gap around 3.0 eV have been reported in Ref. [161] for Fe-18Cr passive film formed in borate buffer solution. We like to mention that as for chromite ($Fe(II)Cr_2O_4$) compound an optical band gap around 3.0 eV is foreseen according to the correlation previously reported, whilst for mixed Fe(III)-Cr(III) oxides at the different composition near those reported in Ref. [187] band gap values ranging between 2.5 eV and 3.0 eV can be estimated on the basis of the correlation for mixed d-metal oxides. It is also interesting that lower $E_g$ values pertain to the passive film grown on Fe-Cr alloys at highest electrode potential. By extending the previous considerations to passive films on ferritic SS, the decrease of their optical band gap with increasing electrode potential can be analogously explained by an increase of the iron content in the passive film.

According to Eqs. (60), (61) and (64), band gap around $2.1 \pm 0.1$ eV strongly point toward the presence on the surface of Cr-Fe alloys of passive films constituted of mixed iron-chromium hydroxides richer in iron, with larger $E_g$ values attributable to passive film at higher Cr content.

As for passive films on austenitic SS (Fe-18Cr-8Ni) and Ni-Cr alloys (alloy 600 series), formed at high anodic potential ($U_E = 0.8$ V(SCE)) in borate buffer solution at pH = 9.2, a rather constant optical band gap ($E_g = 2.35 \pm 0.1$eV) has been reported in Refs. [181–183], with the highest values for passive films on SS at higher Ni content or on Ni-Cr alloys. The $E_g$ value suggests that also in this case a chromium hydroxide rich phase is the main component of the passive films. At lower potentials ($U_E < 0.1$ V(SCE)) higher band gap values ($E_g = 2.75 \pm 0.1$ eV) have been found.[178,180] However, passive films on austenitic SS with higher molybdenum content (> 4% in weight) usually displayed a more limited variation in the $E_g$ values as a function of electrode potentials.[174,178] This finding was particularly evident in the case of the superaustenitic commercial alloy 254-SMO (20Cr-18Ni-6Mo), which displayed an almost constant $E_g$ value ($2.45 \pm 0.1$ eV) in a rather large range of electrode potential ($-0.1 \leq U_E \leq 0.9$ V(SCE)) at pH = 6.5 and in presence of chloride.[174] The beneficial action of Mo in improving the pitting resistance of SS is well known[191] and, according to PCS data, it could be ascribed to its ability to keep on the surface of such an alloy a passive film having an almost constant composition (very close to $Cr(OH)_3$, according to the $E_g$ values) at different potentials. This interpretative hypothesis is in qualitative agreement with the results of STM and XPS analysis reported in Ref. [185] showing that the passive film grown on Fe-18Cr-13Ni in 0.5 M $H_2SO_4$ at 0.5V(SHE) keeps a stable thicker outer layer containing almost chromium hydroxide also at longer polarization times. It seems reasonable to conclude that in the case of 254-SMO alloy the external layer of the passive film formed at pH = 6.5 and in presence of chloride ions in solution has a composition close to $Cr(OH)_3$, as reported for the outer layer of the Fe-Cr-Ni alloy discussed in Ref. [185].

Higher $E_g$ values (> 2.40 eV) are expected for passive films on austenitic SS at higher potentials in presence of a dehydration process and/or for the formation of an anhydrous Cr(III) richer mixed oxides.[185,192] However, very recently, in a series of papers dealing with the PCS characterization of passive film grown in borate solution on different SS (254 SMO, AISI 304L and AISI 316L) Blackwood and coworkers have suggested the possible presence of a surface layer of $Fe_2O_3$ at higher anodic potential and a duplex structure with a chromite-like inner phase underlying the external $Fe_2O_3$ layer at intermediate electrode potentials. At still more cathodic potential the possible formation of a p-type FeO oxide on the surface of the alloy 254SMO austenitic SS was suggested. These results indicate that the nature of passive films could be strongly dependent also from the nature of ionic species present in solution apart from the solution pH. Further studies are necessary for reaching final conclusions on particular systems.

However, all the above mentioned results highlight both the ability of PCS technique to scrutinize very complex systems as well as the possibility to extract semi-quantitative information on the surface layer composition not conflicting with experimental data obtained by other very sophisticated techniques. We are confident that future PCS investigations on carefully chosen systems could provide further evidence in favour of a widespread use of such a technique in passivity and corrosion studies.

## IV.     CONCLUSIONS

A critical analysis of the experimental results on the electronic properties of passive film, obtained by using a traditional approach based on the Mott-Schottky theory, has shown some aspects seriously conflicting with the theoretical hypothesis underlying the validity of the M-S analysis. In this review it has been shown that a better unifying framework for getting out information on electronic properties of passive films is provided by the theory of a-SC Schottky barrier. In the frame of this theory the differential admittance behavior of passive film/electrolyte junction can be qualitatively and quantitatively analyzed providing important information useful to locate the energy levels of the passive film/electrolyte junction as well as new insight on the DOS distribution inside the film.

As for the location of energy levels of the junction, in the case of amorphous or strongly disordered material, they can be seriously in error if the analysis of measured capacitance is carried out by means of traditional M-S approach. Moreover such an approach is not able to explain, neither qualitatively, the changes in the capacitance behavior observed at different *ac* frequencies of electrical signal.

It has been suggested that in presence of very thin passive film, like those reported for base metal and alloys, a different approach, as that suggested by Gerischer for crystalline graphite electrodes, could offer a coherent approach to the study of the highly doped crystalline passive film formed on these material. It appears however rather surprising that, in the case of well documented systems having strongly disordered or amorphous nature, many researchers still employ the classical M-S analysis in spite of the fact that a well developed theory of amorphous semiconductor Schottky barrier has been developed more than 25 years ago, at least, for bulk amorphous semiconductors. Some further refinements to this last theory appear necessary for a quantitative analysis of the electronic properties of amorphous thin passive film in order to account for the limited thickness of the passive film and for the possible metal capacitance contribution to the measured capacitance of the film/electrolyte junction. Once such refinements will be carried out, it is our opinion that we could become able to get more insight also on the influence of anodizing process and post-anodizing treatment on the electronic properties of passive film grown on base-metal and valve-metal too. This last class of materials, which are of large interest in the electronic industry owing to their high k values, and specially anodic niobia and tantala deserve more investigation owing to the fact that they display some attractive features making them ideal candidates for testing further improvements in the theory of a-SC Schottky barriers.

In this review we presented also a relatively simple approach to the fundamentals of PCS technique currently used for characterizing passive films and corrosion layers on metal and alloys. Some of the interpretative models suggested in the past by the present authors to rationalize the experimental PCS data, gathered from many investigated systems, have been validate by different authors but some of them are still under scrutiny and they should be used with some caution. This is particularly true for the proposed correlation between optical band gap of hydroxides and difference of electronegativity of the constituents owing to the still limited, although increasing, number of investigated systems.

Nevertheless, it is in our opinion that, PCS technique remains one of the most versatile techniques at our disposal, for getting out information on the solid-state properties of passive layers in corrosion studies.

The combined use of DA and PCS techniques in the characterization of thin and thick passive films on metal and alloys it has, already, provided important information, on electronic and solid state properties of films, which cannot be accessible by other techniques. If we consider that both techniques are able to test in situ real corroding systems at different time scales and electrode potential window, without any special procedure, it appears reasonable to predict a still wider use of both techniques in corrosion and electrochemical material science.

We are aware that, in many cases, other in situ and ex situ scanning probes having very high spatial resolution or vacuum techniques providing more precise information on the morphology or chemical composition of thin passive layers will be mandatory, but we have no doubts that DA and PCS techniques, together with other electrochemical techniques, will play an unchallenged role for a complete physico-chemical characterization of passive films and corrosion layers.

## REFERENCES


[1] D. Landolt, *Corrosion and Surface Chemistry of Metals*, EPFL Press, Lausanne, CH, 2007.

[2] H.-H. Streblow, in *Advances in Electrochemical and Engineering Science*, R.C. Alkire, D.M. Kolb, Editors, Vol. 8, Wiley-VCH, Weinheim, FRG, 2003, p. 271.

[3] *Passivity of Metals*, Proceedings of the 4th International Symposium on Passivity, R.P. Frankental, J. Kruger, Editors, The Electrochemical Society Inc., Pennington, NJ, 1978.

[4] *Passivity of Metals and Semiconductors*, Proceedings of the 5th International Symposium on Passivity, M. Froment, Editor, Elsevier, Oxford, 1983.

[5] *Passivity of Metals and Semiconductors*, Proceedings of the 6th International Symposium on Passivity, N. Sato, K. Hashimoto, Editors, Pergamon Press, Oxford, 1990.

[6] *Oxide Films on Metals and Alloys*, B.R. MacDougall, R.S. Alwitt, T.A. Ramanarayanan, Editors, PV 92-22, The Electrochemical Society Inc., Pennington, NJ, 1992; *Oxide Films on Metals and Alloys*, K. Hebert, G.E. Thompson, Editors, PV 94-25, The Electrochem Society Inc., Pennington, NJ, 1994.

[7] *Passivity of Metals and Semiconductors*, Proceedings of the 7th International Symposium on Passivity, K.E. Heusler, Editor, Trans Tech. Publications Ltd., Zurich, 1995.

[8] *H.H. Uhlig Memorial Symposium*, F. Mansfeld, A. Asphahani, H. Böhni, R. Latanision, Editors, PV 94-26, The Electrochem Society Inc., Pennington, NJ, 1995.

[9] *Passivity and its Breakdown*, P. Natishan, H. S. Isaacs, M. Janik-Czackor, V. A. Macagno, P. Marcus, M. Seo, Editors, **PV 97-26**, The Electrochemical Society Inc., Pennington, NJ, 1998.

[10] *Critical Factors in Localized Corrosion III*, R. G. Kelly, G. S. Frankel, P. M. Natishan, R.C. Newman, Editors, **PV 98-17**, The Electrochemical Society Inc., Pennington, NJ, 1999.

[11] *Passivity and Localized Corrosion*, M. Seo, B. MacDougall, H. Takahashi, R.G. Kelly, Editors, **PV-99-27**, The Electrochemical Society Inc., Pennington, NJ, 1999.

[12] *Passivity of Metals and Semiconductors*, Proceedings of the 8th International Symposium on Passivity, M. B. Ives, J.L. Luo, J.R. Rodda, Editors, **PV-99-42,** The Electrochemical Society Inc., Pennington, NJ, 2001.

[13] *Surface Oxide Films*, V. Birss, L.D. Burke, A.R. Hillmann, R.S. Lillard, Editors, **PV-2003-25**, The Electrochemical Society Inc., Pennington, NJ, 2004.

[14] *Passivity of Metals and Semiconductors, and properties of Thin Oxide Layers*, Proceedings of the 9th International Symposium on Passivity, P. Marcus, V. Maurice, Editors, Elsevier, Amsterdam, 2006.

[15] W. H. Brattain, C.G.B. Garrett, *Bell System Tech. J.*, **34** (1955) 129; *Phys. Rev.*, **99** (1955) 177.

[16] M. Green, in *Modern Aspects of Electrochemistry*, Ed. by B.E. Conway, J.O'M. Bockris, Butterworths, London, 1959, vol. **2**, p. 343.

[17] H. Gerischer, in *Advances in Electrochemistry and Electrochemical Engineering*, Ed. by P. Delahay, Interscience Publishers, New York, 1961, Vol.1, p. 139.

[18] V. A. Myamlin, Yu. V. Pleskov, *Electrochemistry of Semiconductors*, Plenum Press, New York, 1967.

[19] H. Gerischer, in *Physical Chemistry. An Advanced Treatise*, Ed. by H. Eyring, D. Henderson, W. Jost, Academic Press, New York, 1970, vol. **IXA**, p. 463.

[20] S.R. Morrison, *Electrochemistry at Semiconductor and Oxidized Metal Electrodes*, Plenum Press, New York , 1980.

[21] Yu. V. Pleskov, Yu. Ya. Gurevich, *Semiconductor Photoelectrochemistry*, Consultants Bureau, New York, 1986.

[22] A. Hamnett, in *Comprehensive Chemical Kinetics*, Ed. by R.G. Compton, Elsevier Science, Oxford, 1987, vol. **27**, p. 61.

[23] J.O'M. Bockris, S.U.M. Khan, *Surface Electrochemistry*, Plenum Press, New York, 1993.

[24] N. Sato, *Electrochemistry at Metals and Semiconductor Electrodes*, Elsevier Science B.V., Amsterdam, 1999.

[25] R. Memming, *Semiconductor Electrochemistry*, Wiley-VCH, Weinheim, 2001.

[26] L. Young, *Anodic Oxide Films*, Academic Press, London, 1961.

[27] D.A. Vermilyea, in *Advances in Electrochemistry and Electrochemical Engineering*, Ed. by P. Delahay, Interscience Publishers, New York, 1963, Vol. **3**, p. 211.

[28] K.E. Heusler, K.S. Yun, *Electrochim. Acta*, **22** (1977) 977.

[29] H.J. Engell, *Electrochim. Acta*, **22** (1977) 987.

[30] H. Gerischer, *Corrosion Sci.*, **29** (1989) 257; *Corrosion Sci.*, **31** (1990) 81.

[31] H. Gerischer, *Electrochim. Acta*, **35** (1990) 1677.

[32] J.W. Schultze, M.M. Lohrengel, *Electrochim. Acta*, 45 (2000) 2499.

[33] F. Di Quarto, M. Santamaria, *Corrosion Eng. Sci. Tech.*, **39** (2004) 71.

[34] F. Mott: *Proc. Roy. Soc.(A)*, **171** (1939) 27.

[35] W. Shottky: *Zeits. Physik*, **113** (1939) 367.

[36] J. F. Dewald, *Bell Syst. Tech. J.*, **39** (1960) 615; *J. Phys. Chem. Solids*, **14** (1960) 155.

[37] U. Stimming, *Electrochim. Acta*, **31** (1986) 415.

[38] L. Peter, in *Comprehensive Chemical Kinetics*, Ed. by R.G. Compton, Elsevier Science, Oxford, 1989, vol. **29**, p. 382.

[39] F. Di Quarto, C. Sunseri, S. Piazza, M. Santamaria, in *Handbook of Thin Films*, Ed. by H.S. Nalwa, Academic Press, San Diego, CA, 2002, **vol.2**, p. 373.

[40] F. Di Quarto, M. Santamaria, C. Sunseri, in *Analytical Methods in Corrosion Science and Technology*, Ed. by P. Marcus, F. Mansfeld, Taylor, Francis, Boca Raton, 2005, p. 697.

[41] M. Pourbaix, *Atlas of Electrochemical Equilibria in Aqueous Solutions*, Pergamon Press, Oxford, 1966.

[42] K. Uosaki, H. Kita, *J. Electrochem. Soc.*, **130** (1983) 895.

[43] W.H. Laflère, R.L. Van Meirhaeghe, F. Cardon, W.P. Gomes, *Surface Sci.*, **59** (1976) 401;ibidem, **74**, (1978) 125.

[44] W.H. Laflère, R.L. Van Meirhaeghe, F. Cardon, W.P. Gomes, *J. Appl. Phys.D*, **13** (1980) 2135.

[45] M. Tomkiewicz, *J. Electrochem. Soc.*, **126** (1979) 1505.

[46] G. Cooper, J. A. Turner, A.J. Nozik, *J. Electrochem. Soc.*, **129** (1982) 1973.

[47] H.O. Finklea, *J. Electrochem. Soc.*, **129** (1982) 2003.

[48] W.P. Gomes, F. Cardon, in *Progress in Surface Science*, Pergamon Press, Oxford, 1982, vol. **12**, p. 155.

[49] H. Gerischer, R. McIntyre, *J. Chem. Phys.*, **83** (1985) 1363.

[50] D.S. Ginley, M.A. Butler, in *Semiconductor Electrodes*, Ed. by H.O. Finklea, Elsevier Science, Oxford, 1988, p. 335.

[51] N. Sato, K. Kudo, T. Noda, *Z. Physik. Chem. N.F.*, **98** (1975) 271.

[52] P. Marcus, V. Maurice, in *Interfacial Electrochemistry*, Ed. by A. Wieckowski, M. Dekker, New York, 1999, p. 541.

[53] U. Konig, J. W. Shultze, in *Interfacial Electrochemistry*, Ed. A. Wieckowski, M. Dekker, New York, 1999, p. 649.

[54] H. Streblow, P. Marcus in *Analytical Methods in Corrosion Science and Technology*, Ed. by P. Marcus, F. Mansfeld, Taylor and Francis, Boca Raton, 2005, p. 1.

[55] V. Maurice, P. Marcus, in *Analytical Methods in Corrosion Science and Technology*, Ed. by P. Marcus, F. Mansfeld, Taylor and Francis, Boca Raton, 2005, p. 133.

[56] C.-O.A. Olsson, D. Landolt, in *Analytical Methods in Corrosion Science and Technology*, Ed. by P. Marcus, F. Mansfeld, Taylor and Francis, Boca Raton, 2005, p. 733.

[57] S.U.M. Khan, W. Shmickler, *J. Electroanal. Chem.*, **108** (1980) 329.

[58] P. Schmuki, H. Böhni, *Electrochim. Acta*, **40** (1995) 775.

[59] M. Büchler, P. Schmuki, H. Böhni, T. Stenberg, T. Mäntylä: *J. Electrochem. Soc.*, **145** (1998) 378.

[60] E. Sikora, D. D. MacDonald, *J. Electrochem. Soc.*, **147** (2000), 4087.

[61] V.A. Alves, C.M.A. Brett, *Electrochim. Acta*, **47** (2002) 2081.

[62] Y.M. Zeng, J. L. Luo, *Electrochim. Acta*, **48** (2003) 35.

[63] I. Diez-Perez, P. Gorostiz, F. Sanz: *J. Electrochem. Soc.*, **150** (2003) B348.

[64] F. Gaben, B. Vullemin, R. Oltra, *J. Electrochem. Soc.*, **151** (2004) B595.

[65] D.G. Li, Y.R. Feng, Z.Q. Bai, J.W. Zhu, M.S. Zheng, *Electrochim. Acta*, **52** (2007) 7877.

[66] S. Fujimoto, H. Tsuchiya, *Corr. Sci.*, **49** (2007) 195.

[67] J. Amri, T. Souier, B. Maliki, B. Baroux, ibidem, **50** (2008) 431.

[68] J.S.Blakemore, *Semiconductor Statistics*, Pergamon Press, Oxford (1962).

[69] H. Gerischer, *J. Phys. Chem.*, **89** (1985) 4249.

[70] H. Gerischer, R. McIntyre, D. Scherson, W. Stork, *J. Phys. Chem. Phys.*, **91** (1987) 1930.

[71] D.L. Losee, *J. Appl. Phys.*, 46 (1975) 2204.

[72] G. Nogami, *J. Electrochem. Soc.*, **129** (1982) 2219.

[73] E.C. Dutoit, R.L. Van Meirhage, F. Cardon, W. P. Gomes, *Ber. Bunsenges. Phys. Chem.*, **79** (1975) 1206.

[74] V. Macagno, J.W. Schultze, *J. Electroanal. Chem.*, **180** (1984) 157.

[75] F. Di Quarto, A. Di Paola, C. Sunseri, *Electrochim. Acta*, **26**, (1981) 1177.

[76] F. Di Quarto, C. Sunseri, S. Piazza, *Ber. Bunsenges. Phys. Chem.*, **90** (1986) 549.

[77] F. Di Quarto, S. Piazza, C. Sunseri, *Electrochim. Acta*, **35** (1990) 99.

[78] F. Di Quarto, V.O. Aimiuwu, S. Piazza, C. Sunseri, *Electrochim. Acta*, **36** (1991) 1817.

[79] M.H. Cohen, H. Fritzsche, S.R. Ovishinsky, *Phys. Rev. Lett.*, **22** (1969) 1065.

[80] D.Adler, *Amorphous Semiconductors*, CRS Press, Cleveland, 1971.

[81] N.F. Mott, E.A. Davis, *Electronic Processes in Non-crystalline Materials*, 2$^{nd}$ Ed., Clarendon Press, Oxford, 1979.

[82] J. Tauc, *Amorphous and Liquid Semiconductors*, Plenum Press, London, 1974.

[83] *Amorphous Semiconductors*, edited by M.H. Brodsky, Springer Verlag, Berlin, 1979.

[84] *Physical Properties of Amorphous Materials*, edited by D. Adler, B.B. Schwartz, M.C. Steele, Plenum Press, New York, 1985.

[85] W.E. Spear, P.G. Le Comber, A.J. Snell, *Philos. Mag. B*, **38** (1978) 303.

[86] R.A. Abram, P.J. Doherty, *Philos. Mag. B*, **45** (1982) 167.

[87] J.D. Cohen, D.V. Lang, *Phys. Rev. B*, **25**, (1982) 5321; D.V. Lang, J.D. Cohen, J.P. Harbison, *Phys. Rev. B*, **25**, (1982) 5285.

[88] I. W. Archibald, R. A. Abram, *Philos. Mag. B*, **48**, (1983) 111.

[89] I. W. Archibald, R. A. Abram, *Philos. Mag. B*, **54**, (1986) 421.

[90] W. E. Spear, S. H. Baker, *Electrochim. Acta*, **34**, (1989) 1691.

[91] M. H. Dean, U. Stimming, *J. Electroanal. Chem.*, **228**, (1987) 135.

[92] M. H. Dean, U. Stimming, *J. Phys. Chem.*, **93**, (1989) 8053.

[93] S. Piazza, C. Sunseri, F. Di Quarto, A.I.Ch.E. Journal, **38** (1992) 219.

[94] F. Di Quarto, M. Santamaria, in *Surface Oxide Films*, V. Birss, L. Burke, A. R. Hillman, R. S. Lillard, Editors, PV 2003-25, The Electrochemical Society Proceeding Series, Pennington, NJ, 2004, p. 86.

[95] F. Di Quarto, F. La Mantia, M. Santamaria, *Electrochim. Acta*, **50**, (2005) 5090.

[96] F. Di Quarto, F. La Mantia, M. Santamaria, *Corr. Sci.*, **49**, (2007) 186.

[97] F. La Mantia, M. Santamaria, F. Di Quarto, in *Passivation of Metals and Semiconductors, and properties of Thin Oxide Layers*, Ed. by P. Marcus, V. Maurice, Elsevier, Amsterdam, 2006, p. 343.

[98] C. da Fonseca, M. G. Ferreira, M. da Cunha Belo, *Electrochim. Acta*, **39**, (1994) 2197.

[99] E. Becquerel, *C. R. Hebd. Séan. Acad. Sci.*, **9** (1839) 561.

[100] A.W. Copeland, O.D. Black, A.B. Garrett, *Chem. Rev.*, **31** (1942) 177.

[101] E. K. Oshe, I. L. Rozenfel'd, Elektrokhimiya, **4** (1968) 1200.

[102] R. Williams: *J. Chem. Phys.*, **32** (1960) 1505.

[103] W. Paatsch: *J. Phys.*, **38** (1977) C5-151

[104] E. Angelini, M. Maja, P. Spinelli: *J. Phys.*, **38** (1977) C5-261.

[105] S. M. Wilhelm, Y. Anizawa, Chaung-Yi Liu, N. Hackermann, *Corros. Sci.*, **22** (1982) 791.

[106] A. Fujishima, K. Honda, *Nature*, 238 (1972) 37.

[107] M. A. Butler, *J. Appl. Phys.*, **48** (1977) 1914.

[108] Faraday Discussion No.70, The Royal Society of Chemistry, London, 1980.

[109] Photoeffects at Semiconductor-Electrolyte Interfaces, ACS Symposium Series 146 Ed. A. J. Nozik, American Chemical Society, Washington D.C., 1981.

[110] F. Di Quarto, G. Russo, C. Sunseri, A. Di Paola: *J. Chem. Soc. Faraday Trans.1*, **78** (1982) 3433.

[111] R.H. Wilson. *J. Appl. Phys.*, **48** (1977) 4292.

[112] H. Reiss. *J. Electrochem. Soc.*, **125** (1978) 937.

[113] W.J. Albery, P.N. Bartlett, A. Hamnett, M.P. Dare-Edwards. *J. Electrochem. Soc.*, **128** (1981) 1492.

[114] F. El Guibaly, K. Colbow. *J. Appl. Phys.*, **53** (1982) 1737.

[115] W.W. Gärtner. *Phys. Rev.*, **116** (1959) 84.

[116] J. Reichman, *Appl. Phys. Lett.*, **36** (1980) 574.

[117] R.U. E't Lam, D.R. Franceschetti, *Mater. Res. Bull.*, **17** (1982) 1081.

[118] F. Di Quarto, A. Di Paola, S. Piazza, C. Sunseri, *Solar Energy Mater.*, **11** (1985) 419.

[119] S.U.M. Khan, J.O'M. Bockris *J. Appl. Phys.*, 52 (1981) 7270.

[120] D.M. Pai, R.C. Enck. *Phys. Rev. B*, **11** (1975) 5163.

[121] F. Di Quarto, S. Piazza, C. Sunseri. *Electrochim. Acta*, **38** (1993) 29.

[122] M. Santamaria, F. Di Quarto, H. Habazaki, *Corr. Sci.*, **50** (2008) 2012.

[123] F. Di Quarto, S. Piazza, R. D'Agostino, C. Sunseri. *J. Electroanal. Chem.*, **228** (1987) 119.

[124] F. Di Quarto, S. Piazza, C. Sunseri. *J. Chem. Soc., Faraday Trans. 1*, **85** (1989) 3309.

[125] S. Piazza, C. Sunseri, F. Di Quarto. *J. Electroanal. Chem.*, **293** (1990) 69.

[126] N.F. Mott, E.A. Davis, *Electronic Processes in Non-crystalline Materials*, 2$^{nd}$ Ed., Clarendon Press, Oxford, 1979, p.275.

[127] D.L. Wood, J. Tauc, *Phys. Review B*, **5** (1972) 3144.

[128] Di Quarto, S. Piazza, A. Splendore, C. Sunseri. "Oxide Films on Metals and Alloys", p. 311, B.R. McDougall, R.S. Alwitt, T.A. Ramanarayanan, Editors, PV 92-22, The Electrochemical Society Proceeding Series, Pennington, NJ, 1992

[129] T. Watanabe, H. Gerischer. *J. Electroanal. Chem.*, **122** (1981) 73.

[130] Di Quarto, G. Tuccio, A. Di Paola, S. Piazza, C. Sunseri, "Oxide Films on Metals and Alloys" p. 25, K. Hebert, G.E. Thompson, Editors, PV 94-25, The Electrochemical Society Proceeding Series, Pennington, NJ, 1994.

[131] C. Sunseri, S. Piazza, F. Di Quarto, *Mater. Sci. Forum*, **185-188** (1995) 435.

[132] F. Di Quarto, S. Piazza, C. Sunseri, M. Yang, S.M. Cai, *Electrochim. Acta*, **41** (1996) 2511.

[133] S. Piazza, A. Splendore, A. Di Paola, C. Sunseri, F. Di Quarto, *J. Electrochem. Soc.*, **140** (1993) 3146

[134] Yu.Ya. Gurevich, Yu. V. Pleskov, Z.A. Rotenberg, *Photoelectrochemistry*, Plenum Press, New York, 1980.

[135] M. Santamaria, F. Di Quarto, S. Zanna, P. Marcus, *Electrochim. Acta*, **53** (2007) 1314.

[136] P. Han, D.M. Bartels, *J. Phys. Chem.*, **94** (1990) 5824.

[137] F. Di Quarto, C. Sunseri, S. Piazza, M.C. Romano. *J. Phys. Chem. B*, **101** (1997) 2519.

[138] J.C. Phillips, *Bonds and Bands in Semiconductors*, Academic Press, New York, 1973.

[139] L. Pauling, *The Nature of Chemical Bond*, chapter 3, Cornell University Press, Ithaca, NY, 1960.

[140] W. Gordy, W.J.O. Thomas., *J. Phys. Chem.*, **24** (1956) 439.

[141] L. Allred, *J. Inorg. Nucl. Chem.*, **17** (1961) 215.

[142] W. Teng, J. F. Muth, U. Ozgur, M. J. Bergman, H. O. Everitt, A. K. Sharma, C. Jin, J. Narayan. *Appl. Phys. Letters*, **76** (2000) 979.

[143] M. Santamaria, F. Di Quarto, P. Skeldon, G.E. Thompson, *J. Electrochem. Soc.*, **153** (2006) B518.

[144] M. Santamaria, D. Huerta, S. Piazza, C. Sunseri, F. Di Quarto. *J. Electrochem. Soc.*, **147** (2000) 1366.

[145] H. Kim, N. Hara, K. Sugimoto, . *J. Electrochem. Soc.*, **146** (1999) 955.

[146] M. Santamaria, F. Di Quarto, H. Habazaki, *Electrochim. Acta*, **53** (2008) 2272.

[147] F. Di Quarto, M. Santamaria, P. Skeldon, G.E. Thompson, *Electrochim. Acta*, **48** (2003) 1143

[148] A. R. Newmark, U. Stimming, *Langmuir*, **3** (1987) 905.

[149] F. Di Quarto, M.C. Romano, M. Santamaria, S. Piazza, C. Sunseri, *Russian Journal of Electrochemistry*, **36** (2000) 1203.

[150] M. Santamaria, E. Adragna, F. Di Quarto, *Electrochem. and Solid State Letters*, **8** (2005) B12.

[151] U. Stimming, *Electrochim. Acta*, **3** (1986) 415.

[152] M. K. Carpenter, D.A. Corrigan, *J. Electrochem. Soc.*, **136** (1989) 1022.

[153] L. Anicai, R. Masi, M. Santamaria, F. Di Quarto, *Corros. Sci,.* **47** (2005) 2883.

[154] F. Criminisi. Chem. Eng. Degree Thesis, Università di Palermo, April, 2002

[155] Di Quarto, M. Santamaria, N. Mallandrino, V. Laget, R. Buchheit, K. Shimizu, *J. Electrochem. Soc.*, **150** (1989) B462.

[156] F. Di Quarto, S. Piazza, C. Sunseri, Current Topics in Electrochemistry, Vol. 3, p. 357, edited by J.C. Alexander, Council of Scientific Research Integration, Trivarandum, 1994

[157] F. Di Quarto, S. Piazza, C. Sunseri, *Mater. Sci. Forum*, **192-194** (1995) 633.

[158] L. Anicai, S. Piazza, M. Santamaria, C. Sunseri, F. Di Quarto, *ATB Metallurgie*, (2000) 175.

[159] F. Di Quarto, S. Piazza, C. Sunseri, *Corros. Sci.*, **31** (1990) 721.

[160] H.Tsuchiya, S. Fujimoto, O. Chihara, T. Shibata, *Electrochim. Acta*, **47** (2002) 4357.

[161] H.Tsuchiya, S. Fujimoto,T. Shibata, *J. Electrochem. Soc.*, **151** (2004) B39.

[162] S. Fujimoto, H.Tsuchiya, Corr. Science, **49** (2007) 195.

[163] T.L. Sudesh, L. Wijesinghe, D. J. Blackwood, *J. of Physics: Conference Series*, **28** (2006) 74.

[164] T.L. Sudesh, L. Wijesinghe, D. J. Blackwood, *Appl. Surf. Sci.*, **253** (2006) 1006.

[165] T.L. Sudesh, S.L. Wijesinghe, D. J. Blackwood, *J. Electrochem. Soc.*, **154** (2007) C16.

[166] T.L. Sudesh, L. Wijesinghe, D. J. Blackwood, *Corros. Sci.*, **50** (2008) 23.

[167] S Virtanen, P Schmuki, H. Bohni , P. Vuoristo, T. Mantyla, *J. Electrochem. Soc.*, **142** (1995) 3067.

[168] P. Schmuki, M. Buchler, S Virtanen, H Bohni, R. Muller, L.J. Gauchler, *J. Electrochem. Soc.*, **142** (1995) 3336.

[169] M. Buchler, P. Schmuki, H Bohni, T. Stenberg, T. Mantyla, *J. Electrochem. Soc.*, **145** (1998) 378.

[170] A.J. Davenport, L.J. Oblonsky, M.P. Ryan, M.F. Toney, *J.Electrochem. Soc.*, 147 (2000) 2162.

[171] S. Piazza, M. Sperandeo, C. Sunseri, F. Di Quarto, *Corros. Sci.*, **46** (2004) 831.

[172] C. Sunseri, S. Piazza, A. Di Paola, F. Di Quarto, *J. Electrochem. Soc.*, **134** (1987) 2410.

[173] G. Dagan, W-M. Shen, M. Tomkiewicz, *J. Electrochem. Soc.*, **139** (1992) 1855.

[174] A. Di Paola, F. Di Quarto, C. Sunseri, *Corros. Sci.*, **26** (1986) 935.

[175] T. D. Burleigh, R.M. Latanision. *J. Electrochem. Soc.*, **134** (1987) 137.

[176] A. Di Paola, D. Shukla, U. Stimming, *Electrochim. Acta*, **36** (1991) 345.

[177] M.J. Kloppers, F. Bellucci, R.M. Latanision, *Corrosion*, **48** (1992) 229.

[178] P. Schmuki, H. Bohni, *J. Electrochem. Soc.*, **139** (1992) 1908.

[179] P. Schmuki, H. Bohni, F. Mansfeld. *J. Electrochem. Soc.*, **140** (1993) L119.

[180] A.M.P. Simoes, M.G.S. Ferreira, B. Rondot, M. Da Cunha Belo, *J. Electrochem. Soc.*, **137** (1990) 82.

[181] N.E. Hakiki, M. Da Cunha Belo, A.M. P. Simoes, M.G.S. Ferreira, *J. Electrochem. Soc.*, **145** (1998) 3821.

[182] N.E. Hakiki, M. Da Cunha Belo, M.G.S. Ferreira, *Electrochim. Acta*, **44** (1999) 2473.

[183] M.F. Montemor, M.G.S. Ferreira, N.E. Hakiki, M. Da Cunha Belo, *Corros. Sci.*, **42** (2000) 1635.

[184] L.J. Oblonsky, M.P. Ryan, H.S. Isaacs, *J. Electrochem. Soc.*, **145** (1998) 1922.

[185] P. Marcus, V. Maurice, *Interfacial Electrochemistry*, Ed. A. Wieckowski, M. Dekker Inc., NY, 1999, p. 541.

[186] D. Hamm, C-O. A. Olsson, D. Landolt, *Corros. Sci.*, **44** (2002) 1009.

[187] Keller, H-H. Streblow, *Corros. Sci.*, **46** (2004) 1939.

# Index